Arc Welding Safety

Guide for Safe Arc Welding

LINCOLN ELECTRIC

Safety Practices in Welding

INTRODUCTION

Arc welding is a safe occupation when sufficient measures are taken to protect the welder from potential hazards. When these measures are overlooked or ignored, however, welders can encounter such dangers as electric shock, overexposure to fumes and gases, arc radiation, and fire and explosion; which may result in serious, or even fatal injuries.

This bulletin is written with the arc welding operator in mind, containing both mandatory safety practices and those based on shop experience. Be sure to read ANSI Z49.1, and refer to the other publications listed at the end of the bulletin for more detailed information on specific topics of arc welding safety, as well as the manufacturers' instructions and material safety data sheets (MSDS's).

Important Note:
So that you can protect yourself against these hazards, every welder should be familiar with American National Standard ANSI Z49.1, "Safety in Welding and Cutting," and should follow the safety practices in that document. Z49.1 is now available for download at no charge at:

http://www.lincolnelectric.com/community/safety/ or at the AWS website http://www.aws.org.

Download and read it!

PERSONAL PROTECTIVE EQUIPMENT

Protective Clothing

Welders, like firemen, must wear clothing to protect them from being burned. Of all injuries to welders, burns are the most common due to sparks landing on bare skin. Welding arcs are very intense and can cause burns to skin and eyes with just a few minutes of exposure.

The actual gear varies with the job being performed, but generally **protective clothing** must allow freedom of movement while providing adequate coverage against burns from sparks, weld spatter, and arc radiation. Many types of clothing will protect you from ultra-violet radiation exposure, which appears as a skin burn (much like sunburn). Under the worst conditions, however, severe burns and skin cancer may result from excessive radiation.

Because of its durability and resistance to fire, wool clothing is suggested over synthetics (which should never be worn because it melts when exposed to extreme heat) or cotton, unless it is specially treated for fire protection. If possible, keep your clothes clean of grease and oil, as these substances may ignite and burn uncontrollably in the presence of oxygen.

Avoid rolling up your sleeves and pant-cuffs, because sparks or hot metal could deposit in the folds; also, wear your trousers outside your work boots, not tucked in, to keep particles from falling into your boots. While we're on the subject, we suggest leather high-tops with steel toes (especially when doing heavy work).

Other protective wear for heavy work or especially hazardous situations includes: flame-resistant suits, aprons, leggings, leather sleeves/shoulder capes, and caps worn under your helmet.

Heavy, flame-resistant gloves, such as leather, should **always** be worn to protect your hands from burns, cuts, and scratches. In addition, as long as they are dry and in good condition, they will offer some insulation against electric shock.

As to preventing electric shock, the key word is **dry**! We'll have more on the subject later, but for now keep in mind that moisture can increase the potential for and severity of electric shock. When working in wet conditions, or when perspiring heavily, you must be even more careful to insulate your body from electrically "live" parts and work on grounded metal.

ARC RAYS can burn.
- **Wear eye, ear and body protection.**

Note To Arc Welding Educators and Trainers:
This Arc Welding Safety brochure may be freely copied for educational purposes if distributed to welders and welding students at no additional charge.

ARC RAYS

It is essential that your **eyes are protected** from radiation exposure. Infrared radiation has been known to cause retinal burning and cataracts. And even a brief exposure to ultraviolet (UV) radiation can cause an eye burn known as "welder's flash." While this condition is not always apparent until several hours after exposure, it causes extreme discomfort, and can result in swelling, fluid excretion, and temporary blindness. Normally, welder's flash is temporary, but repeated or prolonged exposure can lead to permanent injury of the eyes.

Other than simply not looking at an arc, the primary preventive measure you can take is to use the proper shade lens in your helmet. Refer to the lens shade selector chart in Supplement 1 for the recommended shade numbers for various arc welding processes. The general rule is to choose a filter too dark to see the arc, then move to lighter shades without dropping below the minimum rating. The filters are marked as to the manufacturer and shade number, the impact-resistant variety are marked with an "H."

Helmets and hand-held face shields (see Figure A) offer the most complete shading against arc radiation. The shade slips into a window at the front of the shield so that it can be removed and replaced easily. The shields are made from a hard plastic or fiberglass to protect your head, face, ears, and neck from electric shock, heat, sparks, and flames. You should also use safety glasses with side shields or goggles to protect your eyes from flying particles.

Visible light can also be harmful, but it is easy to tell if the light is dangerous: if it hurts to look at, then it's too bright. The same is true for infrared radiation: it can usually be felt as heat. However, there's no real way for you to tell if you're being over exposed to UV radiation, so just don't take chances: always wear eye protection (see Supplement 1 for recommended lens shade numbers).

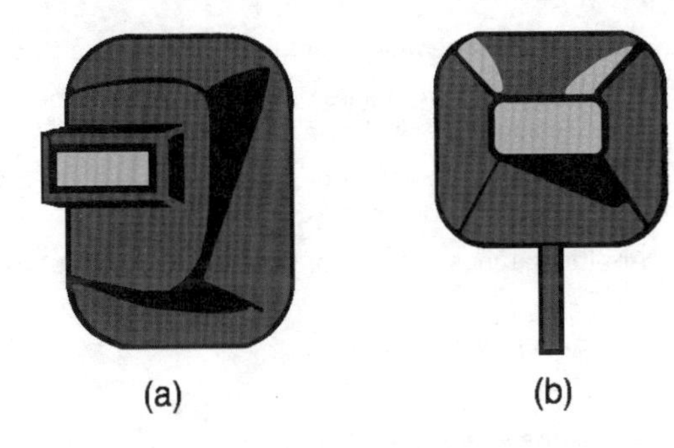

Figure A. A helmet (a) required for protecting the welder's eyes and face and (b) a hand-held face shield that is convenient for the use of foremen, inspectors, and other spectators.

NOISE

There are two good reasons to **wear ear muffs or plugs**:

a) to keep flying sparks or metal out of your ears; and

b) to prevent hearing loss as a result of working around noisy arc welding equipment, power sources, and processes (like air carbon arc cutting or plasma arc cutting).

As with radiation exposure to the eyes, the length and number of times that you are exposed to high levels of noise determine the extent of the damage to your hearing, so be sure to avoid repeated exposure to noise. If it is not possible to reduce the level of noise at the source (by moving either yourself or the equipment, utilizing sound shields, etc.), then you should wear adequate ear protection.

If the noise in your work area becomes uncomfortable, causing a headache or discomfort of the ears, you could be damaging your hearing and should immediately put on ear muffs or plugs.

In fact, the use of ear protection at all times is a good idea, as hearing loss is both gradual and adds up over time. Damage to your hearing may not be noticed until you have a complete hearing test, and then it could be too late.

INSPECTION AND MAINTENANCE OF EQUIPMENT AND WORK

Before starting any arc welding operation, you should make a **complete inspection** of your equipment. All it takes on your part is 5-10 minutes before you turn on your welder; is that too much to spend in preventing injury to yourself or your co-workers?

To begin with:

- Have you read the instruction manual and do you understand the instructions? The instruction manual for your welder is available upon request to your welding distributor or the manufacturer. Manuals for Lincoln Electric welders may be downloaded from lincolnelectric.com at no charge.
- Have you read the warnings and instructions on the equipment nameplates and decals as well as the consumables labels and material safety data sheets? (For older equipment see Supplement 5 to request a FREE Warning Label.)

For the welder:

- Are all the connections tight, including the earth ground?
- OSHA regulations require output terminals to be insulated. Rubber boots are available for that purpose.
- Are the electrode holder and welding cable well insulated and in good condition?
- Are the settings correct for the job you're about to begin?

For an engine-driven welder:

- Is it running OK?
- Are all the hoses on tight?
- Is the fuel cap on tight?
- Is the engine leaking fuel or oil? Some jobsites look for this and may refuse entry if your engine is leaking.
- Is the original enclosure and fan guarding in place? Check with your welding equipment distributor if you are unsure. (See Supplement 6.)

For the work in general: (See also Supplements 4 and 7)

- Are the work area conditions such that normal safety precautions can be observed or must special equipment (i.e., ventilation, exhaust, or respirator, welding equipment, protective equipment, safety equipment) or procedures be used?
- Many jobsites require permits for any welding or cutting. Be sure you have any permits you will need.
- If you will be working in a confined space, many special OSHA regulations and jobsite requirements may apply in addition to the arc welding precautions in this brochure. Understand which of these apply to your jobsite and comply with them.
- Are the cables the right size for your job? Be sure any damaged cable insulation is repaired.
- Are they spread out and run neatly to prevent overheating?
- Is the gas cylinder connected properly?
- Is the cylinder secure?

- Is the work stable and easy to reach from where you're standing?
- Is the Work Lead connected securely?
- Is there enough dry insulation between your body and the work piece?
- Is there adequate ventilation in your work area?

Take some personal responsibility for your own safety. Notify your supervisor if equipment is in need of repair or not working properly or any unsafe condition. You have the most to lose if you get hurt. Don't allow yourself to work in a hazardous situation without taking appropriate safety precautions.

If the hazard is serious and cannot be corrected readily, the machine should be shut down until the needed repairs are made. If the problem is limited to the outside of the welder, such as a loose connection or a damaged cable that needs to be replaced, **disconnect power to the welder** and correct the problem per the manufacturers instructions in the operating/service manual. If the hazard requires repairs to the inside of the welder or to the electrical input supply lines, call a service technician or an electrician. Never attempt to make these repairs if you are untrained.

Important Safety Note:
Consider whether the area in which you will be working creates or increases the level of hazard to you thus requiring special procedures or equipment. Factors such as electrical safety, fume ventilation/exhaust and risk of fire or explosion may be affected. See later sections on those topics and other documents in "Bibliography and Suggested Reading" for further information.

CARE AND CLEANING OF THE WORK AREA

Keeping the area around your work neat is as important as maintaining your equipment. Perhaps even more-so, as the risk of injury is amplified by the larger group of people involved. You may have already inspected your equipment and found it to be OK, but all your caution won't matter when, for example, a co-worker trips over your cable, causing you, and/or the people around you, to be injured by shock, hot metal, or from falling.

Keep all your equipment, cables, hoses, cylinders, etc. out of any traffic routes such as doors, hallways, and ladders. A good practice is to avoid clutter ... and **clean up your work area** when you're done! Not only will it help to protect yourself and others, you'll find it much easier for you to work efficiently.

Also, bear in mind that while you're paying attention to your work, other welders may be preoccupied with their own tasks and not watching where they're going. So be sure that there are protective screens in place, just in case somebody happens to be passing into your work area or walks into a shower of sparks or spatter.

GAS CYLINDERS

Because of the high pressure gas in cylinders, you must pay particularly close attention to their storage and use. Examine the cylinders as you did the rest of your equipment; check the cylinder label to make sure it is the correct shielding gas for the process, and that the regulators, hoses, and fittings are the right ones for that gas and pressure, and are in good condition.

Cylinders must be secured in an upright position, with the valve caps in place, in an area away from combustibles and fuels, and safeguarded from damage, heat, and flames. When in use, keep them out of traffic routes and flying sparks, with all hoses run neatly to the welding area. Never allow the electrode or other "electrically hot" parts of your welder to touch a cylinder. "Crack" the valve open to prevent dirt from entering the regulator; open the cylinder valve only when standing to one side of the cylinder, away from welding or other sources of ignition. Return damaged cylinders to the supplier. Refer to the Compressed Gas Association pamphlet P-1, "Safe Handling of Gas Cylinders," for further information.

CYLINDER may explode if damaged.
- **Keep cylinder upright and chained to support.**
- **Never allow welding electrode to touch cylinder.**

Electric and Magnetic Fields

Electric current flowing through any conductor causes localized Electric and Magnetic Fields (EMF). Welding current creates EMF fields around welding cables and welding machines. EMF fields may interfere with some pacemakers, and welders having a pacemaker should consult their physician before welding. Exposure to EMF fields in welding may have other health effects which are now not known. All welders should use the following procedures in order to minimize exposure to EMF fields from the welding circuit:

- Route the electrode and work cables together – Secure them with tape when possible.
- Never coil the electrode lead around your body.
- Do not place your body between the electrode and work cables. If the electrode cable is on your right side, the work cable should also be on your right side.
- Connect the work cable to the workpiece as close as possible to the area being welded.
- Do not work next to welding power source.

SPECIFIC CONCERNS

Possible Shock Hazards

The hazard of electric shock is one of the most serious and immediate risks facing you as a welder. Contact with metal parts which are "electrically hot" can cause injury or death because of the effect of the shock upon your body or a fall which may result from your reaction to the shock. The electric shock hazard associated with arc welding may be divided into two categories which are quite different:

- Primary Voltage Shock (i.e., 230, 460 volts); and
- Secondary Voltage Shock (i.e., 20-100 volts).

⚠ WARNING

HIGH VOLTAGE can kill.
- **Do not operate with covers removed.**
- **Disconnect input power before servicing.**
- **Do not touch electrically live parts.**

The **primary voltage shock** is very hazardous because it is much greater voltage than the welder secondary voltage. You can receive a shock from the primary (input) voltage if you touch a lead inside the welder with the **power to the welder** "on" while you have your body or hand on the welder case or other grounded metal. Remember that turning the welder power switch "off" does not turn the power off **inside the welder**. To turn the power inside the welder "off", the input power cord must be unplugged or the power disconnect switch turned off. You should never remove fixed panels from your welder; in fact, always have a qualified technician repair your welder if it isn't working properly. Also, your welder should be installed by a qualified electrician so it will be correctly wired for the primary voltage which supplies it power and so the case will be connected to an earth ground. When electrical supply lines are connected to a welder, check the welder capacity nameplate and connection instructions to be sure the input is the correct phase (single phase or three phase) and voltage. Many welders may be set up for single phase or three phase and for multiple input voltages. Be certain the welder is set up for the electrical supply to which it is connected. **Only a qualified electrician should connect input power**. The case must be grounded so that if a problem develops inside the welder a fuse will blow, disconnecting the power and letting you know that repair is required. Never ignore a blown fuse because it is a warning that something is wrong.

⚠ WARNING

ELECTRIC SHOCK can kill.
- **Do not touch electrically live parts or electrode with skin or wet clothing.**
- **Insulate yourself from work and ground.**

If welding must be performed under electrically hazardous conditions (in damp locations or while wearing wet clothing; on metal structures such as floors, gratings or scaffolds; when in cramped positions such as sitting, kneeling or lying, if there is a high risk of unavoidable or accidental contact with the work piece or ground) use the following equipment:

- **Semiautomatic DC Constant Voltage Welder**
- **DC Manual (Stick) Welder**
- **AC Welder with Reduced Voltage Control**

A **secondary voltage shock** occurs when you touch a part of the **electrode circuit** — perhaps a bare spot on the electrode cable — at the same time another part of your body is touching the metal upon which you're welding (work). To receive a shock your body must touch both sides of the welding circuit — electrode and work (or welding ground) — at the same time. To prevent secondary voltage shock, you must develop and use safe work habits. Remember the voltage at the electrode is highest when you are **not** welding (open circuit voltage).

- Wear **dry** gloves in good condition when welding.
- Do not touch the electrode or metal parts of the electrode holder with skin or wet clothing.
- Keep **dry** insulation between your body (including arms and legs) and the metal being welded or ground (i.e., metal floor, wet ground).
- Keep your welding cable and electrode holder in good condition. Repair or replace any damaged insulation.

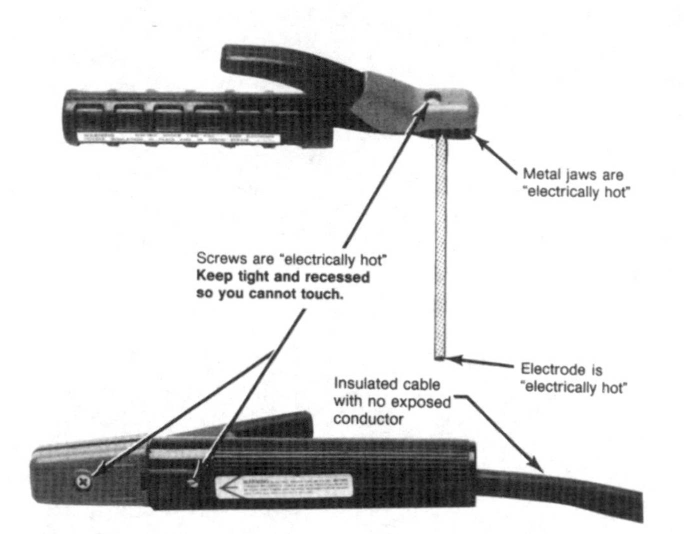

Figure B. Always inspect your electrode holder before turning the welder on.

These rules are basic to welding and you should already know them. Check out the warning on your welder or electrode box next time you weld. You will probably not have a shock while welding if you follow these rules.

Though it may be more difficult to follow the rules under some conditions, the rules still apply. Keep your gloves **dry** even if you have to keep an extra pair. Use plywood, rubber mats, or some other **dry** insulation to stand or lie upon. Insulate your body from the metal you are welding. Don't rest your body, arms, or legs on the workpiece, especially if your clothing is wet or bare skin is exposed (and it should not be if you are dressed properly). In addition to the normal safety precautions, if welding must be performed under electrically hazardous conditions (in damp locations or while wearing wet clothing; on metal structures such as floors, gratings or scaffolds; when in cramped positions such as sitting, kneeling or lying, if there is a high risk of unavoidable or accidental contact with the work piece or ground) use the following equipment:

- Semiautomatic DC Constant Voltage Welder
- DC Manual (Stick) Welder
- AC Welder with Reduced Voltage Control

The condition of your electrode holder and electrode cable is also very important. The plastic or fiber insulation on the electrode holder protects you from touching the metal "electrically hot" parts inside. Always inspect your electrode holder before turning the welder on. Replace the holder if it is damaged — don't try to repair it unless you have replacement parts.

The same is true of the electrode cable except that when not replaced it may be repaired using good electrical tape. If your cable has been repaired, be sure to check and see that the tape is secure before you turn the welder on.

Remember, a stick electrode is always "electrically hot" when the welder is on — treat it with respect. If you do experience a shock, think of it as a warning — check your equipment, work habits and work area to see what is wrong before continuing to weld.

⚠ WARNING

WELDING SPARKS can cause fire or explosion.
• Keep flammable material away.

Fire Hazards
Because of the extreme temperatures associated with any arc welding process, you should always be aware of fire hazards. The heat of the welding arc can reach temperatures of 10,000°F, but this heat in itself is not generally a fire hazard. The danger of fire actually results from the effects of this intense heat upon your work and in the form of **sparks and molten metals**. Because these can spray up to 35 feet from your work, you must recognize and protect combustible materials from the welding arc, sparks and spatter. It is also important to be sure the work is not in contact with any combustible which it may ignite when heated. These materials fall into three categories: liquid (gasoline, oil, paints, and thinners); solid (wood, cardboard, and paper); and gaseous (acetylene and hydrogen).

Watch where the sparks and metals are falling from your work: if there are flammable materials including fuel or hydraulic lines in your work area and you can't move either your work or the combustible substances, put a fire-resistant shield in place. If you're welding above the ground or off a ladder, make sure that there are no combustibles underneath. Also, don't forget about your co-workers, and everybody else who may be in the work area, as they probably wouldn't appreciate being hit with slag or sparks from your work.

Particular care must be taken when welding or cutting in dusty locations. Fine dust particles may readily oxidize (burn) and without warning result in a flash fire or even an explosion when exposed to the welding arc or even sparks.

If you are not sure of the combustible or volatile nature of residue or dust in the work area, no welding or cutting should take place until a responsible person has inspected the area and given approval for the work.

Before you start welding, inspect the surface of your work, looking for flammable coatings or any unknown substances that would ignite when heated. Because of the extreme fire and explosion hazards inherent to welding on or around containers and piping that may have combustible materials, such work should be handled only by experienced welders who review and follow the safety practices recommended in the American Welding Society document F4.1, "Recommended Safe Practices for the Preparation for Welding and Cutting of Containers and Piping Which Had Held Hazardous Substances."

Know where the fire alarms and fire extinguishers are located, and check the pressure gauges so you don't rely upon one that's empty. If there are none in the area, make sure that you have access to fire hoses, sand buckets, fire-resistant blankets, or other fire fighting equipment. If you're welding within 35 feet or so of flammable materials, you should have a fire watcher to see where your sparks are flying, and to grab an extinguisher or alarm if needed. Both you and the fire watcher should wait for a half hour after all welding is finished to find and put out any smoldering fires that may have resulted from your welding.

As with other emergencies that may result from welding accidents, the first rule is: don't panic. Depending on the size of the fire, sound the fire alarm to warn others and call the fire department; shut off your welder; and get to the fire exits as quickly as possible.

⚠ WARNING

FUMES & GASES can be dangerous to your health

- **Keep fumes and gases from your breathing zone and general area.**
- **Keep your head out of the fumes.**
- **Use enough ventilation or exhaust at the arc, or both, to keep fumes and gases from your breathing zone and general area.**

Fumes and Gases

Because of the variables involved in fume and gas generation from arc welding, cutting and allied processes (such as the welding process and electrode, the base metal, coatings on the base metal, and other possible contaminants in the air), we'll have to treat the subject in a rather general way, lumping all but the more hazardous situations together. The precautions we describe will hold true for all arc welding processes.

The **fume plume** contains solid particles from the consumables, base metal, and base metal coating. For common mild steel arc welding, depending on the amount and length of exposure to these fumes, most immediate or short term effects are temporary, and include symptoms of burning eyes and skin, dizziness, nausea, and fever. For example, zinc fumes can cause metal fume fever, a temporary illness that is similar to the flu.

Long-term exposure to welding fumes can lead to siderosis (iron deposits in the lungs) and may affect pulmonary function. Bronchitis and some lung fibrosis have been reported.

Some consumables contain certain compounds in amounts which may require special ventilation and/or exhaust. These Special Ventilation products can be identified by reading the labels on the package. If Special Ventilation products are used indoors, use local exhaust. If Special Ventilation products are used outdoors, a respirator may be required. Various compounds, some of which may be in welding fume, and reported health effects, in summary, are:

Barium: Soluble barium compounds may cause severe stomach pain, slow pulse rate, irregular heart beat, ringing of the ears, convulsions and muscle spasms. In extreme cases can cause death.

Cadmium also requires extra precautions. This toxic metal can be found on some steel and steel fasteners as a plating, or in silver solder. Cadmium fumes can be fatal even under brief overexposures, with symptoms much like those of metal fume fever. These two conditions should **not** be confused. Overexposure to cadmium can be enough to cause fatalities, with symptoms appearing quickly, and, in some circumstances, death a few days later.

Chromium: Chromium is on the IARC (International Agency for Research on Cancer) and NTP (National Toxicology Program) lists chromium as posing a carcinogenic risk to humans. Fumes from the use of stainless steel, hardfacing and other types of consumables contain chromium and/or nickel. Some forms of these metals are known or suspected to cause lung cancer in processes other than welding and asthma has been reported. Therefore, it is recommended that precautions be taken to keep exposures as low as possible. OSHA recently adopted a lower PEL (Permissible Exposure Limit) for chromium (see Supplement 3). The use of local exhaust and/or an approved respirator may be required to avoid overexposure.

Coatings on the metal to be welded, such as paint, may also contain toxic substances, such as lead, chromium and zinc. In general, it is always best to remove coatings from the base metal before welding or cutting.

Cobalt: Exposure to cobalt can cause respiratory disease and pulmonary sensitization. Cobalt in metallic form has been reported to cause lung damage.

Copper: Prolonged exposure to copper fume may cause skin irritation or discoloration of the skin and hair.

Manganese: Manganese overexposure may affect the central nervous system, resulting in poor coordination, difficulty in speaking, and tremor of arms or legs. This condition is considered irreversible.

Nickel: Nickel and its compounds are on the IARC (International Agency for Research on Cancer) and NTP (National Toxicology Program) lists as posing a carcinogenic risk to humans.

Silica: Crystalline silica is present in respirable dust form submerged arc flux. Overexposure can cause severe lung damage (silicosis).

Zinc: Overexposure to zinc (from galvanized metals) may cause metal fume fever with symptoms similar to the common flu.

The **gases** that result from an arc welding process also present potential hazard. Most of the shielding gases (argon, helium, and carbon dioxide) are non-toxic, but, as they are released, they **displace oxygen** in your breathing air, causing dizziness, unconsciousness, and death, the longer your brain is denied the oxygen it needs. Carbon monoxide can also be developed and may pose a hazard if excessive levels are present.

The heat and UV radiation can cause irritation to the eyes and lungs. Some degreasing compounds such as trichlorethylene and perchlorethylene can decompose from the heat and ultraviolet radiation of an arc. Because of the chemical breakdown of vapor-degreasing materials under ultraviolet radiation, arc welding should not be done in the vicinity of a vapor-degreasing operation. Carbon-arc welding, gas tungsten-arc welding and gas metal arc welding should be especially avoided in such areas, because they emit more ultraviolet radiation than other processes. Also, keep in mind that ozone and nitrogen oxides are formed when UV radiation passes through the air. These gases cause headaches, chest pains, irritation of the eyes, and an itchiness in the nose and throat.

There is one easy way to **reduce the risk** of exposure to hazardous fumes and gases: **keep your head out of the fume plume!** As obvious as this sounds, the failure to follow this advice is a common cause of fume and gas overexposure because the concentration of fume and gases is greatest in the plume. Keep fumes and gases from your breathing zone and general area using natural ventilation, mechanical ventilation, fixed or moveable exhaust hoods or local exhaust at the arc. Finally, it may be necessary to wear an approved respirator if adequate ventilation cannot be provided (see Ventilation section).

As a rule of thumb, for many mild steel electrode, if the air is visibly clear and you are comfortable, then the ventilation is generally adequate for your work.The most accurate way to determine if the worker exposure does not exceed the applicable exposure limit for compounds in the fumes and gases is to have an industrial hygienist take and analyze a sample of the air you are breathing. This is particularly important if you are welding with stainless, hardfacing or Special Ventilation products. All Lincoln MSDS have a maximum fume guideline number. If exposure to total fume is kept below that number, exposure to all fume from the electrode (not coatings or plating on the work) will be below the TLV.

There are also steps that you can take to identify hazardous substances in your welding environment. First, read the product label and material safety data sheet for the electrode posted in the work place or in the electrode or flux container to see what fumes can be reasonably expected from use of the product and to determine if special ventilation is needed. Secondly, know what the base metal is, and determine if there is any paint, plating, or coating that could expose you to toxic fumes and/or gases. Remove it from the metal being welded, if possible. If you start to feel uncomfortable, dizzy or nauseous, there is a possibility that you are being overexposed to fumes and gases, or suffering from oxygen deficiency. Stop welding and get some **fresh air** immediately. Notify your supervisor and co-workers so the situation can be corrected and other workers can avoid the hazard. Be sure you are following these safe practices, the consumable labeling and MSDS and improve the ventilation in your area. Do not continue welding until the situation has been corrected.

NOTE: The MSDS for all Lincoln consumables is available on Lincoln's website: www.lincolnelectric.com

Before we turn to the methods available to control welding fume exposure, you should understand a few basic terms:

Natural Ventilation is the movement of air through the workplace caused by natural forces. Outside, this is usually the wind. Inside, this may be the flow of air through open windows and doors.

Mechanical Ventilation is the movement of air through the workplace caused by an electrical device such as a portable fan or permanently mounted fan in the ceiling or wall.

Source Extraction (Local Exhaust) is a mechanical device used to capture welding fume at or near the arc and filter contaminants out of the air.

The ventilation or exhaust needed for your application depends upon many factors such as:

- workspace volume
- workspace configuration
- number of welders
- welding process and current
- consumables used (mild steel, hardfacing, stainless, etc.)
- allowable levels (TLV, PEL, etc.)
- material welded (including paint or plating)
- natural airflow

Your work area has **adequate ventilation** when there is enough ventilation and/or exhaust to control worker exposure to hazardous materials in the welding fumes and gases so the applicable limits for those materials is not exceeded. See Supplement 2 for the legal limits, the OSHA PEL (Permissible Exposure Limit), and the recommended guideline, the ACGIH TLV (Threshold Limit Value), for many compounds found in welding fume.

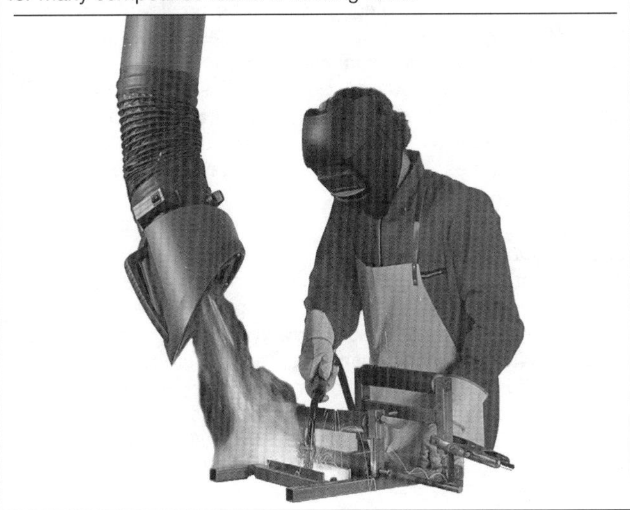

Ventilation

There are many methods which can be selected by the user to provide adequate ventilation for the specific application. The following section provides general information which may be helpful in evaluating what type of ventilation equipment may be suitable for your application. When ventilation equipment is installed, you should confirm worker exposure is controlled within applicable OSHA PEL and/or ACGIH TLV. According to OSHA regulations, when welding and cutting (mild steels), natural ventilation is usually considered sufficient to meet requirements, provided that:

1. The room or welding area contains at least 10,000 cubic feet (about 22' x 22' x 22') for each welder.
2. The ceiling height is not less than 16 feet.
3. Cross ventilation is not blocked by partitions, equipment, or other structural barriers.
4. Welding is not done in a confined space.

Spaces that do not meet these requirements should be equipped with mechanical ventilating equipment that exhausts at least 2000 cfm of air for each welder, except where local exhaust hoods or booths, or air-line respirators are used.

Important Safety Note:
When welding with electrodes which require special ventilation such as stainless or hardfacing (see instructions on container or MSDS) or on lead or cadmium plated steel and other metals or coatings which produce hazardous fumes, keep exposure as low as possible and below exposure limit values (PEL and TLV) for materials in the fume using local exhaust or mechanical ventilation. In confined spaces or in some circumstances, for example outdoors, a respirator may be required if exposure cannot be controlled to the PEL or TLV. (See MSDS and Supplement 3 of this brochure.) Additional precautions are also required when welding on galvanized steel.

Source Extraction Equipment

Mechanical ventilation is an effective method of fume control for many welding processes. Because it captures fume near the arc or source of the fume, which is more efficient in most cases, local exhaust, also called "source extraction", is a very effective means to control welding fume.

Source extraction of welding fumes can be provided by mobile or stationary, single or multi-station, exhaust and/or filtration equipment designed with adjustable fume extraction arms nozzles or guns, by fixed enclosures, booths or tables with extraction canopies also known as down-draft, or by back-draft or cross-draft tables/booths. Source extraction of weld fume falls into two categories: low vacuum/high volume, or high vacuum/low volume.

Low Vacuum/High Volume

Mobile or stationary, single or multi-station, large centralized exhaust and/or filtration equipment designed with adjustable fume extraction arms are usually low vacuum/high volume systems. When correctly positioned, the capture rate of adjustable fume extraction arms is suitable for all position welding and cutting. For more difficult to reach work areas, flexible hose may be used in place of adjustable fume extraction arms.

Mobiflex™ 200-M low vacuum mobile fume extraction unit.

These arms generally move between 560 and 860 cubic feet per minute (CFM) (900 – 1400 m^3/hr) of air, but use low vacuum levels (3 to 5 inches water gauge [750 – 1250 Pa]) to minimize power requirements. Water gauge (WG) is a measure of negative pressure: higher numbers mean more negative pressure (more "suction"). With this volume of airflow, the end of the arm can be placed 6 to 15 inches (160 – 375 mm) away from the arc and still effectively capture weld fume.

Fume extraction arms generally use a 6 or 8 inch diameter hose, or hose and tubing combinations. Arm lengths are typically 7, 10, or 13 feet (2, 3, or 4 m), with boom extensions available. The arms may be wall mounted, attached to mobile units, or incorporated into a centralized system.

In general, the farther the extraction hose is from the arc, the more volume of air movement is required to effectively capture welding fume. Overhead hoods (canopies), for example, capture most of the fume, but care must be taken to be sure fume is not pulled through the breathing zone of the operator.

Fixed enclosures, booths or tables with extraction canopies also known as **down-draft, back-draft or cross-draft booths/tables** are a variation of overhead hood technology and can be used as source extraction equipment. A booth is a fixed enclosure that consists of a top and at least two sides that surround the welding operation. These systems use a plenum with openings to the side, back or bottom of the work space rather than above it to capture the weld fume. The weld fume is extracted through the plenum and away from the breathing zone of the operator that is welding or cutting. Down-draft or back-draft booths/tables can be mobile or stationary, single or multi-station, exhaust and/or filtration systems. They are particularly suitable for in-position bench welding or cutting jobs and can be effective when small parts are being welded. The airflow required for effectiveness varies depending upon the installation design, but may be 1,000 CFM or higher.

There are advantages and limitations associated with low vacuum/high volume source extraction systems.

Advantages	***Limitations***
Source extraction with large volume of air being extracted from welder breathing zone.	If not using filtration unit, exhausting air to outside requires make-up air systems and make-up heaters (ie. large volumes of displaced air need to be replaced, resulting in increased utility costs).
Auto-stop delay assists with removal of residual fumes.	Welder must stop to reposition arm over weld area(s).
Low noise level.	Filtration systems larger due to volume of air flow.
Flexible arm for repositioning.	Depending on design, ductwork can be large.
Low installation costs (ductwork).	
Low energy consumption (small fan unit with low rpm).	
Adjustable arms suitable for all-position welding.	

High Vacuum/Low Volume

High vacuum/low volume fume extraction systems are designed for close proximity (2 to 4 inches) positioning. High vacuum/low volume weld fume extraction is achieved with lower airflow rates than those encountered when utilizing low vacuum/high volume systems. There are two methods of high vacuum extraction: welding guns with built-in extraction (fume extraction guns), or separate suction nozzles of various designs.

Fume extraction guns use fume capture nozzles built into the gun tube and handle. The extraction airflow is approximately 35 to 60 CFM (60 – 100 m^3/hr) for integrated fume extraction guns. Therefore, no repositioning is required, since the suction automatically follows the arc. The vacuum level is high (40 to 70 inches WG [9.96 X 103 to 1.74 X 104 Pascal]) permitting the use of hose featuring longer lengths (10 to 25 feet) and smaller diameters (1.25 to 1.75 inches). Fume extraction gun designs have been improved to be more ergonomic and user friendly. Depending upon the type of welding, particularly "in position" welding, extraction guns may be a good solution.

Suction nozzles are positioned near the weld, and commonly use capture distances of less than four inches. Depending upon the design, airflow of suction nozzles is typically between 80 to 100 CFM (135 – 170 m^3/hr). Suction nozzles must be kept near the arc to be used effectively.

The capture rate for fume extraction guns or nozzles is highest when used in flat and horizontal welding positions. High vacuum equipment ranges from small, portable, mobile units to stationary, single or multi-station, large centralized filtration systems.

Miniflex™ high vacuum portable fume extraction unit.

There are advantages and limitations associated with high vacuum/low volume source extraction systems.

Advantages	***Limitations***
When using a fume extraction gun, welder does not need to stop and reposition extraction device.	Required when using a suction nozzle. Welder may need to stop to reposition extraction device.
Low volume of air is displaced-results in energy efficiency and conservation.	High noise level due to increased air velocity and high motor rpm of the fan unit.
Ductwork smaller in diameter (3 to 10 inches) vs. low vacuum systems.	Possible removal of shielding gases affecting weld integrity if nozzle or gun placed too close to source.
Low obstruction of welder vision.	Greater energy consumption (large fan unit with high rpm).
Suitable for heavier particulate (ie. grinding dust).	Residual fumes not extracted.
Suitable option for confined, difficult to reach work spaces.	Less effective in out-of-position welding.
Smaller filter systems due to less volume of airflow.	

Fume extraction is only one component in reducing welding fume. Users should also consider the selection of the welding process, welding procedure, or consumable. Many times a combination of fume extraction, training, process change, and/or consumable change is needed to reduce the amount of fume to acceptable levels. Solutions to a particular application may involve one or all of these factors and the user must determine which solution best fits their application.

OSHA regulations include specific requirements for exhaust systems which should be reviewed when selecting fume extraction systems (see Supplement 2).

Exhaust vs. Filtration

Source extraction exhaust equipment captures and extracts weld fumes from the source and exhausts the fumes to the outside atmosphere. This technique removes welding fume from the breathing zone of the welder but can also displace large volumes of conditioned air which may lead to increased utility and heating costs.

Source extraction filtration equipment captures and extracts weld fumes from the source and filters the fumes by passing them through a cellulose and/or polyester filter cartridge or electrostatic filter. Depending on the weld application, environment, federal or local regulations, and filtration efficiency levels, filtered air may be re-circulated back into the facility or exhausted to the outside atmosphere. By re-circulating filtered air back into the work environment compared to exhausting to the outside, source extraction filtration equipment can be more economical to operate. Particularly in winter months, substantially lower heating costs may be recognized, as less replacement air is required with filtration versus exhaust systems.

Using a cellulose or polyester filter cartridge or electrostatic filter will depend upon the weld application. Electrostatic filters may also be used however, they lose efficiency if they are not frequently washed.

Regardless of the type of mechanical ventilation (exhaust or filtration) source extraction system used, the important factor is that it is a tool designed to control exposure to welding fume and its constituents. All forms of mechanical ventilation or source extraction equipment require routine maintenance. In addition, when using weld fume source extraction equipment, sparks from welding, cutting or grinding processes can cause fire within the equipment. To control this potential fire hazard, operation, service and maintenance instructions for source extraction equipment should be followed.

Note:

It is the equipment owner and operator's responsibility to comply with Occupational Safety, Health Administration (OSHA) Permissable Exposure Limits (PELs) or American Conference of Governments Industrial Hygienists (ACGIH) TLVs for welding fume. It is the responsibility of the equipment owner to research, test and comply with regulations which may apply to filtered air recirculated inside the facility or unfiltered air is exhausted outside of the facility.

Working in Confined Spaces

When arc welding in a confined area, such as a boiler, tank, or the hold of a ship, bear in mind that all the hazards associated with normal arc welding are amplified, so the precautions mentioned here are even more important. This subject is very complicated and only these precautions related to arc welding will be discussed in this brochure. Per OSHA document 29 CFR 1910.146, a particular area is considered a confined space if it:

1) Is large enough and so configured that an employee can bodily enter and perform assigned work; and
2) Has limited or restricted means for entry or exit (for example, tanks, vessels, silos, storage bins, hoppers, vaults, and pits are spaces that may have limited means of entry.); and
3) Is not designed for continuous employee occupancy.

There is a greater danger that enough **flammable gases** may be present in the confined space to cause an explosion. The metal of the enclosure can become part of the welding circuit, so any metal you touch (the walls, floor, ceiling) is **electrically "hot"**. Welding fumes can accumulate more rapidly, with a higher concentration; gases can force out the **breathable air**, suffocating you in the process.

Per OSHA document 29 CFR 1910.146(d)(5)(iii); after an area has been deemed a confined space, the existence of the following atmospheric hazards are to be determined:

1) Test for oxygen
2) Test for combustible gases and vapors
3) Test for toxic gases and vapors

The workplace and OSHA rules regarding confined spaces must be followed. Make sure that your body is insulated from the work-piece using dry insulation. Wear dry gloves and only use a well-insulated electrode holder. Semiautomatic constant voltage welders with cold electrode or stick welders equipped with a device to lower the no-load voltage are recommended, especially when the work area is wet. Make sure that there is adequate ventilation and exhaust (a respirator or an air-supplied respirator may be necessary depending on the application), and that there are no flammable coatings, liquids or gases nearby.

Lastly, you must have someone outside the enclosure trained to handle emergencies, with rescue procedures and a means to disconnect power to your equipment and pull you out if danger arises. We cannot stress this strongly enough: however experienced you are, do not attempt work of this nature without constant communication with the person outside the confined area. When welding within a confined area, problems which arise can immediately become very serious and, in some cases, life-threatening. It is for that reason that OSHA regulations and workplace procedures for confined space work must be followed.

GUIDE FOR SHADE NUMBERS

OPERATION	ELECTRODE SIZE 1/32 in. (mm)	ARC CURRENT (A)	MINIMUM PROTECTIVE SHADE	SUGGESTED(1) SHADE NO. (COMFORT)
Shielded metal arc welding	Less than 3 (2.5) 3-5 (2.5–4) 5-8 (4–6.4) More than 8 (6.4)	Less than 60 60-160 160-250 250-550	7 8 10 11	– 10 12 14
Gas metal arc welding and flux cored arc welding		Less than 60 60-160 160-250 250-500	7 10 10 10	– 11 12 14
Gas tungsten arc welding		Less than 50 50-150 150-500	8 8 10	10 12 14
Air carbon Arc cutting	(Light) (Heavy)	Less than 500 500-1000	10 11	12 14
Plasma arc welding		Less than 20 20-100 100-400 400-800	6 8 10 11	6 to 8 10 12 14
Plasma arc cutting	(Light)(2) (Medium)(2) (Heavy)(2)	Less than 300 300-400 400-800	8 9 10	9 12 14
Torch brazing		–	–	3 or 4
Torch soldering		–	–	2
Carbon arc welding		–	–	14
	PLATE THICKNESS in.	PLATE THICKNESS mm		
Gas welding Light Medium Heavy	 Under 1/8 1/8 to 1/2 Over 1/2	 Under 3.2 3.2 to 12.7 Over 12.7		 4 or 5 5 or 6 6 or 8
Oxygen cutting Light Medium Heavy	 Under 1 1 to 6 Over 6	 Under 25 25 to 150 Over 150		 3 or 4 4 or 5 5 or 6

(1) As a rule of thumb, start with a shade that is too dark, then go to a lighter shade which gives sufficient view of the weld zone without going below the minimum. In oxyfuel gas welding or cutting where the torch produces a high yellow light, it is desirable to use a filter lens that absorbs the yellow or sodium line the visible light of the (spectrum) operation

(2) These values apply where the actual arc is clearly seen. Experience has shown that lighter filters may be used when the arc is hidden by the workpiece.

Data from ANSI Z49.1-1999

BIBLIOGRAPHY AND SUGGESTED READING

ANSI Z87.1, *Practice for Occupational and Educational Eye and Face Protection*, American National Standards Institute, 11 West 42nd Street, New York, NY 10036.

Arc Welding and Your Health: A Handbook of Health Information for Welding. Published by The American Industrial Hygiene Association, 2700 Prosperity Avenue, Suite 250, Fairfax, VA 22031-4319.

NFPA Standard 51B, *Cutting and Welding Processes*, National Fire Protection Association, 1 Batterymarch Park, P.O. Box 9146, Quincy, MA 02269-9959.

OSHA General Industry Standard 29 CFR 1910 Subpart Q. OSHA Hazard Communication Standard 29 CFR 1910.1200. Available from the Occupational Safety and Health Administration at http://www.osha.org or contact your local OSHA office.

The following publications are published by The American Welding Society, P.O. Box 351040, Miami, Florida 33135. AWS publications may be purchased from the American Welding society at http://www.aws.org or by contacting the AWS at 800-854-7149.

ANSI, Standard Z49.1, *Safety in Welding, Cutting and Allied Processes*. Z49.1 is now available for download at no charge at http://www.lincolnelectric.com/community/safety/ or at the AWS website http://www.aws.org.

AWS F1.1, *Method for Sampling Airborne Particulates Generated by Welding and Allied Processes.*

AWS F1.2, *Laboratory Method for Measuring Fume Generation Rates and Total Fume Emission of Welding and Allied Processes*.

AWS F1.3, *Evaluating Contaminants in the Welding Environment: A Strategic Sampling Guide*.

AWS F1.5, *Methods for Sampling and Analyzing Gases from Welding and Allied Processes*.

AWS F3.2, Ventilation Guide for Welding Fume Control

AWS F4.1, *Recommended Safe Practices for the Preparation for Welding and Cutting of Containers and Piping That Have Held Hazardous Substances*.

AWS SHF, *Safety and Health Facts Sheets*. Available free of charge from the AWS website at http://www.aws.org.

LISTED BELOW ARE SOME TYPICAL INGREDIENTS IN WELDING ELECTRODES AND THEIR TLV (ACGIH) GUIDELINES AND PEL (OSHA) EXPOSURE LIMITS

INGREDIENTS	CAS No.	TLV mg/m³	PEL mg/m³
Aluminum and/or aluminum alloys (as Al)*****	7429-90-5	10	15
Aluminum oxide and/or Bauxite*****	1344-28-1	10	5**
Barium compounds (as Ba)*****	513-77-9	****	****
Chromium and chromium alloys or compounds (as Cr)*****	7440-47-3	0.5(b)	.005(b)
Fluorides (as F)	7789-75-5	2.5	2.5
Iron	7439-89-6	10*	10*
Limestone and/or calcium carbonate	1317-65-3	10	15
Lithium compounds (as Li)	554-13-2	10*	10*
Magnesite	1309-48-4	10	15
Magnesium and/or magnesium alloys and compounds (as Mg)	7439-95-4	10*	10*
Manganese and/or manganese alloys and compounds (as Mn)*****	7439-96-5	0.2	5.0(c)
Mineral silicates	1332-58-7	5**	5**
Molybdenum alloys (as Mo)	7439-98-7	10	10
Nickel*****	7440-02-0	1.5	1
Silicates and other binders	1344-09-8	10*	10*
Silicon and/or silicon alloys and compounds (as Si)	7440-21-3	10*	10*
Strontium compounds (as Sr)	1633-05-2	10*	10*
Zirconium alloys and compounds (as Zr)	12004-83-0	5	5

Supplemental Information:

(*) Not listed. Nuisance value maximum is 10 milligrams per cubic meter. PEL value for iron oxide is 10 milligrams per cubic meter. TLV value for iron oxide is 5 milligrams per cubic meter.

(**) As respirable dust.

(*****) Subject to the reporting requirements of Sections 311, 312, and 313 of the Emergency Planning and Community Right-to-Know Act of 1986 and of 40CFR 370 and 372.

(b) The PEL for chromium (VI) is .005 milligrams per cubic meter as an 8 hour time weighted average. The TLV for water-soluble chromium (VI) is 0.05 milligrams per cubic meter. The TLV for insoluble chromium (VI) is 0.01 milligrams per cubic meter.

(c) Values are for manganese fume. STEL (Short Term Exposure Limit) is 3.0 milligrams per cubic meter. PEL of 1.0 milligrams per cubic meter proposed by OSHA in 1989. Present PEL is 5.0 milligrams per cubic meter (ceiling value).

(****) There is no listed value for insoluble barium compounds. The TLV for soluble barium compounds is 0.5 mg/m³.

TLV and PEL values are as of April 2006. Always check Material Safety Data Sheet (MSDS) with product or on the Lincoln Electric website at http://www.lincolnelectric.com

WARNING

CALIFORNIA PROPOSITION 65 WARNINGS

Diesel engine exhaust and some of its constituents are known to the State of California to cause cancer, birth defects, and other reproductive harm.

The Above For Diesel Engines

The engine exhaust from this product contains chemicals known to the State of California to cause cancer, birth defects, or other reproductive harm.

The Above For Gasoline Engines

ARC WELDING CAN BE HAZARDOUS. PROTECT YOURSELF AND OTHERS FROM POSSIBLE SERIOUS INJURY OR DEATH. KEEP CHILDREN AWAY. PACEMAKER WEARERS SHOULD CONSULT WITH THEIR DOCTOR BEFORE OPERATING.

Read and understand the following safety highlights. For additional safety information, it is strongly recommended that you purchase a copy of "Safety in Welding & Cutting - ANSI Standard Z49.1" from the American Welding Society, P.O. Box 351040, Miami, Florida 33135 or CSA Standard W117.2-1974. A Free copy of "Arc Welding Safety" booklet E205 is available from the Lincoln Electric Company, 22801 St. Clair Avenue, Cleveland, Ohio 44117-1199.

BE SURE THAT ALL INSTALLATION, OPERATION, MAINTENANCE AND REPAIR PROCEDURES ARE PERFORMED ONLY BY QUALIFIED INDIVIDUALS.

FOR ENGINE powered equipment.

1.a. Turn the engine off before troubleshooting and maintenance work unless the maintenance work requires it to be running.

1.b. Operate engines in open, well-ventilated areas or vent the engine exhaust fumes outdoors.

1.c. Do not add the fuel near an open flame welding arc or when the engine is running. Stop the engine and allow it to cool before refueling to prevent spilled fuel from vaporizing on contact with hot engine parts and igniting. Do not spill fuel when filling tank. If fuel is spilled, wipe it up and do not start engine until fumes have been eliminated.

1.d. Keep all equipment safety guards, covers and devices in position and in good repair.Keep hands, hair, clothing and tools away from V-belts, gears, fans and all other moving parts when starting, operating or repairing equipment.

1.e. In some cases it may be necessary to remove safety guards to perform required maintenance. Remove guards only when necessary and replace them when the maintenance requiring their removal is complete. Always use the greatest care when working near moving parts.

1.f. Do not put your hands near the engine fan. Do not attempt to override the governor or idler by pushing on the throttle control rods while the engine is running.

1.g. To prevent accidentally starting gasoline engines while turning the engine or welding generator during maintenance work, disconnect the spark plug wires, distributor cap or magneto wire as appropriate.

1.h. To avoid scalding, do not remove the radiator pressure cap when the engine is hot.

ELECTRIC AND MAGNETIC FIELDS may be dangerous

2.a. Electric current flowing through any conductor causes localized Electric and Magnetic Fields (EMF). Welding current creates EMF fields around welding cables and welding machines

2.b. EMF fields may interfere with some pacemakers, and welders having a pacemaker should consult their physician before welding.

2.c. Exposure to EMF fields in welding may have other health effects which are now not known.

2.d. All welders should use the following procedures in order to minimize exposure to EMF fields from the welding circuit:

2.d.1. Route the electrode and work cables together - Secure them with tape when possible.

2.d.2. Never coil the electrode lead around your body.

2.d.3. Do not place your body between the electrode and work cables. If the electrode cable is on your right side, the work cable should also be on your right side.

2.d.4. Connect the work cable to the workpiece as close as possible to the area being welded.

2.d.5. Do not work next to welding power source.

ELECTRIC SHOCK can kill.

3.a. The electrode and work (or ground) circuits are electrically "hot" when the welder is on. Do not touch these "hot" parts with your bare skin or wet clothing. Wear dry, hole-free gloves to insulate hands.

3.b. Insulate yourself from work and ground using dry insulation. Make certain the insulation is large enough to cover your full area of physical contact with work and ground.

In addition to the normal safety precautions, if welding must be performed under electrically hazardous conditions (in damp locations or while wearing wet clothing; on metal structures such as floors, gratings or scaffolds; when in cramped positions such as sitting, kneeling or lying, if there is a high risk of unavoidable or accidental contact with the workpiece or ground) use the following equipment:

- **Semiautomatic DC Constant Voltage (Wire) Welder.**
- **DC Manual (Stick) Welder.**
- **AC Welder with Reduced Voltage Control.**

3.c. In semiautomatic or automatic wire welding, the electrode, electrode reel, welding head, nozzle or semiautomatic welding gun are also electrically "hot".

3.d. Always be sure the work cable makes a good electrical connection with the metal being welded. The connection should be as close as possible to the area being welded.

3.e. Ground the work or metal to be welded to a good electrical (earth) ground.

3.f. Maintain the electrode holder, work clamp, welding cable and welding machine in good, safe operating condition. Replace damaged insulation.

3.g. Never dip the electrode in water for cooling.

3.h. Never simultaneously touch electrically "hot" parts of electrode holders connected to two welders because voltage between the two can be the total of the open circuit voltage of both welders.

3.i. When working above floor level, use a safety belt to protect yourself from a fall should you get a shock.

3.j. Also see Items 6.c. and 8.

ARC RAYS can burn.

4.a. Use a shield with the proper filter and cover plates to protect your eyes from sparks and the rays of the arc when welding or observing open arc welding. Headshield and filter lens should conform to ANSI Z87. I standards.

4.b. Use suitable clothing made from durable flame-resistant material to protect your skin and that of your helpers from the arc rays.

4.c. Protect other nearby personnel with suitable, non-flammable screening and/or warn them not to watch the arc nor expose themselves to the arc rays or to hot spatter or metal.

FUMES AND GASES can be dangerous.

5.a. Welding may produce fumes and gases hazardous to health. Avoid breathing these fumes and gases.When welding, keep your head out of the fume. Use enough ventilation and/or exhaust at the arc to keep fumes and gases away from the breathing zone. **When welding with electrodes which require special ventilation such as stainless or hard facing (see instructions on container or MSDS) or on lead or cadmium plated steel and other metals or coatings which produce highly toxic fumes, keep exposure as low as possible and below Threshold Limit Values (TLV) using local exhaust or mechanical ventilation. In confined spaces or in some circumstances, outdoors, a respirator may be required. Additional precautions are also required when welding on galvanized steel.**

5.b. Do not weld in locations near chlorinated hydrocarbon vapors coming from degreasing, cleaning or spraying operations. The heat and rays of the arc can react with solvent vapors to form phosgene, a highly toxic gas, and other irritating products.

5.c. Shielding gases used for arc welding can displace air and cause injury or death. Always use enough ventilation, especially in confined areas, to insure breathing air is safe.

5.d. Read and understand the manufacturer's instructions for this equipment and the consumables to be used, including the material safety data sheet (MSDS) and follow your employer's safety practices. MSDS forms are available from your welding distributor or from the manufacturer.

5.e. Also see item 1.b.

WELDING SPARKS can cause fire or explosion.

6.a. Remove fire hazards from the welding area. If this is not possible, cover them to prevent the welding sparks from starting a fire. Remember that welding sparks and hot materials from welding can easily go through small cracks and openings to adjacent areas. Avoid welding near hydraulic lines. Have a fire extinguisher readily available.

6.b. Where compressed gases are to be used at the job site, special precautions should be used to prevent hazardous situations. Refer to “Safety in Welding and Cutting” (ANSI Standard Z49.1) and the operating information for the equipment being used.

6.c. When not welding, make certain no part of the electrode circuit is touching the work or ground. Accidental contact can cause overheating and create a fire hazard.

6.d. Do not heat, cut or weld tanks, drums or containers until the proper steps have been taken to insure that such procedures will not cause flammable or toxic vapors from substances inside. They can cause an explosion even though they have been “cleaned”. For information, purchase “Recommended Safe Practices for the Preparation for Welding and Cutting of Containers and Piping That Have Held Hazardous Substances”, AWS F4.1 from the American Welding Society (see address above).

6.e. Vent hollow castings or containers before heating, cutting or welding. They may explode.

6.f. Sparks and spatter are thrown from the welding arc. Wear oil free protective garments such as leather gloves, heavy shirt, cuffless trousers, high shoes and a cap over your hair. Wear ear plugs when welding out of position or in confined places. Always wear safety glasses with side shields when in a welding area.

6.g. Connect the work cable to the work as close to the welding area as practical. Work cables connected to the building framework or other locations away from the welding area increase the possibility of the welding current passing through lifting chains, crane cables or other alternate circuits. This can create fire hazards or overheat lifting chains or cables until they fail.

6.h. Also see item 1.c.

CYLINDER may explode if damaged.

7.a. Use only compressed gas cylinders containing the correct shielding gas for the process used and properly operating regulators designed for the gas and pressure used. All hoses, fittings, etc. should be suitable for the application and maintained in good condition.

7.b. Always keep cylinders in an upright position securely chained to an undercarriage or fixed support.

7.c. Cylinders should be located:

- Away from areas where they may be struck or subjected to physical damage.
- A safe distance from arc welding or cutting operations and any other source of heat, sparks, or flame.

7.d. Never allow the electrode, electrode holder or any other electrically “hot” parts to touch a cylinder.

7.e. Keep your head and face away from the cylinder valve outlet when opening the cylinder valve.

7.f. Valve protection caps should always be in place and hand tight except when the cylinder is in use or connected for use.

7.g. Read and follow the instructions on compressed gas cylinders, associated equipment, and CGA publication P-l, “Precautions for Safe Handling of Compressed Gases in Cylinders,” available from the Compressed Gas Association 1235 Jefferson Davis Highway, Arlington, VA 22202.

FOR ELECTRICALLY powered equipment.

8.a. Turn off input power using the disconnect switch at the fuse box before working on the equipment.

8.b. Install equipment in accordance with the U.S. National Electrical Code, all local codes and the manufacturer's recommendations.

8.c. Ground the equipment in accordance with the U.S. National Electrical Code and the manufacturer's recommendations.

PRÉCAUTIONS DE SÛRETÉ

Pour votre propre protection lire et observer toutes les instructions et les précautions de sûreté specifiques qui parraissent dans ce manuel aussi bien que les précautions de sûreté générales suivantes:

Sûreté Pour Soudage A L'Arc

1. Protegez-vous contre la secousse électrique:

   a. Les circuits à l'électrode et à la piéce sont sous tension quand la machine à souder est en marche. Eviter toujours tout contact entre les parties sous tension et la peau nue ou les vétements mouillés. Porter des gants secs et sans trous pour isoler les mains.
   b. Faire trés attention de bien s'isoler de la masse quand on soude dans des endroits humides, ou sur un plancher metallique ou des grilles metalliques, principalement dans les positions assis ou couché pour lesquelles une grande partie du corps peut être en contact avec la masse.
   c. Maintenir le porte-électrode, la pince de masse, le câble de soudage et la machine à souder en bon et sûr état defonctionnement.
   d. Ne jamais plonger le porte-électrode dans l'eau pour le refroidir.
   e. Ne jamais toucher simultanément les parties sous tension des porte-électrodes connectés à deux machines à souder parce que la tension entre les deux pinces peut être le total de la tension à vide des deux machines.
   f. Si on utilise la machine à souder comme une source de courant pour soudage semi-automatique, ces precautions pour le porte-électrode s'applicuent aussi au pistolet de soudage.

2. Dans le cas de travail au dessus du niveau du sol, se protéger contre les chutes dans le cas ou on recoit un choc. Ne jamais enrouler le câble-électrode autour de n'importe quelle partie du corps.

3. Un coup d'arc peut être plus sévère qu'un coup de soliel, donc:

   a. Utiliser un bon masque avec un verre filtrant approprié ainsi qu'un verre blanc afin de se protéger les yeux du rayonnement de l'arc et des projections quand on soude ou quand on regarde l'arc.
   b. Porter des vêtements convenables afin de protéger la peau de soudeur et des aides contre le rayonnement de l'arc.
   c. Protéger l'autre personnel travaillant à proximité au soudage à l'aide d'écrans appropriés et non-inflammables.

4. Des gouttes de laitier en fusion sont émises de l'arc de soudage. Se protéger avec des vêtements de protection libres de l'huile, tels que les gants en cuir, chemise épaisse, pantalons sans revers, et chaussures montantes.

5. Toujours porter des lunettes de sécurité dans la zone de soudage. Utiliser des lunettes avec écrans lateraux dans les zones où l'on pique le laitier.

6. Eloigner les matériaux inflammables ou les recouvrir afin de prévenir tout risque d'incendie dû aux étincelles.

7. Quand on ne soude pas, poser la pince à une endroit isolé de la masse. Un court-circuit accidental peut provoquer un échauffement et un risque d'incendie.

8. S'assurer que la masse est connectée le plus prés possible de la zone de travail qu'il est pratique de le faire. Si on place la masse sur la charpente de la construction ou d'autres endroits éloignés de la zone de travail, on augmente le risque de voir passer le courant de soudage par les chaines de levage, câbles de grue, ou autres circuits. Cela peut provoquer des risques d'incendie ou d'echauffement des chaines et des câbles jusqu'à ce qu'ils se rompent.

9. Assurer une ventilation suffisante dans la zone de soudage. Ceci est particuliérement important pour le soudage de tôles galvanisées plombées, ou cadmiées ou tout autre métal qui produit des fumeés toxiques.

10. Ne pas souder en présence de vapeurs de chlore provenant d'opérations de dégraissage, nettoyage ou pistolage. La chaleur ou les rayons de l'arc peuvent réagir avec les vapeurs du solvant pour produire du phosgéne (gas fortement toxique) ou autres produits irritants.

11. Pour obtenir de plus amples renseignements sur la sûreté, voir le code "Code for safety in welding and cutting" CSA Standard W 117.2-1974.

PRÉCAUTIONS DE SÛRETÉ POUR LES MACHINES À SOUDER À TRANSFORMATEUR ET À REDRESSEUR

1. Relier à la terre le chassis du poste conformement au code de l'électricité et aux recommendations du fabricant. Le dispositif de montage ou la piece à souder doit être branché à une bonne mise à la terre.

2. Autant que possible, l'installation et l'entretien du poste seront effectués par un électricien qualifié.

3. Avant de faires des travaux à l'interieur de poste, la debrancher à l'interrupteur à la boite de fusibles.

4. Garder tous les couvercles et dispositifs de sûreté à leur place.

WARNING LABEL/OPERATING MANUAL REQUEST FORM

NOTE: S18494 WARNING LABELS, SUCH AS THE ONE BELOW FOR LINCOLN ELECTRIC WELDERS, ARE AVAILABLE FREE OF CHARGE to update your welding equipment. Operating manuals are also available upon request. PLEASE write to The Lincoln Electric Company, 22801 St. Clair Ave., Cleveland, Ohio 44117-1199 or visit www.lincolnelectric.com and make the request online.

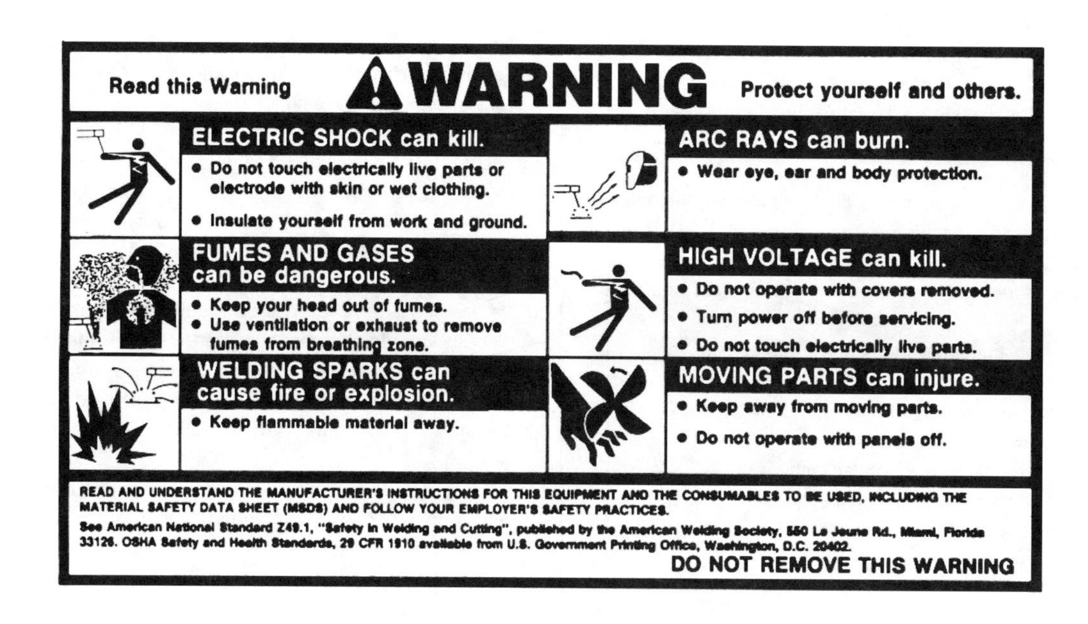

ENGINE WELDER FAN GUARDS

In order to determine whether your engine welder has the proper fan guards, compare your welder with the photo below. If your welder lacks the guards shown, contact your nearest Lincoln Field Service Shop or Distributor for assistance.

NOTE: On some engine welders, including the SA-200, the original fan guard design shown below has been modified to provide added protection and/or to make it more likely to be replaced after maintenance. Check with a Lincoln Field Service Shop or Distributor to determine if updated guarding is available for your welder.

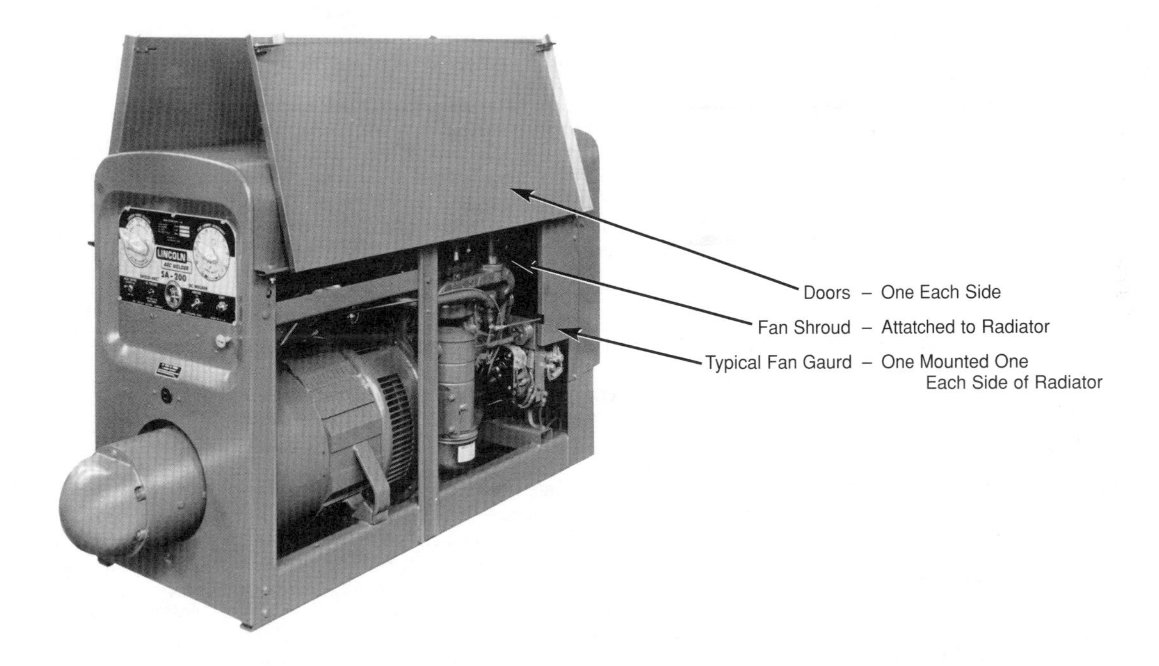

WELDING SAFETY CHECKLIST

HAZARD	FACTORS TO CONSIDER	PRECAUTION SUMMARY
Electric shock can kill	**· Wetness** **· Welder in or on workpiece** **· Confined space** **· Electrode holder and cable insulation**	• Insulate welder from workpiece and ground using *dry* insulation. Rubber mat or dry wood. • Wear *dry, hole-free* gloves. (Change as necessary to keep dry.) • Do not touch electrically "hot" parts or electrode with bare skin or wet clothing. • If wet area and welder cannot be insulated from workpiece with dry insulation, use a semiautomatic, constant-voltage welder or stick welder with voltage reducing device. • Keep electrode holder and cable insulation in good condition. Do not use if insulation damaged or missing.
Fumes and gases can be dangerous	**· Confined area** **· Positioning of welder's head** **· Lack of general ventilation** **· Electrode types, i.e., manganese, chromium, etc. See MSDS** **· Base metal coatings, galvanize, paint**	• Use ventilation or exhaust to keep air breathing zone clear, comfortable. • Use helmet and positioning of head to minimize fume in breathing zone. • Read warnings on electrode container and material safety data sheet (MSDS) for electrode, • Provide additional ventilation/exhaust where special ventilation requirements exist. • Use special care when welding in a confined area. • Do not weld unless ventilation is adequate.
Welding sparks can cause fire or explosion	**· Containers which have held combustibles** **· Flammable materials**	• Do not weld on containers which have held combustible materials (unless strict AWS F4.1 procedures are followed). Check before welding. • Remove flammable materials from welding area or shield from sparks, heat. • Keep a fire watch in area during and after welding. • Keep a fire extinguisher in the welding area. • Wear fire retardant clothing and hat. Use earplugs when welding overhead.
Arc rays can burn eyes and skin	**· Process: gas-shielded arc most severe**	• Select a filter lens which is comfortable for you while welding. • Always use helmet when welding. • Provide non-flammable shielding to protect others. • Wear clothing which protects skin while welding.
Confined space	**· Metal enclosure** **· Wetness** **· Restricted entry** **· Heavier than air gas** **· Welder inside or on workpiece**	• Carefully evaluate adequacy of ventilation especially where electrode requires special ventilation or where gas may displace breathing air. • If basic electric shock precautions cannot be followed to insulate welder from work and electrode, use semiautomatic, constant-voltage equipment with cold electrode or stick welder with voltage reducing device. • Provide welder helper and method of welder retrieval from outside enclosure.
General work area hazards	**· Cluttered area**	• Keep cables, materials, tools neatly organized.
	· Indirect work (welding ground) connection	• Connect work cable as close as possible to area where welding is being performed. Do *not* allow alternate circuits through scaffold cables, hoist chains, ground leads.
	· Electrical equipment	• Use only double insulated or properly grounded equipment. • Always disconnect power to equipment before servicing.
	· Engine-driven equipment	• Use in only open, well ventilated areas. • Keep enclosure complete and guards in place. • See Lincoln service shop if guards are missing. • Refuel with engine off. • If using auxiliary power, OSHA may require GFI protection or assured grounding program (or isolated windings if less than 5KW).
	· Gas cylinders	• Never touch cylinder with the electrode. • Never lift a machine with cylinder attached. • Keep cylinder upright and chained to support.

NOTES

Publication E205 | Issue Date 06/14

THE LINCOLN ELECTRIC COMPANY
22801 Saint Clair Avenue • Cleveland, OH • 44117 • U.S.A.
Phone: +1 216.481.8100 • www.lincolnelectric.com

ARC WELDED PROJECTS

Volume IV

The JAMES F. LINCOLN ARC WELDING FOUNDATION

Foreword

A number of hazards are associated with arc welding that can cause death or serious injury. However, these hazards can be avoided by use of the safe welding practices described in the first section of this document.

The serviceability of a product or structure utilizing the type of information contained in this document is and must be the sole responsibility of the builder/user. Many variables beyond the control of The James F. Lincoln Arc Welding Foundation affect the results obtained in applying this type of information. These variables may include, but are not limited to, welding procedure, plate chemistry and temperature, weldment design, fabrication methods and service requirements.

The JAMES F. LINCOLN ARC WELDING FOUNDATION

ARC WELDED PROJECTS, Volume IV
Published by
James F. Lincoln Arc Welding Foundation
P.O. Box 17188
Cleveland, Ohio 44117-1199
www.jflf.org

ISBN 0-937390-01-1

Important—Read Before Using These Plans

The plans, suggestions and instructions for making the projects included in this book were prepared by students as entries in The James F. Lincoln Arc Welding Foundation Award Programs. The material has been edited and some of the drawings improved, but neither the projects nor drawings and instructions have been reviewed for accuracy or safety. The Foundation will welcome the comments of persons who, in making a project, uncover an error or fault. This will permit correction in the next printing of the book.

The projects are included because they appear to be interesting, and, in some respects, propose novel applications. However, since the Foundation has not tested the material, nor verified the computations or other aspects described, The James F. Lincoln Arc Welding Foundation cannot, and does not, assume responsibility for the accuracy of the plans or safety of the projects. The projects are submitted for such use as may appear feasible, but those making the projects must assume full responsibility for the results of their efforts to make or use the projects described.

The name of the student, school and teacher that produced the material is shown, if available, for each project. Further information about a project, if needed, should be sought from the teacher involved.

Award Programs Participation

The projects described in this book represent the range of entries submitted in The James F. Lincoln Arc Welding Foundation's Award Programs both as to type and size of project, and the nature of the descriptive information included. Students make many similar projects each year, yet relatively few of these are submitted to the Foundation in competition for cash awards. The reason for this may be that teachers have not used the opportunities afforded by the Program as incentives to student endeavor.

The planning, design and execution of a project, and then the preparation of a written and illustrated description of how it was accomplished provide a student with: (1) involvement with something of his or her own choosing and desire; (2) a challenge to complete the work successfully, using many different shop skills that are applicable to industry and business; (3) an experience in report writing and communications that is as important in business as other skills; (4) the satisfaction and recognition that come from participating in a national competition.

The Foundation annually sponsors Award Programs for secondary and post-secondary students. To learn more about current Award Programs and obtain a copy of the appropriate rules brochure, go to *jflf.org*.

ARC WELDED PROJECTS

Table of Contents

Agriculture

Table of Contents

Hydraulic Trench Digger

Author: Justin Brown
Instructor: James L. King
School: Burlington High
City & State: Burlington, IA

Bill of Materials

Tubing

Quantity	*Description*	*Length*
1	4″x 4″x 1/4″	62-1/2″
1	3″x 3″ x 1/4″	37″
2	2″x 2″ x1/4″	24″
1	3-1/2″ x 3-1/2″ x 3/16″	10-1/2″
1	3-1/2″ x 3-1/2″ x 3/16″	31″
2	3-1/2″ x 3-1/2″ x 3/16″	8″
1	3″x 3″ x 3/16″	67″
2	3″x 3″ x 3/16″	36″
2	3″x 3″ x 3/16″	19″
1	2-1/2″ x 2-1/2″ x 3/16″	14″

Plate A36

Quantity	*Thickness*	*Size*
2	1/2″	12″ x 12″
2	1/2″	8″ x 8″
2	1/2″	8″ x 10″
2	1/2″	60″ x 96″
4	1/2″	2″ x 9″
2	1/2″	2″ x 11″
2	1/2″	3″ x 14″
2	1/2″	14″ x 9-1/2″
2	1/2″	4-1/2″ x 10″
2	1/2″	4-1/2″ x 31-1/2″
2	3/8″	2″ x 5-1/2″
2	5/16″	12″ x 14″
2	5/16″	16″ x 10″
2	1/4″	24″ x 14″
2	1/4″	3-1/2″ x 3-1/2″
2	1/4″	22-1/2″ x 14″
1	1/4″	13″ x 30″
1	1/4″	3″ x 5-1/2″
2	1/4″	3″ x 2″

Mechanic Tubing

Quantity	*Size*	*Length*
4	1-1/2″ ID x 2″ OD	2″
4	1″ ID x 1-1/2″ OD	3/4″
4	1″ ID x 2″ OD	1″

Miscellaneous

1 pc. 2-1/2″ x 2-1/2″ x 1/4″ Angle Iron
46″ Cutter Blade
4 pcs. Grease Fitting
2″ Sch 40 Pipe, 13″ long
4 Cylinders
1 Hydraulic Pump
1 Hydraulic Valve
1 Hydraulic Strainer
1 Gas Engine, 5.5 hp or larger
1 Pump Mounting Bracket
Misc. Hydraulic Hose (1/2″ 2 wire)

Procedure

1. Cut all materials to specific size.

2. Lay a piece of 2.5" x 2.5" x 3/16" x 72" tubing on a flat, level floor. Then lay a piece of 3" x 3" x 3/16"x 67" on top of the base piece, with the bottom pivot plate between the two pieces. The pivot plate is made from 12" x 12" x 1/2" plate steel with a 1.5" hole in the center.

3. Attach a piece of 3.5" x 3.5" x 3/16" x 10.5" on top of the upper frame rail. At the rear of the unit, weld a piece of 2.5" x 2.5" angle iron to the top rail.

4. Weld a piece of 3.5" x 3.5" x 3/16" x 31" in front of the bottom frame rail. After that is welded in place, put a piece of 2.5" x 2.5" x 3/16" x 31" in front and weld in place.

5. Weld a piece of 2.5" x 2.5" x 3/16" x 14" on each side of the bottom frame rail, 11.5" from the front.

6. Take a piece of 3" x 3" x 1/4" x 37" and cut two holes in one end. The first will be centered and 1" in; the second will also be centered and 6" in.

7. Take a piece of 14" x 12" x 5/16" plate steel and drill holes in the plate as shown in the drawings. Then weld the two plates, one on each side of the boom extension.

8. Build the main boom out of a piece of 4" x 4" x 1/4" x 62.5" tubing. Cap off both ends with pieces of 16 gauge steel.

9. Cut a piece of 16" x 10" x 6/16" plate steel and drill two one-inch holes where shown on the plans, and weld into place.

10. Take a piece of 3" x 3" x 3/16" x 36" tubing and weld a 3.5" x 3.5" x 8" piece to one end, then weld the two plates together to make the feet.

11. Weld the hubs onto a piece of 2" x 2" x 1/4" x 24" tubing.

12. Finally, trial fit all pieces, make necessary adjustments, chip off all weld splatters, and paint the unit. Install all hydraulic lines and cyclinders to complete the project.

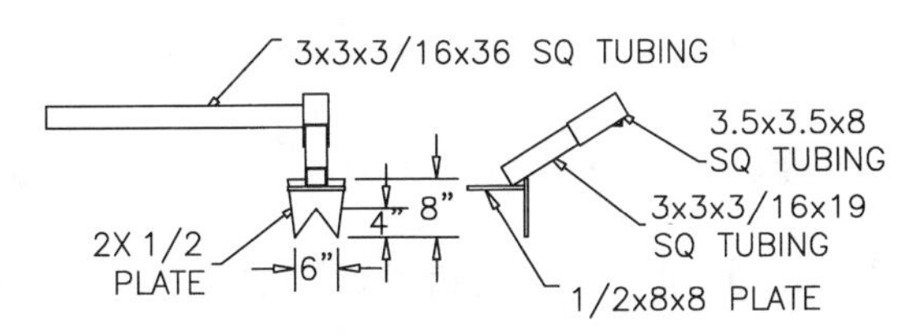

OUTRIGGER DETAIL

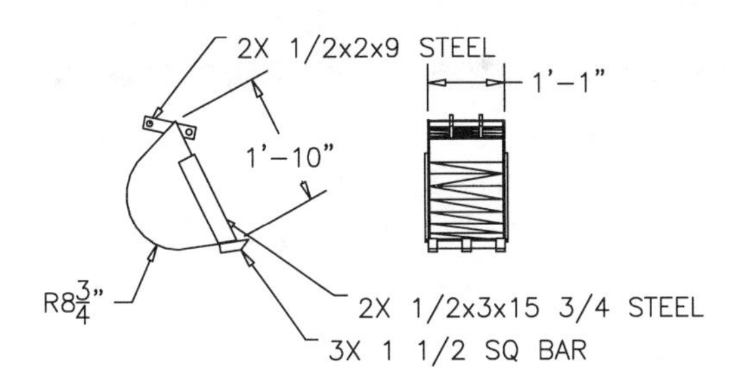

BUCKET DETAIL

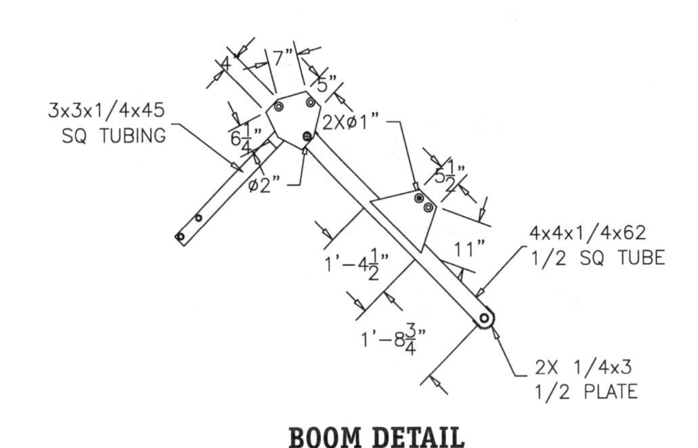

BOOM DETAIL

2.5x2.5x3/16x31 SQ TUBING
3.5x3.5x3/16x31 SQ TUBING
2.5x2.5x3/16x14 SQ TUBING
2'-7"
1'-3"
2'-6 1/2"
5'-7"
R2"
R2"
11"
5"
9"
2x2x1/8 SQ TUBING
SEAT & SEAT POST
1 1/2"X13 1/2" SQ TUBING
9"
1'-8"
OIL TANK
8"X9"X15"
2'-4"
1'-2"
R5 1/2"
11"
5 1/2"
2.5x2.5x1/4x28 ANGLE
6'
3x3x3/16x67 SQ TUBING
2.5x2.5x3/16x72 SQ TUBING

FRAME

10-Foot Dirt Scraper

Authors: Chris Augustine & Caleb Carlson
Instructor: James L. King
School: Burlington High School
City & State: Burlington, IA

Bill of Materials

1	4″ x 8″ x 5/16″ plate steel
10′	2″ x 6″ x 1/4″ tube
24′	2″ x 4″ x 3/16″ tube
20′	4″ x 4″ x 1/4″ tube
10′	4″ SCH 40 pipe
10′	3 1/2″ SCH 40 pipe
20′	3″ x 3″ x 1/4″ angle
10′	2″ x 5″ x 1/4″ tube
12′	1″ x 4″ bar
2	hubs/wheels
1	4′ x 4′ x 1″ plate steel
3	hydraulic cylinders
4′	1″ round stock
1	4′ x 4′ x 1/2″ plate steel
1	3′ x 10′ x 3/16″ plate steel
1	Jack/wheel
4	Gallon paint

Step by Step

Cut out two pieces of 2″ x 4″ square tubing 19″ long. Also cut four pieces of 4″ x 9-1/2″ by 1/4″ stock. Cut 1″ hole in all four pieces with slugger to connect pieces for the scarifier arm.

Cut out two 17″ pieces of 2″ x 4″ square tubing with a 1″ hole in one end to connect to scarifier lift arm and 3 1/2″ pipe to act as a sleeve on the other end.

Fabricate the scarifier bar lift, hydraulic cylinder lift bracket, and the scarifier bar lift sleeves.

Weld backing plates to lift arms and pipe supports.

Cut pipe 83″ long. Then cut 4″ x 6″ square tubing for tongue and lay out lines on lift bracket.

Finish welding pipe to backing plates on both sides.

Weld two pieces of 4″ x 4″ tubing to make one piece 120″ long.

Center 10′ x 6″ x 4″ piece of square tubing for tongue.

Weld sides on inside and weld plates to the tongue.

Weld back side of tongue and grind welds on the sides.

Cut 4″ x 4″ square tubing 95-3/4″ and 22-3/4″ and grind edges to be welded.

Drill holes in 4″ x 5″ and 4″ x 9″ plates, then cut two 2′ x 4′ square tubing 36″ long and cap both ends.

Grind welds of end caps on 2″ x 4″ x 36″ square tubing.

Cut gusset for tongue, then square up and weld to tongue.

Put scarifier bar on side brackets and put pins in lift arms and make placement lines.

Tack scarifier arms to scarifier bar, then tack and weld pipe to top of bar to hold lift arms.

Level up the scarifier bar and lift bottom bar to tongue and weld in gussets.

Make new hydraulic lift support, then cut one piece of square tubing 2″ x 4″ x 32″.

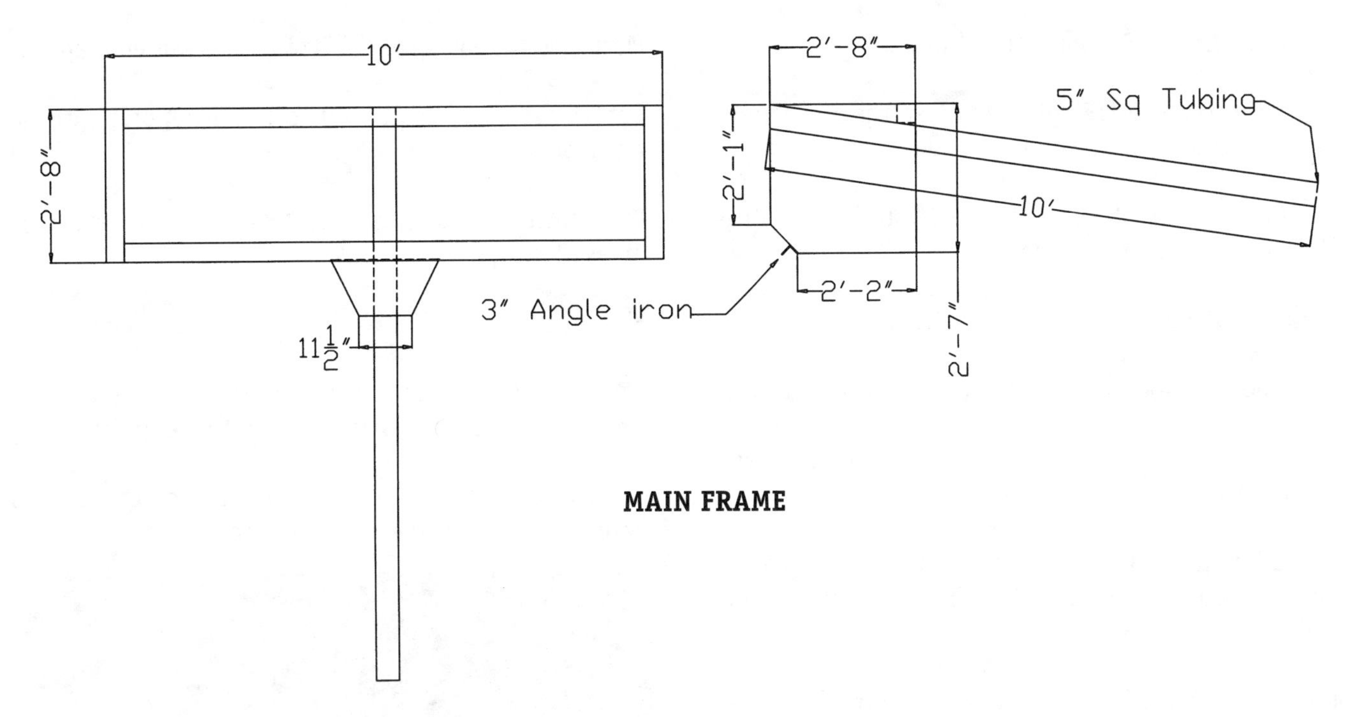

MAIN FRAME

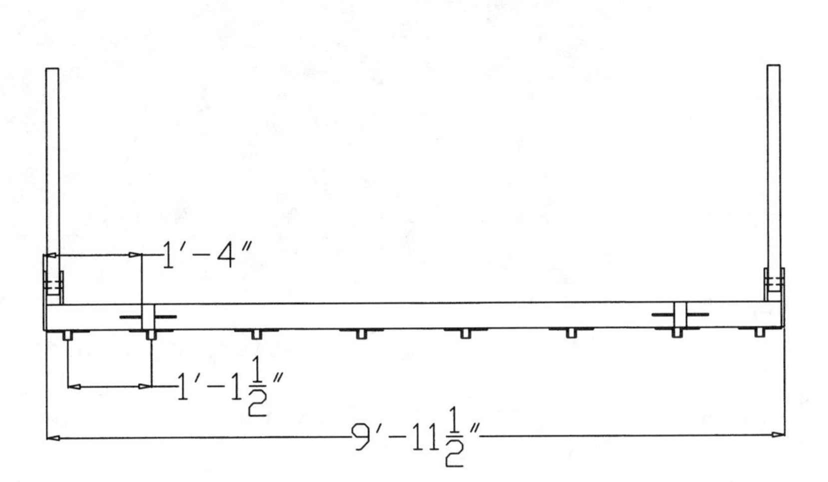

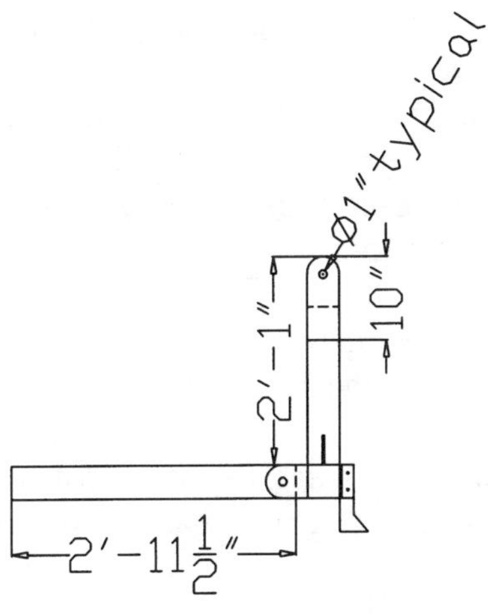

FRONT SCRAPER

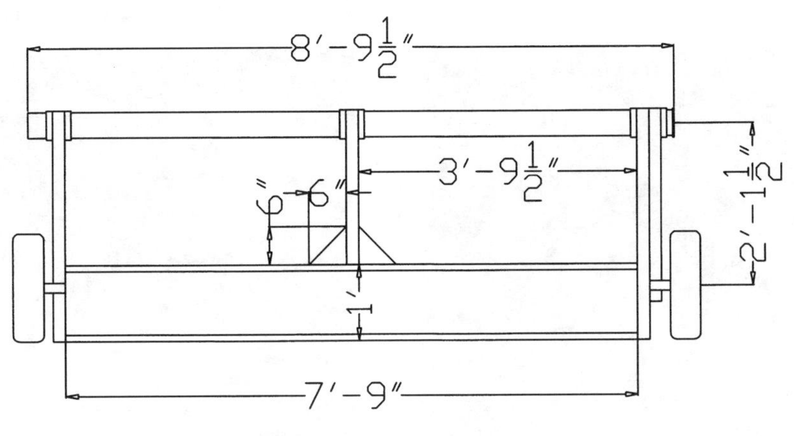

BACK AXLE

Cut two pieces of 5″ channel 32″ long.

Cut two pieces of 3/16″ plate to make the 5″ channel iron into 5″ x 2″ sq. tubing.

Cut out one piece of 3″ x 9-1/2″ x 1″ stock, then drill a line hole for hydraulic mount.

Drimmel out holes so the pins will fit tightly.

Make another pipe mount out of 2″ x 5″ square tubing.

Cut four pieces of 4-1/2″ tubing 3″ long with grease fittings. These are used as sleeves.

Plasma cut three 3-1/2″ holes in axle support plates, then weld end caps to 2″ x 5″ tubing and grind down welds.

Make a 3″ x 9-1/2″ x 1″ hydraulic support, then level up the rest of the axle supports.

Plasma cut holes in 2″ x 5″ tubing for wheel hubs.

Cut a 10′ piece of 3″ x 3″ x 1/4″ angle iron to go along back of scraper.

Cut 3-1/2″ pipe for axle, then square two pieces and weld them together.

Drimmel out holes for hubs, plasma cut off ends of pipe so they won't hit back plate.

Make end caps for 2″ x 5″ tubing, then weld them on.

Tack up axle, then level up hubs and tack them in place.

Weld hubs on all sides, then weld gussets on to support hubs, then make one 2″ x 5″ x 3″ square tubing out of 2″ x 5″ channel and a 38″ long plate 1/4″ thick and weld together.

Plasma cut 4″ hole in one side of tubing. Put end caps on.

Cut 16 pieces of angle iron to help support teeth on front of scraper.

Drill holes in axle supports, then tap them for 1/8″ grease fittings.

Plasma cut a 3-1/2″ hole in axle arm for sleeve.

Weld 2″ x 4″ tube that holds wheel hub to axle, then weld middle bar that is connected to weight box.

Tack up axle, then make an end cap for the axle to hold the floating wheel hub bar on.

Drill holes in end cap so that you can plug weld it on.

Cut weight box back down to size, then tack the bottom together.

Cut middle box support down to size, then weld box together.

Weld scarifier bar, and then weld scarifier support bracket to inside of end plates.

Finish welded back box and make more hydraulic brackets.

Cut eight teeth made of 1" stock for front of scraper.

Drill 1/4" pilot holes, then drill 1/2" holes through teeth and braces — two holes in each.

Cut blade down to size, then cut blade in half to get the bow out.

Lay out holes to bolt blade to edge of scraper.

Clamp up blade and drill holes with slugger.

Insert bolts to hold blade in place.

Bevel edge of teeth to get a good penetrating weld.

Tack weight box up in back, then level it up and weld it complete.

Finish welding everything on back.

Weld scarifier arms to top pipe and weld everything to make the hydraulic box scraper complete.

Finish Schedule
Chip off all slag and spatter with a chisel.

Grind off all welds with a Milwaukee 4-1/2" and 7-1/4" angle grinder.

Power sand all welds with Milwaukee 4-1/2" finishing sander.

Send project to get sand blasted and painted.

Ditch Digger

Author: Brandon Oetken
Instructor: James L. King
School: Burlington High School
City & State: Burlington, IA

Bill of Materials

Quantity	*Description*
2	3/16″ plates, 56″ φ
2	2-1/4″ OD x 1-7/16″ ID sleeves
4	(4) 3/4″ x 6″ x 8″
2	1/4″ x 11″ x 60″ plates, rolled
7″	2″ OD x 1-3/4″ ID sleeves
2	5/8″ x 10″ x 5″
4′	4″ x 6″ x 1/4″ rect. tube
10′	1/2″ x 4″ flat
12′	2″ x 4″ x 1/4″ rect. tube
24′	4″ x 4″ x 1/4″ square tube
100″	1-1/2″ x 1-1/2″ x 3/16″ square tube
2′	2″ x 2″ x 3/16″ square tube
2	1-1/2″ hitch pins
1	1″ pin
4	3/8″ pins
1	Silage blower (used)

Introduction

The ditcher's main frame consists of 4′ x 4′ x 1/2″ square tube that comes up and over the star wheel/blower. The star wheel has six blades and came from a Kools Brothers silage blower, along with the PTO shaft. The rest is all new construction. It will cut a ditch about 6″ deep and 30″ wide, much like the manufactured versions. The ditcher will cut the dirt away and throw it approximately 100′ to 150′ away from the ditch so there will be no berm left on either side of the ditch. This will allow water to drain from the field with minimal erosion and permit the field to dry faster. The ditcher can be used with a 120 horsepower tractor to increase drainage and field productivity on any farm.

Step by Step Construction Procedure

Cut (4) 4″ x 4″ x 1/4″ tube to 44″.

Cut (4) 3-3/4″ x 3-3/4″ x 1/4″ end caps for 4″ x 4″ tube.

Tack and MIG weld end caps on 4″ x 4″ tube, and then grind welds.

Cut (2) 4″ x 4″ x 1/4″ tube to 32″ each.

Tack and MIG weld the two 44″ 4″ x 4″ tubes together with the 32″ 4″ x 4″ into an H frame (make two of these).

Cut, tack, and MIG weld (1) 2″ x 4″ x 1/4″ tube to one of the H frames (this will be the back part of the frame).

Cut a 2″ x 2″ tube to 32″ to keep front frame spread apart while welding 3 pt. arms on.

Grind down all welds on both frames.

Have (4) 6″ x 8″ x 3/4″ plates cut and drill holes in each plate to 1-1/2″ and counter bore them.

Cut (2) 4″ x 6″ 1/4″ tube to 24″ each.

Tack and MIG weld (4) 6″ x 8″ x 3/4″ plates to 4″ x 6″ tube.

Have (2) 10″ x 5″ x 5/8″ cut and have slight bend put in them.

Cut (1) 2″ x 4″ x 1/4″ tube to 24″ and then tack and MIG weld the two 10″ x 5″ plates to the tube.

Cut (2) 2″ x 4″ x 1/4″ tube to 24″ and then tack and MIG weld the two 10″ x 5″ plates to the tube.

Line up and clamp 3 pt. arms on front frame and then tack and MIG weld together.

MIG weld 2 frames together using an 11″ 4″ x 4″ x 1/4″ tube on each side.

Cut (1) 2″ x 4″ x 1/4″ tube to 28″ and then tack and MIG weld to rear part of frame (for future accessory use).

Cut (4) 4″ x 8″ x 1/4″ plates and cap the top of the 2 frames and then grind all welds.

Have (2) 56″ ɸ 3/16″ plates cut and cut a keyway in each of them to get the starwheel out if necessary.

Drill 3/8″ holes in both plates for a 1/4″ cover for keyway and then MIG weld studs in the holes.

Drill 1/2″ holes in plates for bearing and welding in studs.

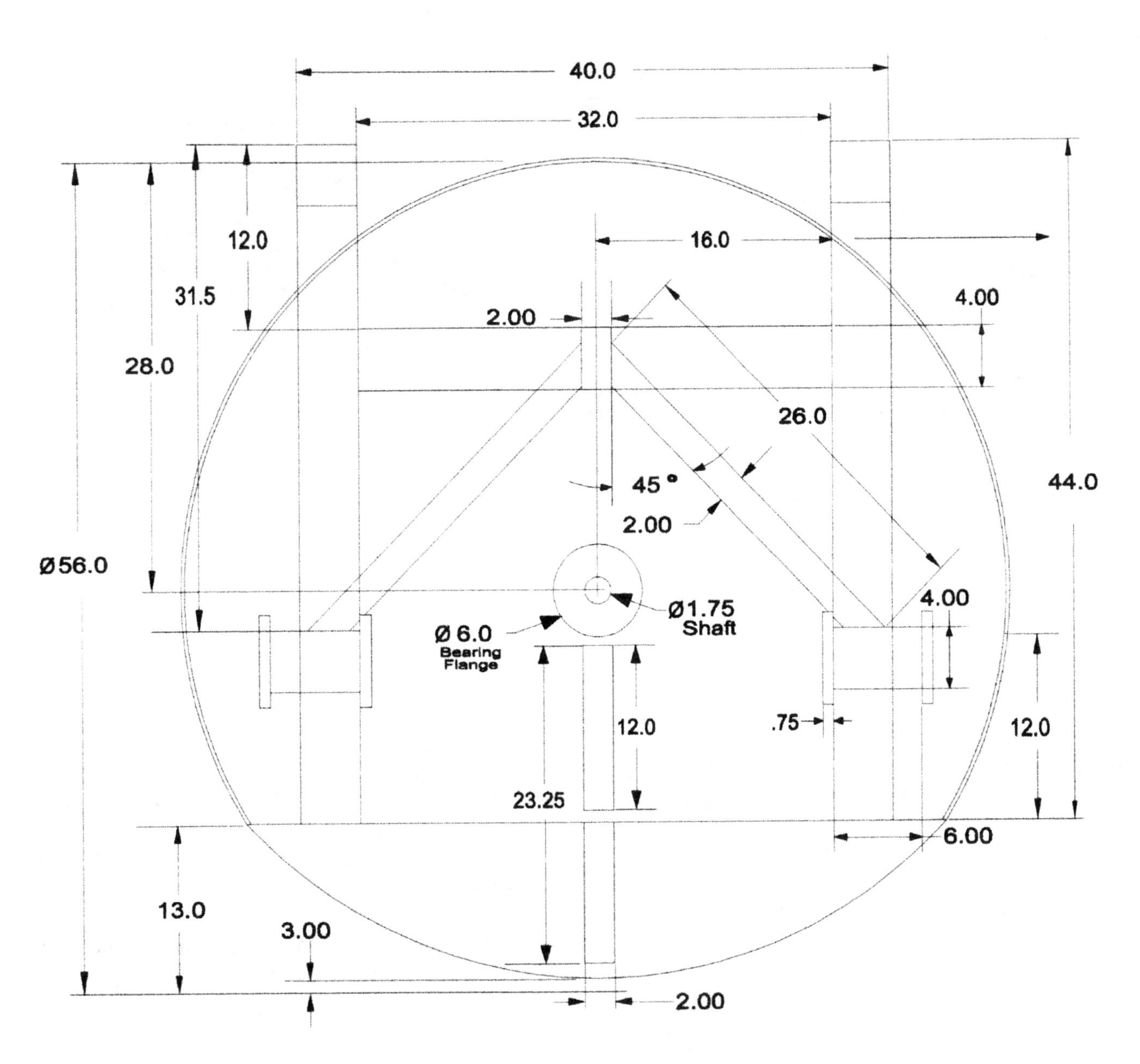

FRONT VIEW

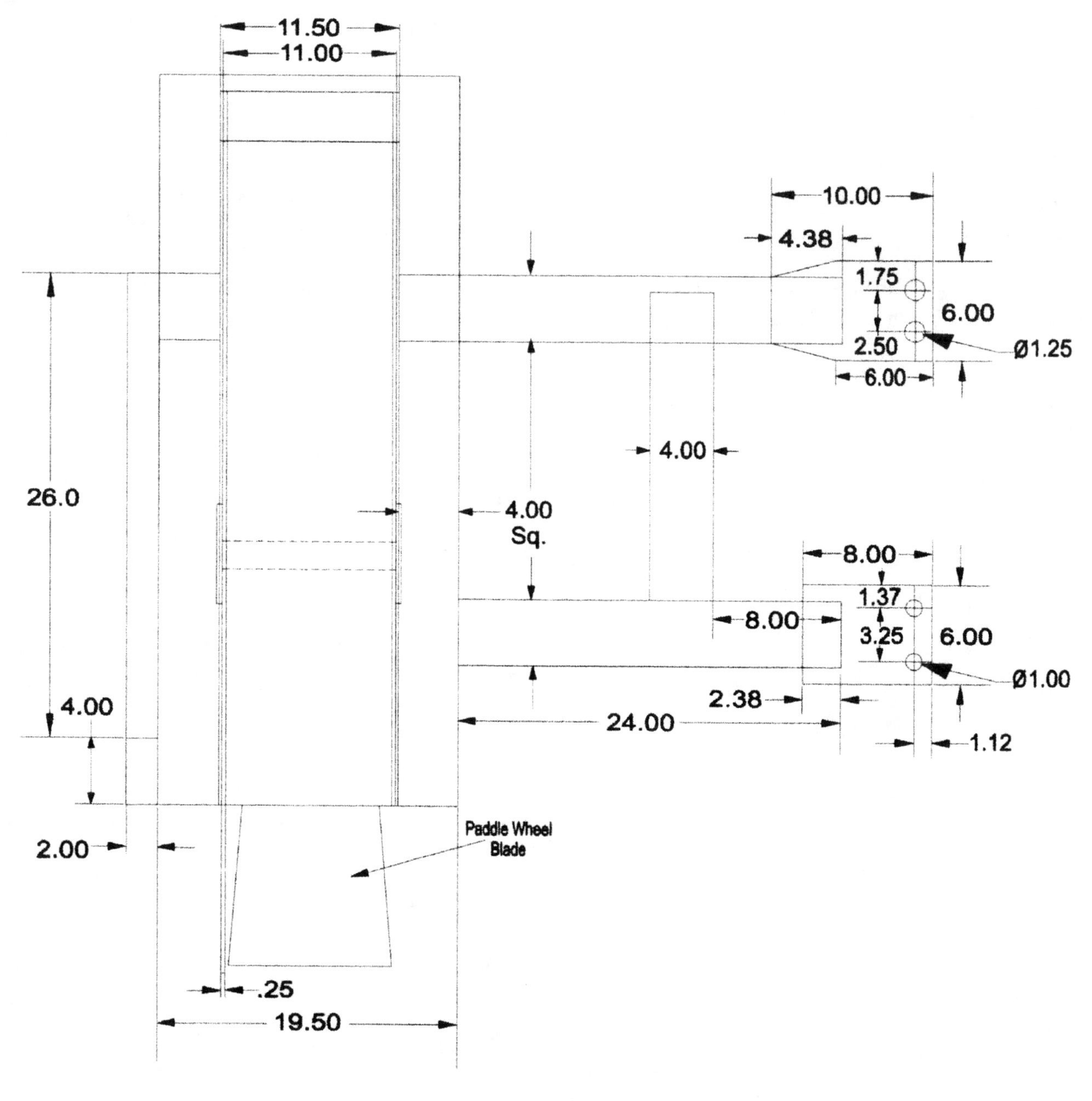

SIDE VIEW

Roll main frame backwards and then forwards to clamp, tack, and weld plates onto the main frame.

Tear a starwheel out of the silage blower.

Install bearings, starwheel, and cover plates on main frame.

Have (2) 11″ x 60″ x 1/4″ plates specially cut and rolled for housing.

Position, tack, and MIG weld rolled plates to the two plates.

Finish Schedule
Wire-wheel all welds to remove spatter.

Hammer and chisel all spatter that the wire wheel doesn't take off.

Smooth out welds with 4-1/2″ and 9″ hand angle grinders with hard grinding wheel and flap wheel.

Spray down with degreaser and power wash.

Prime unit.

Apply a couple of coats of paint.

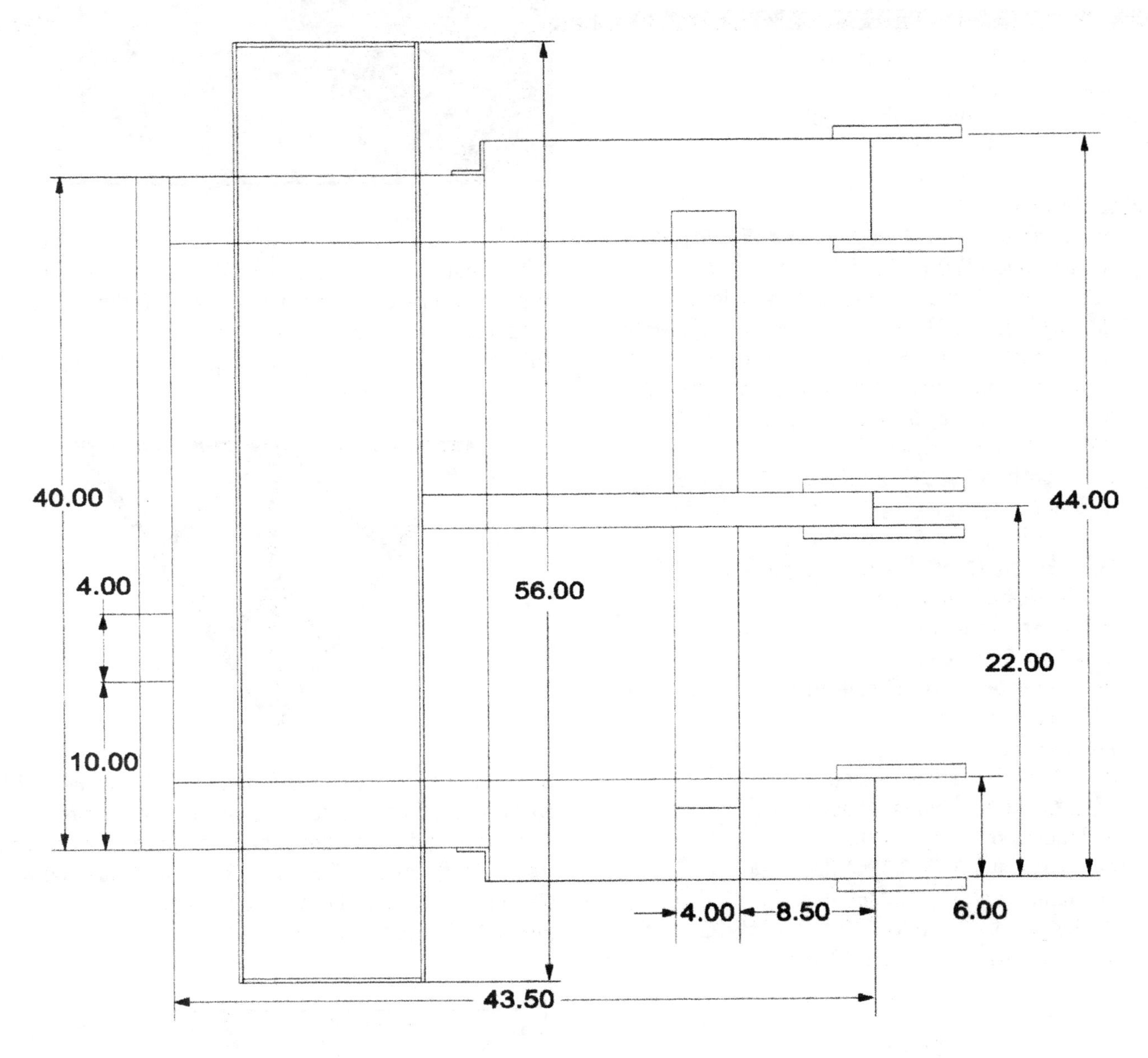

TOP VIEW

24-Foot Plow Roller

Author: Trevor Pitruzzello
Instructor: Lex Godfrey
School: Burley High School
City & State: Burley, ID

Materials List

2	40′ lengths of 4″ x 6″ x 0.25″ cut into (2) 20′ lengths and (1) 16′ length
1	40′ length of 4″ x 6″ x 0.375″ cut to 24′ length
1	20′ length of 6″ x 8″ x 0.375″ cut into (2) 42″ lengths
2	4′ x 8′ sheet of 0.275″ plate cut into (6) 8″ x 8″ x 17″ tire retainers, (2) 6″ x 8″ caps, (2) 5-1/2″ x 7-1/2″ caps, (2) 6″ x 6″ caps, and (1) 33″ x 25″ x 33′ by 6″ trapezoid
1	24″ length of 4″ x 4″ x 0.25″0.250 cut into (6) 4″ lengths
1	31 1/2″ length of 6″ x 6″ x 0.25″
1	36 1/2″ length of 4″ x 0.25″ pipe
1	pre-fabricated 24′ roller
2	0.75″ bearing plates
2	5″ x 7″ x 0.75″ caps
1	pre-fabricated 2 5/16″ ball hitch
4	3″ block bearings
31	24″ used tires
4	8″ x 8″ x 8″ triangle gussets
2	14′ x 14″ x 14″ triangle gussets
6	4″ lengths of 4″ x 4″ x 0.25″
2	40″ lengths of 3″ x 0.375″ flat
4	20″ lengths of 2″ x 0.275″ flat
1	20′ length of 6″ x 0.375″ flat
1	4 1/2′ length of 6″ x 0.375″ flat

Procedure

Frame

To make the frame, cut the lengths of steel as described in the Materials List. Make the two caps that will fit over the top and the two caps that will fit inside out of the 3/8″ plate and weld them on. Place the caps that fit inside the tube about an inch inside and weld them in.

Grind the welds off of the top caps. Next, weld the 6″ x 8″ pieces upright onto the ends of the caps of the 6″ x 4″ so that it is 40″ off of the ground. Following affixing the uprights, cut the 6″ x 4″ x 0.25″ tube into two 20′ lengths and one 16′ length. Angle the 20′ lengths to fit with the 16′ length and the 24′ length. At the front, leave a space for the hitch.

After that is all tacked, weld the frame solid. Next, flip the frame over with the aid of a forklift and an overhead hoist. Then cut the leftover 6″ x 4″ x 0.25″ tube into four cross pieces. Angle these pieces to fit into the frame like so:

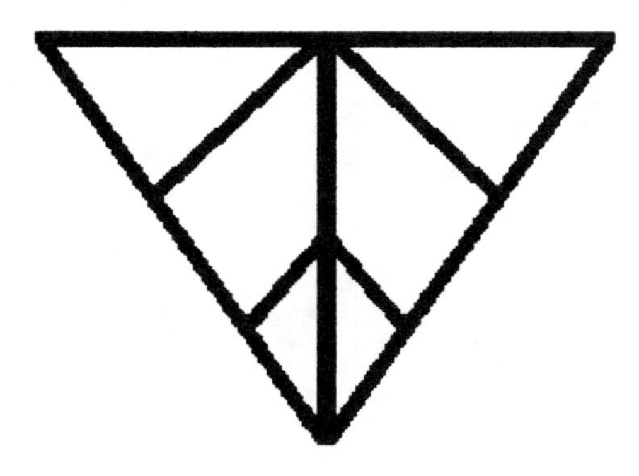

Use wide jaw vise grips and pieces of flat to hold the cross pieces in while tacking them. Then remove the vise grips and weld them solid on top and down the sides. Next, cut a 33″ x 35″ x 33″ x 6″ trapezoid out of the 3/8″ plate. Fit this up top to tie the cross pieces to the main frame.

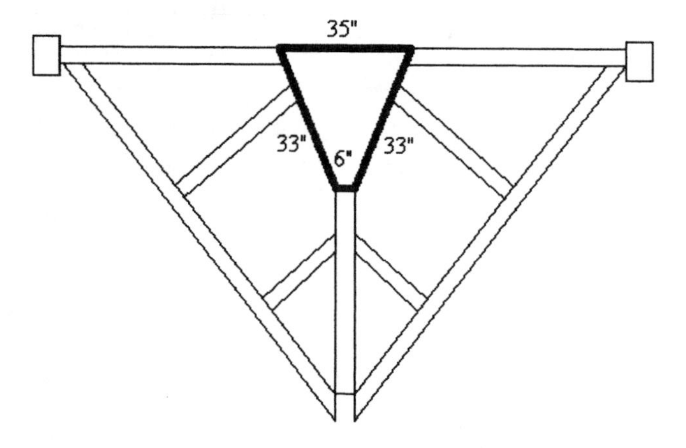

Clamp it down using C-clamps and then weld six 3″ welds all around it to keep it on.

Roller

First, make some plates for bearings to bolt to. Cut them with a computerized plasma cutter. After cutting the plates out, drill some 3/4″ holes on four inch centers in them so that you can bolt the bearings on. Here is a diagram of one bearing plate after it is drilled and welded onto the frame:

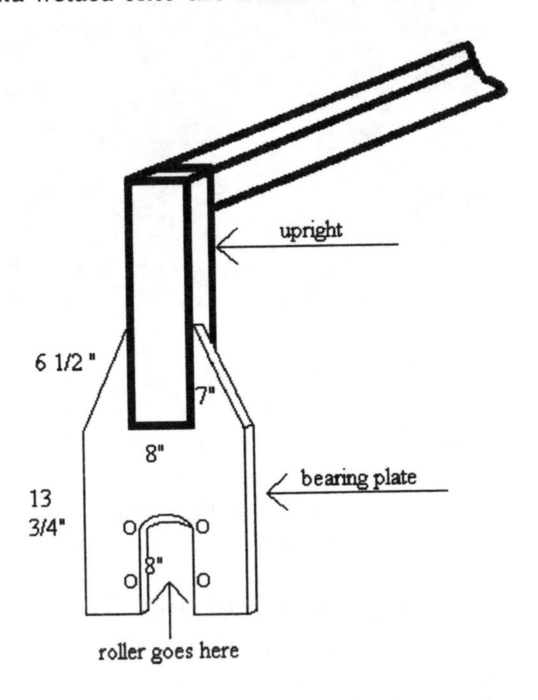

To keep the tires on the roller, cut six tire retainers out of the 3/8″ plate with the computerized plasma cutter. Drill 3/8″ holes in the retainers 4″ over and 4″ up from the corners.

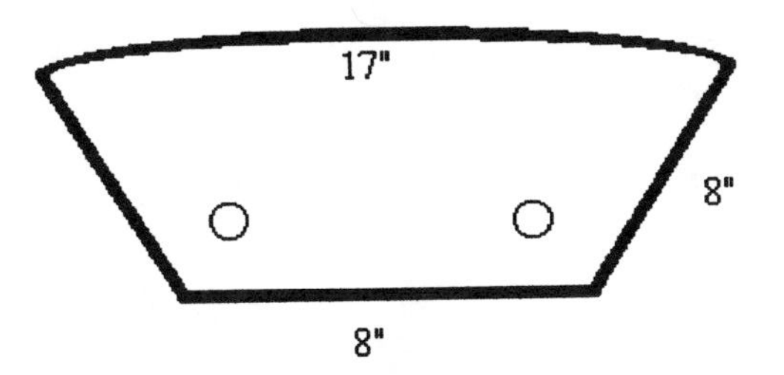

After the retainers are finished, affix them to the roller by first tacking each plate at an equal distance of 16-1/4″ from the next. Then use a magnetic drill to drill some 5/16″ holes into the roller. After breaking the tacks, the roller will look like this:

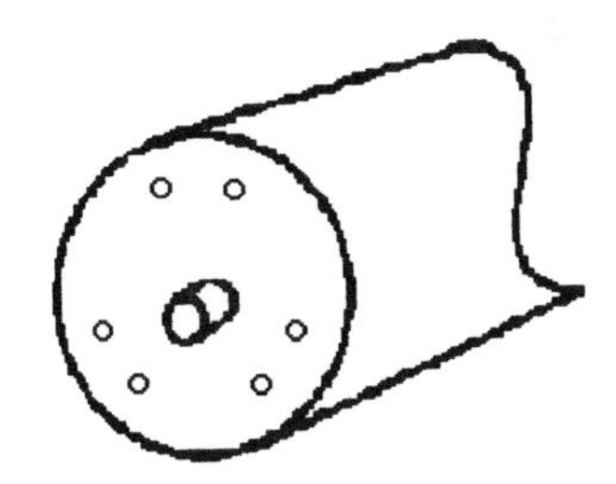

Next, thread the holes for 3/8″ bolts and bolt three plates on one side of the roller so you can push all the tires on the other without having them fall off. The next step is to put tires on the roller. This can be quite an undertaking since the outside measurement of the roller is 24″ and the inside measurement of the tires may be slightly smaller. You may have to shave about 1/2″ out of the inside of the tires with potato knives so that they will fit. After cutting the tires, slick up the roller with oil thoroughly to get them to fit. Next, use the overhead hoist and the forklift to set the frame onto the roller. Once the hoist and forklift are positioned correctly, put the four bearings on and bolt them tightly. Position the lock collars on the bearings and lock them tightly into place.

Hitch

Start the hitch by taking a 31-1/2″ length of 6″ x 6″ x 0.25″ and capping off both sides. Then cut holes big enough to accommodate the 36-1/2″ length of 4″ x 0.25″ pipe. Cut the iron left over from cutting the holes so that it fits inside the pipe to cap it off. Then weld the pipe in solid on both sides.

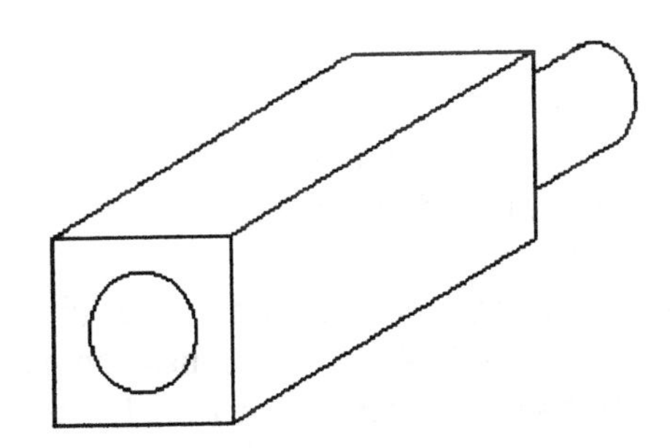

Next, set the ball hitch in the open end of the pipe and weld around it. After that, set the hitch in the open space at the front of the frame, make sure it is level and square, and tack it. After tacking the hitch in, weld the front and top of it to the frame. Here's how the roller looks with the hitch attached:

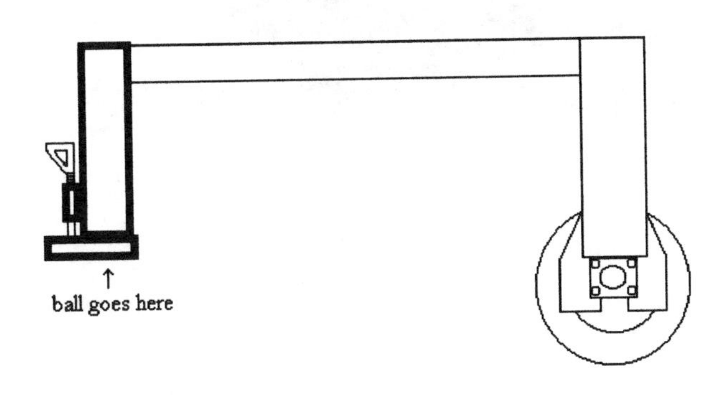

Reinforcement

Using a front-end loader, lift the frame up and flip it over on its back side. Then cut six 4″ pieces of 4″ x 4″ x 0.25″ tube and space them like so on the 24′ piece:

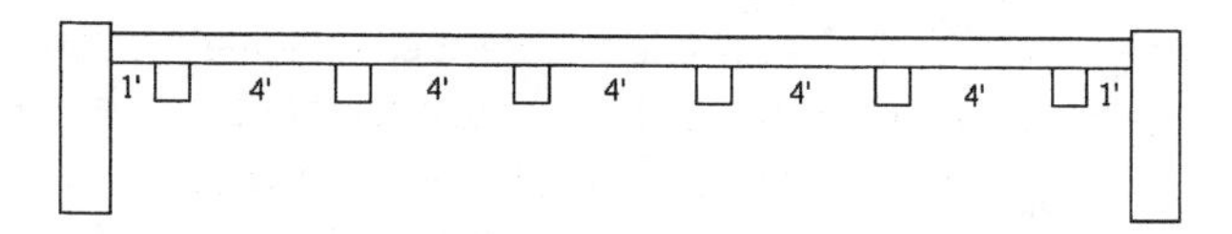

After that, cut an angle the same on the outermost two pieces. Next, take a 4-1/2″′ piece of 6″ x 0.75″ flat and weld it on like this:

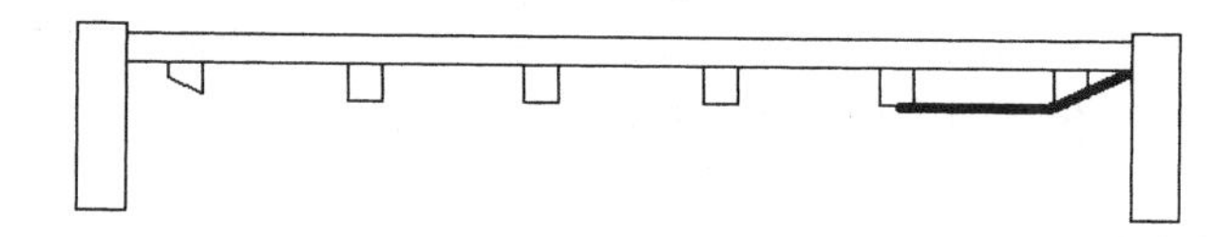

Then take the 20′ piece of 6″ x 0.75″ flat and weld it on like so:

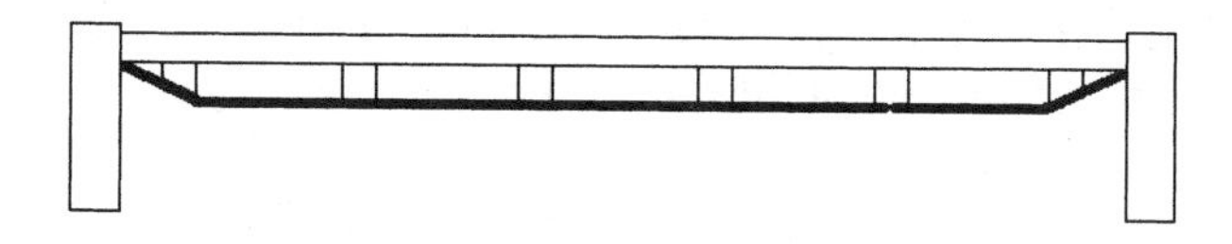

Finally, weld the two pieces of flat together where they meet and around all of the 4″ pieces.

Next, reinforce the bearing plates. Weld four 20″ lengths of 2″ x 0.375″ flat onto the bearing plates. You may have to heat the metal a bit to get it to bend.

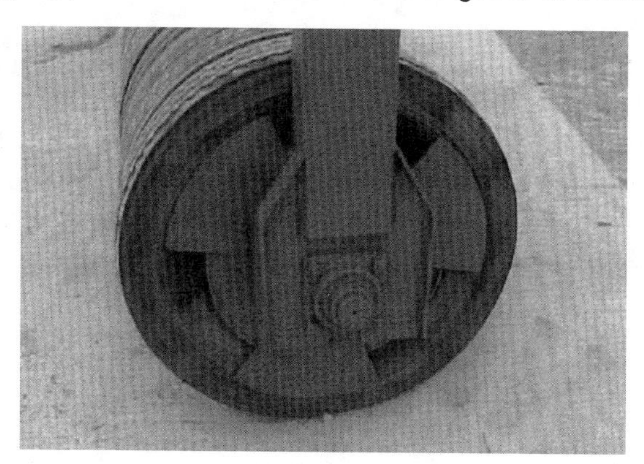

Reinforce the uprights by welding on an 8″ x 8″ x 8″ triangle gusset and a 14″ x 14″ x 14″ triangle gusset on each side of the frame.

Reinforce the hitch even more by welding an 8″ x 8″ x 8″ triangle gusset and a 40″ length of 3″ x 0.375″ on to tie the hitch even better to the frame. Finally, while the roller is on its back side, check and reinforce every weld that was done overhead. This can be done by welding a large weave bead over each weld.

Finishing Touches and Painting

Flip the roller over again with the loader and weld the other sides of the 4″ squares on the bow. Chip all the slag off the welds, and basically clean the assembly up for painting. Spray one coat of primer and one coat of paint.

Fence Line Feeder

Author: Mark A. Brown
Instructor: Thayer Davis
School: Tomahawk High School
City & State: Tomahawk, WI

Materials

Quantity	*Description*
2 24′ length	1-1/2″ x 3″ x 11 ga. tube steel
6 20′ length	1-1/2″ x 1-1/2″ x 11 ga. tube steel
2	4′ x 8′ x 18 ga. sheet metal
1	Primer
1	Paint

Procedure

1. Cut six pieces of 24″ long 1-1/2″ x 3″ tube steel.
2. Cut a 45-degree angle on one end of each of the 24″ pieces.
3. Square these and butt weld the 45-degree ends together, using MIG.
4. Cut three 58.5″ pieces of the 1-1/2″ x 3″ tube steel. Square and weld them to the 24″ welded pieces.
5. Clean up the welds with a hand grinder.
6. Cut two 21″ supports and MIG weld them in, per drawings.
7. Cut six 32″ x 30″ pieces of sheet metal.
8. Roll them up to 24″ diameter on a slip roll former. Spot weld them together and fit them to the inside of the framework.
9. Take all the pieces and MIG weld them together to make the frame, as per drawings. Tack weld and square each assembly before welding solid.
10. Clean up the welds with the hand grinder.
11. MIG weld the trough in (skip welding with 1″ welds to the frame).
12. Clean with solvent, prime and paint.

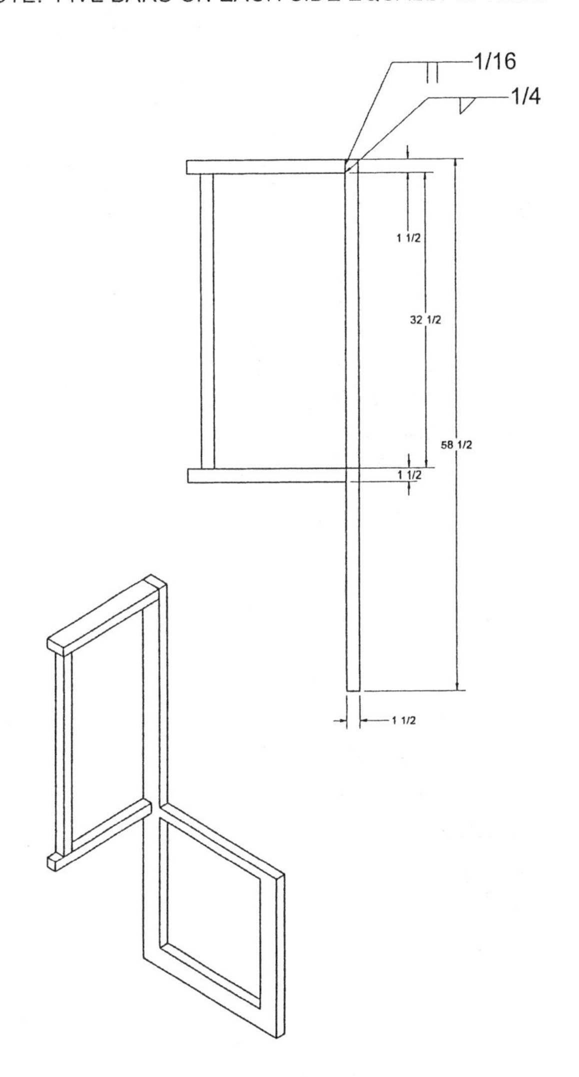

DETAIL A

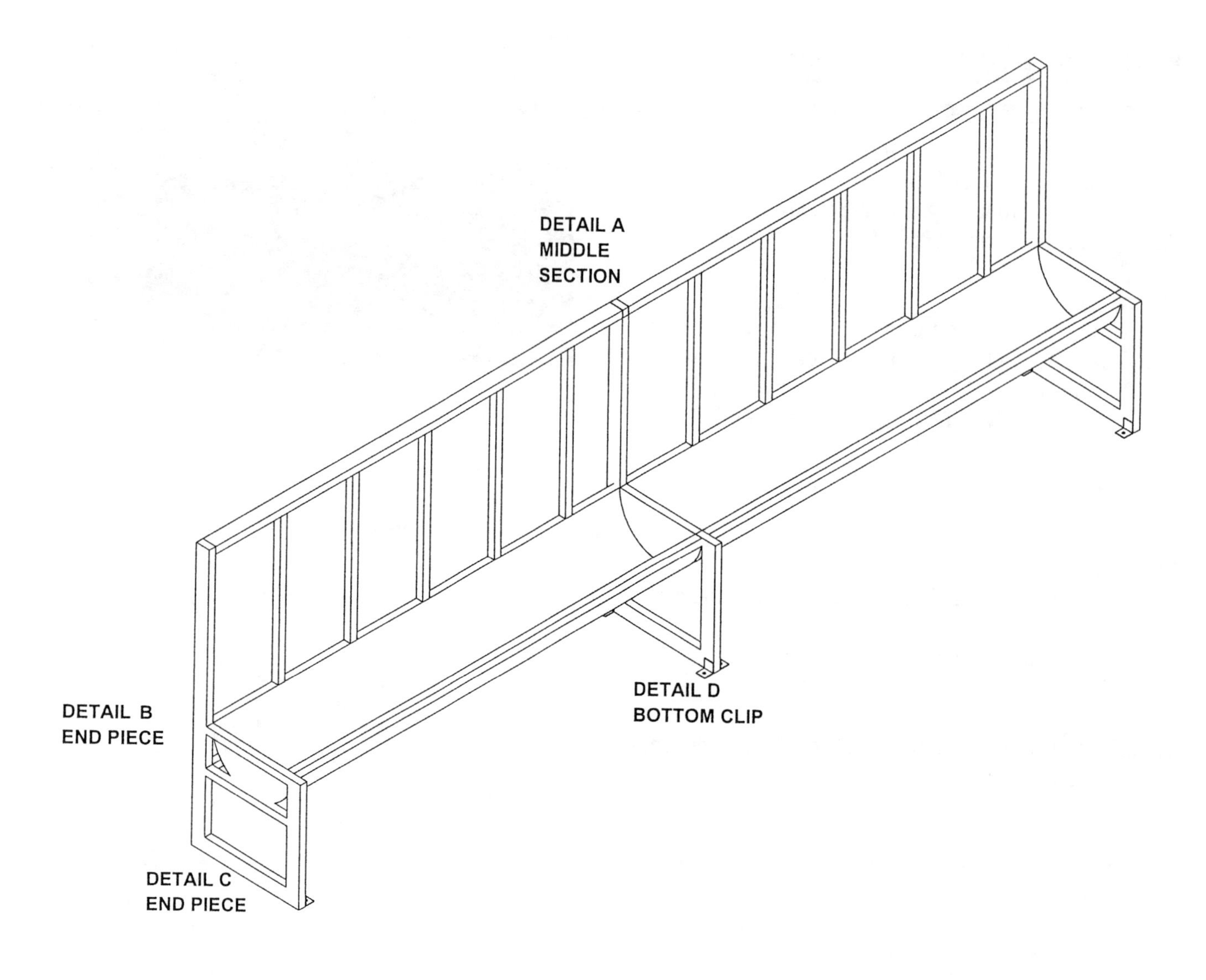
DETAIL A
MIDDLE
SECTION
DETAIL B
END PIECE
DETAIL C
END PIECE
DETAIL D
BOTTOM CLIP

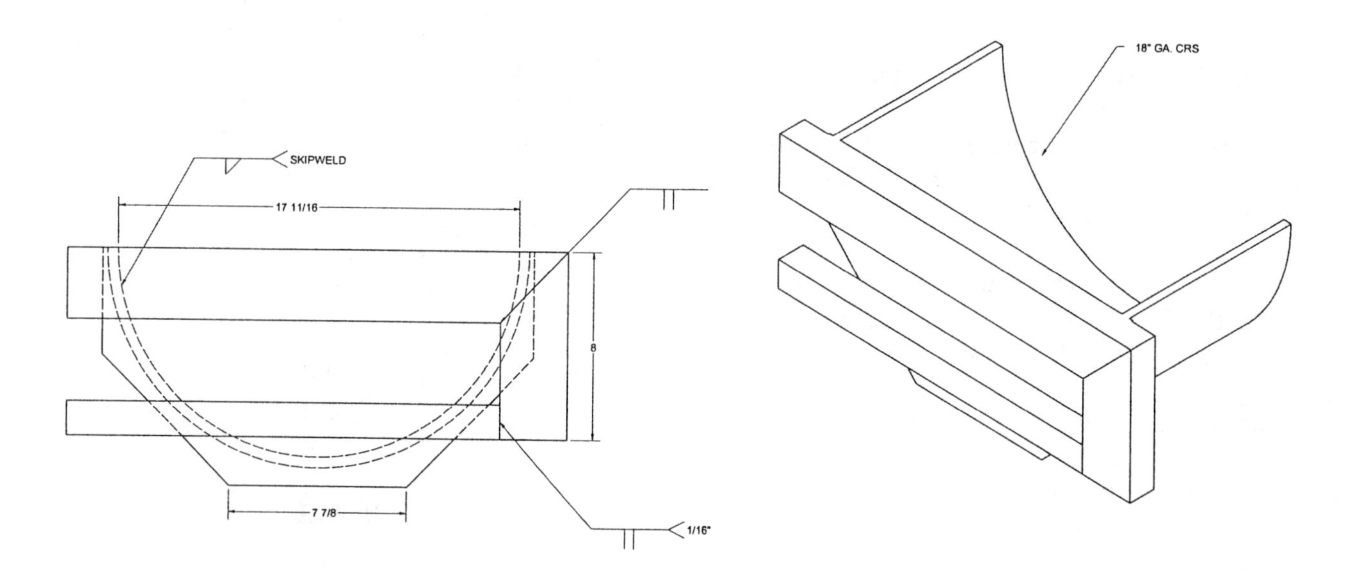
SKIPWELD
17 11/16
8
7 7/8
1/16"
18" GA. CRS

DETAIL B

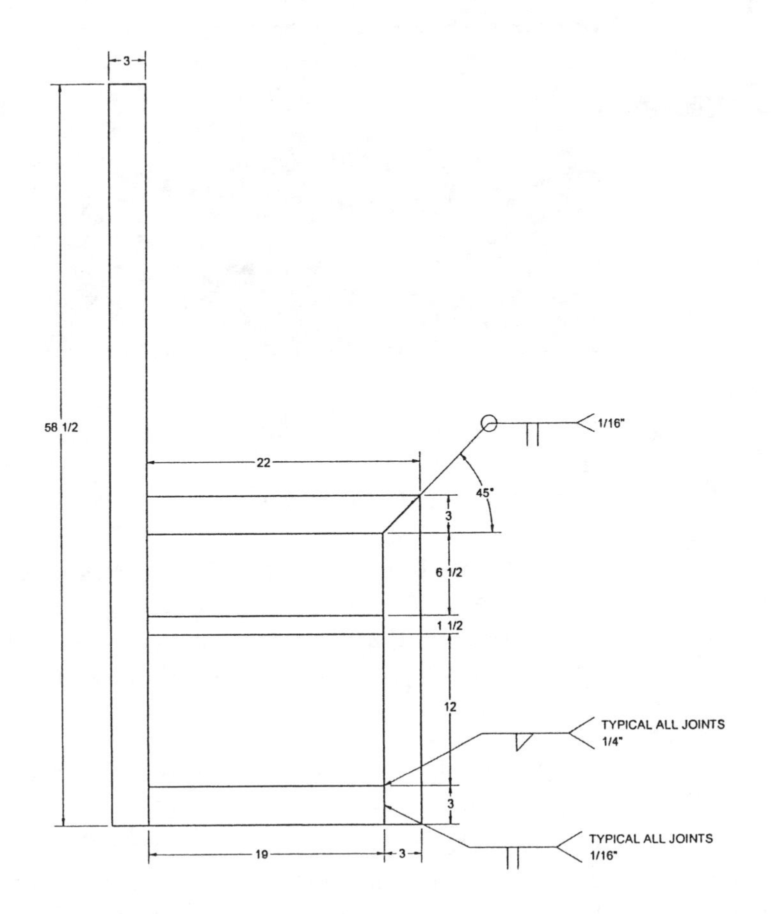

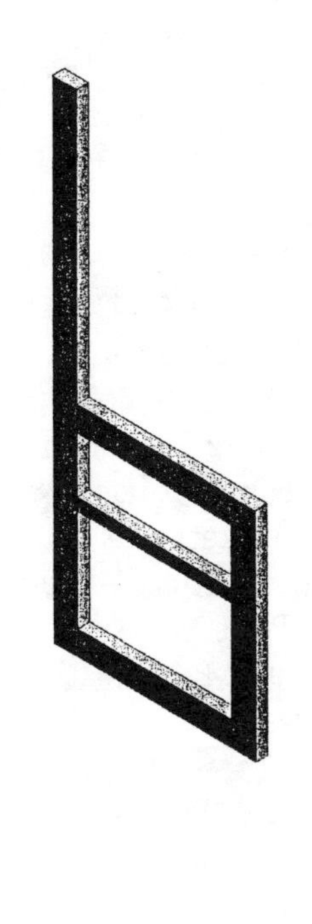

DETAIL C

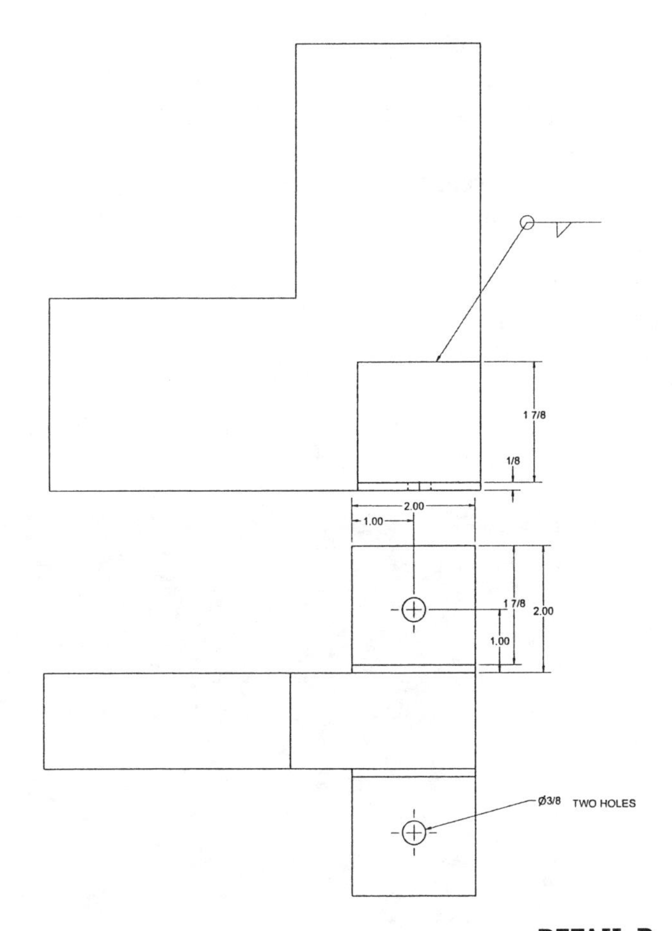

DETAIL D

20-Foot Culti-Packer

Author: Frank Guerrero
Instructor: Lex Godfrey
School: Burley High School
City & State: Burley, ID

Bill of Materials

Quantity	*Description*
2	4 x 4 x 250 – 40
1	4 x 4 x 188 – 40
1	4 x 4 x 188 sq 40′
2	2-1/2 x 12 Hydraulic rams
2	2″ Block Bearings
2	Tire Brackets
	Hydraulic Hose and Fittings
	3/4″ Plate
	Paint and Primer
	Hardware
	Roller
	Tires and Wheels

Procedure

This roller frame is built using 2 pieces of 40′ lengths of 4 x 4 x 1/4 tube and 2 pieces of 4 x 4 x 3/16 for the top frame.

The process begins with carrying the metal over to the band saw. This may require help! Next, measure it to 20′ and cut both sides of the tube at 45 degree angles.

Cut the other 20′ long piece to make the bottom rectangular frame. Then take another piece and cut four pieces at 32″ inches long and also at 45-degree angles so that the four pieces will match with the long pieces. When they are done, square them together so that they are even and ready to tack together. Making sure again that they are square, tack them together. Tell anyone around to be careful and not bump the frame.

Lift the tacked assembly off the floor and put it on some horses. Weld everything you can for now until you are ready to flip it over.

Seven pieces of 24 in. long tube are need for the braces to go in the middle of the top and bottom frame. When they are finished, four more are needed for the bottom. Measuring from the outside of the frame in to 44 in., where one of the pieces will go, and tack it so it won't move. Then measure from the outside of the frame again. (All of these measurements will be from outside of the frame.) For the second tube, measure from the outside to 92 in. Tack it in and then measure to put the third piece in at 144 in., and tack it in also.

For the last piece, go 192 in. from the outside of the frame and then tack it in. Double check the measurements and weld the pieces in.

Now it is time to put the top frame together. The bottom frame used 1/4″ thick tube. For the top, use 3/16″ thick, so that it won't put so much weight on top. Follow the procedure used for the top frame and weld it together.

For the middle pieces of the bottom frame, the first piece will be at 59 in. from the outside. Tack it in and then the second piece will be 118 in. from the outside of the frame. This piece is in the middle of the frame. Finally, the third piece goes to 177 in. from the same side as all the measurements.

When both frames are done, put the lightest frame on top. You will need eight pieces of 15 in. long tubes to put in between the two frames. Measure again to see if everything is level, then weld the four corner tubes in place. For the other tubes, go from the same side and measure in 59 in. for the first two. These two tubes need to line up with a top frame tube that has the same measurement. The next two tubes are the same as the first two tubes. They will be lining up with a tub of the top frame. After welding those in, cut the metal that will hold up the roller. Using 3/4″ thick plate, cut two pieces that are 18 in. on top and 10 in. on the bottom.

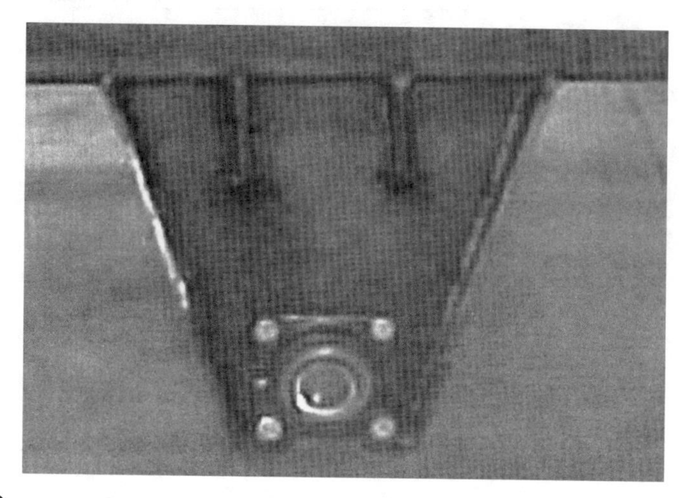

After the roller holders are done and cut out, flip the frame over so that the bottom frame is on top. That way it is easier to weld the holders on. Find the center of the frame and then the center of the holder and line them up and tack them in place. When you are done, use the torch to heat the metal up, because the holder is thick and the frame is thin and the weld won't hold all that weight. You should heat it up until the holder changes color, the put a big bead on it so it would hold. The bead goes on the inside part of the frame. To hold the holder up, put in two big triangle pieces and then two little other pieces to hold it.

Next, flip the frames over to put the tires on. You need two pieces of metal 18 in. long and 1 in. thick. Drill five 3/4" holes for the jack to hold on to. When you are finished, find the center of the tube (2 in.) and then the center of the bar (1/2"). Tack it in place, and then warm it up just like the holder and weld it. When it is in place, put in a piece of flat so it will hold the other end of the hydraulic jack.

To make the tongue of the roller, use a leftover piece of 1/4" tube. After this is all welded together on both sides, you will be ready to put the front of the tongue together. Go from the center of the frame (10') and then go out 11', right to the front point of the tongue. Cut the angles out so they will go together perfectly. The angle should be at 105 degrees for both bars. Put it together and tack it so it won't move.

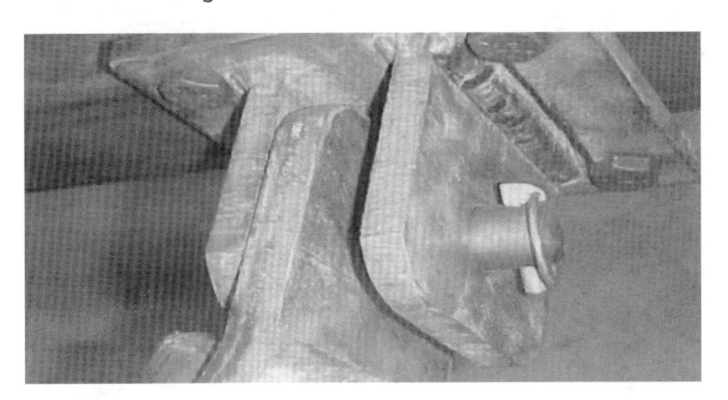

Next, cut two pieces of flat 3/4" thick at 1' long. First it goes straight and then it goes into a wider end.

You will need a brace in the middle of the tongue so it won't bend. Find a piece of scrap that will fit in the gap, mark it, cut it, tack it in, and weld it.

Finally, you will need a hydraulic ram to hold the tongue up. That way it will bounce up and down when it goes down the road. Cut out a 4 x 4 x 1/4 flat and two little pieces at 4 x 2 x 1/4 flat. Then on the first piece of flat, drill four holes in the corners at 1/2 diameter bit and for the two little pieces drill them on top and bottom. Then put it on the top frame around the middle brace and bolt it down. After that is done, cut two pieces of flat so that they will fit around the tongue brace and weld it so that the ram is straight.

Chisel Plow

Author: Reid Bowen
Instructor: Lex Godfrey
School: Burley High School
City & State: Burley, ID

Bill of Materials

Quantity	*Description*
4	4 x 4 x .250 sq. 40′
1	2 x 2 x .188 sq
4	1″ plt
4	1 gallon paint

Procedure

To start the frame, cut two 16 ft. pieces of 4 x 4 to make the width. Next, cut out two 6 ft.-2 in. pieces for the length. Butt weld the long pieces to the shorter, pulling the front bar back 2 in. If extensions are added to the outside, the 2 in. will make it easier to drill the holes for the bolts to go all the way through the bar.

First weld the bars with .035 wire. After the bars are all square, weld over the first pass with 7018. This adds strength to the welds. After welding the whole frame together, cut the center bar to 16 ft. to fit into the outer frame. Centering the bar in the center of the plow, weld the bar to the frame the same as the first.

After the frame is welded together, start cutting the 7 ft. bars that will become the hitch in the front and back. The 7 ft. length will push the front hitch out a foot so that the plow will be far enough that it can be pulled up without hitting the tires of the tractor.

Weld the bars to the frame 43 in. apart. The side on which the bars are welded will become the bottom. Weld one pass with .035 wire, then two passes of 7018. With the bars crossing over, cut the bars that will brace the frame with the center bar, welding them the same as over the cross bars. Do all of this with 4 x 4 steel.

Next, add the bars that will make the top hitch. Cut the bars to 2 ft. long, and on top of the bars, weld a bar cut to 8 ft. long. Weld the end of the bar closed to keep dirt out of it. Next, weld the 2 ft. bar in the center.

Go on welding the bars from the top bar to the back bar of the frame. Two of the bars are of 4 x 4 and the other two bars are of 2 x 2. This is so that the bigger bar will not be in the way of the shank. The bars are to keep the outside of the plow in the ground. Weld the bars with just .035 wire.

Flip the plow to weld the under side. With the under side up, cut the back hitch. Cut the bars to 7 ft. long. The length is to keep the tool behind the plow from hitting the bar. Weld the bar to the rest of the plow very close to the frame.

With the plow upside down, start welding the bottom of the hitch on the front. Cut the plate to go all the way along the extra foot out of the bar. Drill a 1-7/16″ hole in the plate to fit the hitch of the tractor just right. Also weld the plate of the third on. Weld this plate 20 in. apart from the other parts of the bottom hitch. Preheat the 1-in. plate so that the larger metal will not take all the heat. After heating the plate, weld it on using four passes of 7018.

With the hitch welded on, proceed to add the gusset into the corner of the outside. This will help move the force of the plow around, relieving the stress so it will not split the metal. Weld the gussets in with .035 wire

on the top side and the bottom, but not on the side. This is done so that the weld will not weaken the frame. After the gussets are all welded in, check over the whole plow prior to painting.

Wire brush the plow to remove any rust. Spray it with two coats of primer, then add two coats of paint.

One-Ton Bale Feeder

Author: Chad Smith
Instructor: Lex Godfrey
School: Burley High School
City & State: Burley, ID

Bill of Materials

Quantity	*Description*
2	6" channel 40'
16	2 x 2 x 1/8 L
1	3" channel 40'
4	4 x 4 x 0.375 40'
2	4" channel 40'
1	3 x 3 x 0.375 120"
1	14 ga 60 x 120
2	3" C 40'
1	60" 3-1/2" sch 40
1	5 x 3 x 1/4 L 40'
2	Tires and wheels
4	Nap 211-32-L3 2" pillow block
8	NANF 211-32-L3 2" flange
1	Hub's
3	10-3/4" pipe sch.40 102"
1	48 x 96 1/4" plt A-514
1	Bolts (AMI 13625)
1	240" 6 x 4 x .250
1	BLLDG jack Dropleg 12K TW 12-1/2" lift
1	Western Truck –Bearings, Chain, Sprockets
1	120" 1-1/2" CR rd
1	saw
4	1-1/4" 1/2 x 1/8
2	Bearings
1	40" 3/8 key stock
1	AMI 15593 ready rod, nuts
1	Key shaft
1	60" 3/4" key stock
1	Burley Iron Works
1	Bolts
1	Hydraulics
250	Knives
1	Paint / Primer

Background

In this age of agriculture, efficiency is everything. Nothing can be wasted when feeding 10,000 head of cattle. My boss asked me to build an efficient ton bale feeder. In his operation he has to get feed to the cattle on the range during the winter months. In the past, he has used different kinds of feeders. One pushed the bales through choppers on the back and the feed was dropped on the ground. This was wasteful because the cattle used the feed as bedding and not feed. The other type of feeder pushed the bales through the front, and dropped the processed bales on a conveyer, which then dropped the feed into a narrow row. This was the most efficient way to feed, but the machine was built too light for rocky and hilly range ground. In addition, it was hard to work on. I had two challenges: to build a machine that took the benefits of both machines and combine them into one; and to make my design simple and easy to work on if needed.

Procedure

Main Frame

For the frame, use 4" x 4" x 0.375" tubing. After cutting the pieces on the band saw, lay them out on the shop floor, then square the frame and tack weld it. Tack weld each corner at opposite times, so that it will not warp or distort. Since the frame will have to be very strong, used a dual shield process to accomplish this. After the parts are welded, flip the frame and weld the bottom seams.

Sub-Frame

For the sub-frame, use 4" channel iron, cutting the channel at 96". Space the cross members at 30" centers to keep the structure sound while eliminating excess weight. After finishing with the 4" channel

cross members, use 6″ channel iron to make the deck. Many bale choppers use a "push-bar" on the back of the machine to push the ton bales forward. This bar incorporates the 6″ channel. A push-bar is not required, but the option is there if it is needed in the future.

Decking

Now that the sub-frame is complete, begin work on the decking. A front-end loader loads the ton bales; four shafts are pushed through the bale so that it can be maneuvered by the loader operator. It is not always easy to get the bales off; they have a tendency to stick-on the shafts. If the bale is stuck, it can pull the drive chains, bending, or breaking sprockets, drive-lines, or the chain itself. Having the chains run inside channel solves the problem.

The plans call for four separate drive chains. Place 3″ channel, 11″ from the outside of the deck, then 22-1/2″ from that channel. Repeat the process on the other side. Why such different spacing? This is essential because every 12″ on the drive chains there is a pintel attachment, or little spike that digs into the ton bales. These cannot be placed on the outside of the bale because they can't get "traction." Now that the runners for the chain are done, create the rest of the deck. Place 2″ x 2″ angle iron so that the peak or corner is up. This may look different, but for good reason. It reduces the amount of friction the hale bales have on the deck. This makes it easier on the hydraulic motor to drive the bales forward.

Axles

A major complaint is that the axles of hay bale feeders are not built strong enough to support heavy loads. Pre-fabricated axles are too expensive. To fabricate your own, the first step is to buy the right spindles and hubs. They have to be able to support the structure and four tons of feed. 4″ x 4″ x 0.375 tubing will work. To mount the spindles on the tubing while keeping them square cut two pieces of 4″ x 4″ x .0375

at 5 inches long, then find the center and use a hole saw with a 1/16″ bigger diameter than the spindle. Then slide the spindle inside the 5″ piece of tubing, butt it up to the axle and weld it in place. Cap the ends of the 5″ tubing to add strength, making it look cleaner.

Now that the spindles are square, mount them to the frame. Cross-measure from the tip of the spindle to the front inside corners of the frame, measuring from the inside because it is a more e x act measurement than the outside. Now the axles are in place and need to be gusseted solid. Gusset the axles in two places, one running with the frame (a 28″ piece of 4″ x 4″ x 0.375), and the other running in-between the two spindles. Cap all open ends of the 4″ tubing with 3/8″ plate and dual shield weld.

Flipping the Bale Feeder

Between the process of building the axles and the deck, the project will have to be flipped over a few times. This is quite a process because of the weight of the project. Use a forklift and an overhead hoist equipped with heavy chains. Be sure to observe all safety precautions.

Tongue

Fabricate the tongue out of 6″ x 4″ x 0.25 tubing. Cut the first piece at 72″ and run this length of tubing toward the back of the machine, then weld it to the mainframe and sub-frame for strength. The tongue will have to drop somewhat so it will sit level when hooked to a tractor. Calculate the angle needed to make this drop and cut the 6″ x 4″ to fabricate it.

After this is complete, make a 1″ slot in the end of the 6″ x 4″ for the hitch to pin to. Fabricate this out of 1″ plate, and cut a 1-1/2″ hole in it for the hitch pin (1/2″ is the standard pin size for the sizes of tractor that will pull this feeder). Now, make sure that the tongue is square with the bale feeder. Measure from the exact same spot on each side of the axle or the feeder will pull crooked. Once everything is square, weld all the joints solid. Next, install four gussets. Gusset #1 is made of 1/2″ plate and is 13″ wide and 9″ long; there are identical gussets on each side of the tongue. Gusset #2 is located on the back part of the tongue, it runs from the top piece of 6″ x 4″ to the angled piece. It is made of 1/2″ plate and is 9″ x 9″. Gusset #3 gussets the bottom piece of the tongue to the angled piece, it is also made of 1/2″ plate and is 9″ x 9″.

Beaters

To assemble the beaters, begin by constructing the uprights; cut two 69″ lengths of 10″ channel iron. Next, make the holes for the beater bearings. Be careful: if the holes on each side are not drilled exactly right, the beaters will not spin true. A simple way to do this is to make the two pieces of 10″ into one. So make sure both pieces of 10″ are perfectly even and square, and tack the two together so they resemble an I-beam. Use a 3/8″ drill bit and a 2-1/4″ hole saw to do the drilling.

Break the tack welds and tack one piece of 30″ back from the front of the feeder (after making sure that it is square and level). Next comes a unique feature of this project: providing access to the beaters. To do this, put one side on hinges. Use a 1-1/2″ cold rolled shaft and 1-1/2″ Shelby tubing with a 1/2″ wall for the hinges. There will be three pin assemblies in all. One will be at the bottom of the 6″ channel (which the uprights are mounted to) to serve as the main hinge. The second will be placed right above the 6″ channel to act as a "lock," and the last will be put at the top to keep the cross member in between the two uprights locked in place but still let the upright hinge down.

For the actual beaters, use (3) 10-3/4″ pipe sch.40 at 102″ long. Next, make the brackets that will be welded to the pipe, for the knives to be bolted to. The triangular shaped brackets with a 1/4″ hole in the center for the knife to bolt to were designed with a program called Master Cam. Cut the 250 brackets out of 3/8″ plate using an automated plasma cutter.

Drive Chains

The first step is to get the front drive shaft mounted. This can be done by cutting holes in the 6″ channel, right behind the beaters. Place the drive shaft close to the beaters because if the bales are not fed close enough to the beater they won't be chopped, therefore causing the machine to plug. To place the drive shaft, measure from the front and the back on both sides until you have the same measurement, then tack the bearing mounts in position.

With the front drive shaft mounted, lay the chains out where they need to run. A warning: While I was doing my plans, I thought I had calculated enough room for the drive chains to return under the bale feeder. I was wrong. The 1-1/2″ pintel attachments would hit on the tires as the chains returned. The solution was to cut into the 4″ channel of the sub frame and run the chains through the channel.

Adjustments had to be made. It worked out that 2 drive chains ran outside of the mainframe and two ran inside the mainframe. For the two outside chains I fabricated 9″ x 4-1/2″ gussets out of 3/8″ plate. For the chains in between the mainframe I added another length of 3″ channel beneath the existing channel. This proved to strengthen the mainframe just as before, if not better.

Having the chains running, you need only a back idler shaft to make the drive chains complete. This is where

to put the tension adjustment for the chains. Use a 112″ length of 1-1/2″ cold rolled shaft for this idler shaft and 1-1/2″ Shelby tubing with 1/2″ wall for the adjustments. Then weld a 15″ length of 5/8″ bolt stock to the tubing for the adjustment.

Feed Belt

The bearing mounts for the feed belt are crucial because if they are not square, the belt will crawl to one side. To engineer the mounts for the adjustable bearings, build a frame out of 3/8″ plate. On the top and bottom side of the bearings there is a 3/4″ slot that runs the length of the bearing. This is how the bearing slides. So place two 8″ pieces of 3/4″ key stock on the 3/8″ frame. This will allow the bearing to be adjusted. For the adjustment, use 1″ bolt stock, cutting them 12″ long to provide plenty of adjustment. Now that the adjustment bearings are in place, mount the bearings on the opposite end of the bale chopper.

Feed Catch

The beaters spin at about 1200 RPMs. At this speed, feed would be thrown out the front of the bale feeder if there was nothing there to stop it. So fabricate a catch out of 1″ x 1″ x 0.125 tubing and 16 gauge sheet metal. The catch rises vertical 48″ with the front of the bale feeder, and then it breaks 38″ to the top cross member of the beaters. There are 18″ openings at each end. The opening on the left side is for the feed to be discharged, the opening on the right is for easy access to the feed belt adjustment and beaters.

The right opening is covered with 1/4″ rubber belt. This ensures no feed will fall out the wrong side. The feed catch is also removable.

Twine Guard

With the left over sheet metal from the catch, cover the 2″ x 2″ gussets on the beater uprights. This keeps the feed from being thrown over the side. Place a 5″ wide piece of 16 gauge running vertical with the beaters was placed on each side. This will help keep the baling twine out of the bearings.

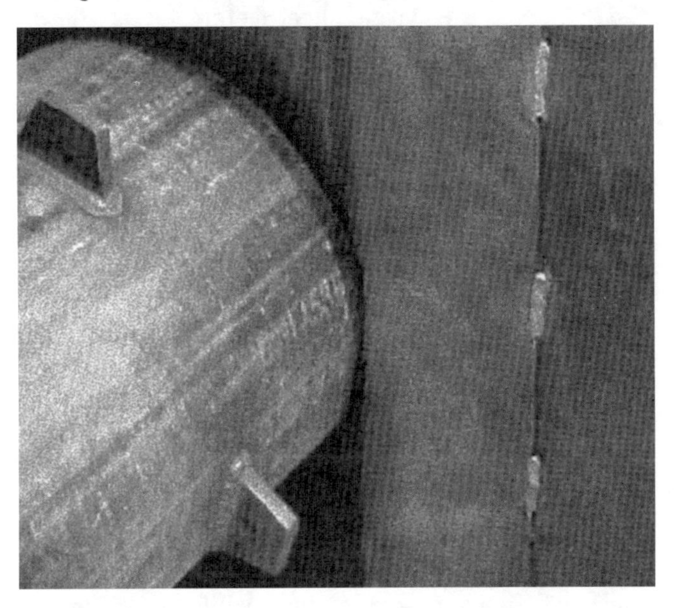

Motor Mounts

The cattleman that I was building the bale feeder for wanted it to be driven by hydraulics. This eliminated the driveline and hopefully some breakdowns. The agreement was for me to mount the motors. The farm mechanics would run the hydraulic lines and install all of the chains and sprockets.

To mount the first motor, which runs the beaters, use 1/2″ plate with precisely drilled holes. Weld the mount to the 10″ channel that extends beneath the 6″ channel of the deck.

Next is the mount for the floor chain motor. Use the same design as the beater motor, except extend out to the side of the 6″ channel. This can be done by using two pieces of 4″ x 5″ x 3/8″ angle iron 15″ long. Extend out the motor to the needed length, then weld the angle to the 6″ channel and the sub frame.

The feed belt motor has to move with the driveshaft's adjustment. To make the mount, use an automated plasma cutter to create two brackets with three long slots the sizes of the motor drive shaft and bolts. The design lets the motor move as the shaft adjusts.

Portable Hay Feeder

Author: Andrew J. Mason and Jesse J. Wooldridge
Instructor: James L. King
School: Burlington High School
City & State: Burlington, IA

Bill of Materials

Quantity	*Description*
192′	1-1/2″ round tube, 13 ga
6	4′ x 8′ 11 ga. sheet metal
72′	2-1/2″ sq 11 ga. tube
72′	3″3/16″ tube
72 lbs.	1-1/2 x 1-1/2 x 3/16″ angle
4	stub axle
1	2-5/16″ hitchball
4	ez hub kit
1	5th wheel coupler, 2-5/16″
4	16 x 6 bolt rim and tire

Step by Step Process

Cut two pieces of 3″ x 3″ square tubing 16 ft. long with 45 degree angles on the end for the frame.

Cut two pieces of 3″ x 3″ square tubing 5′6″ long with 45 degree angles on the end, also for the frame.

Construct frame tacked corners while in 90-degree angle jig. Check diagonals, then weld completely around.

Cut seven pieces of 1-1/2″ by 1-1/2″ angle iron at 60″ long for the cross members in the frame.

Weld cross members at two feet on center on the frame with the vee end flush with the top of the frame.

Lay out four 4′ x 8′ sheets of sheet metal for the bed-pans, sheets of 11 gauge steel, 5′6″ wide at the bottom 6′ 4-5/8″ at the top with a 3″ lip at the top of the pans on both sides.

Bend the pans to the right specifications.

Lay out for end pieces for the bed pans 5′6″ at the bottom and 6′ 4-5/8″ at the top with 1″ flanges on the bottom and the sides of the end pieces.

Align and tack the bed pans to the frame, check for square.

Align and tack the end pieces for the bedpans.

Weld seams inside and outside of the bed pans completely stopping 6″ from the center of the bed to allow for the center divider.

Weld end pieces onto the bed pans inside and outside of pan.

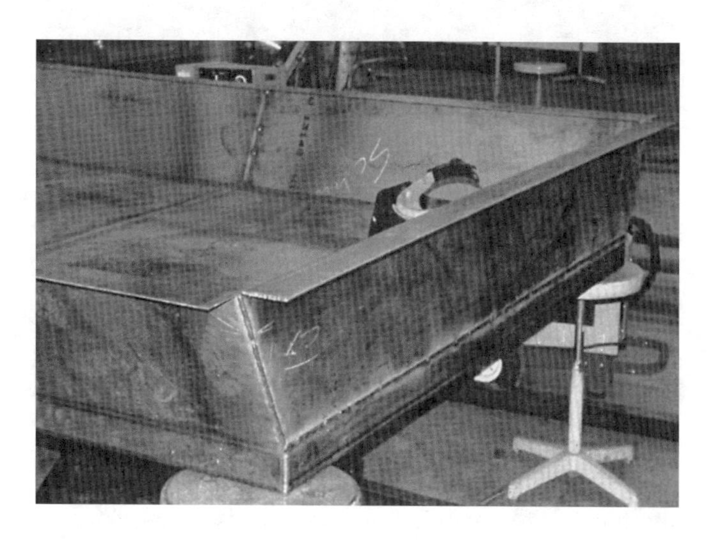

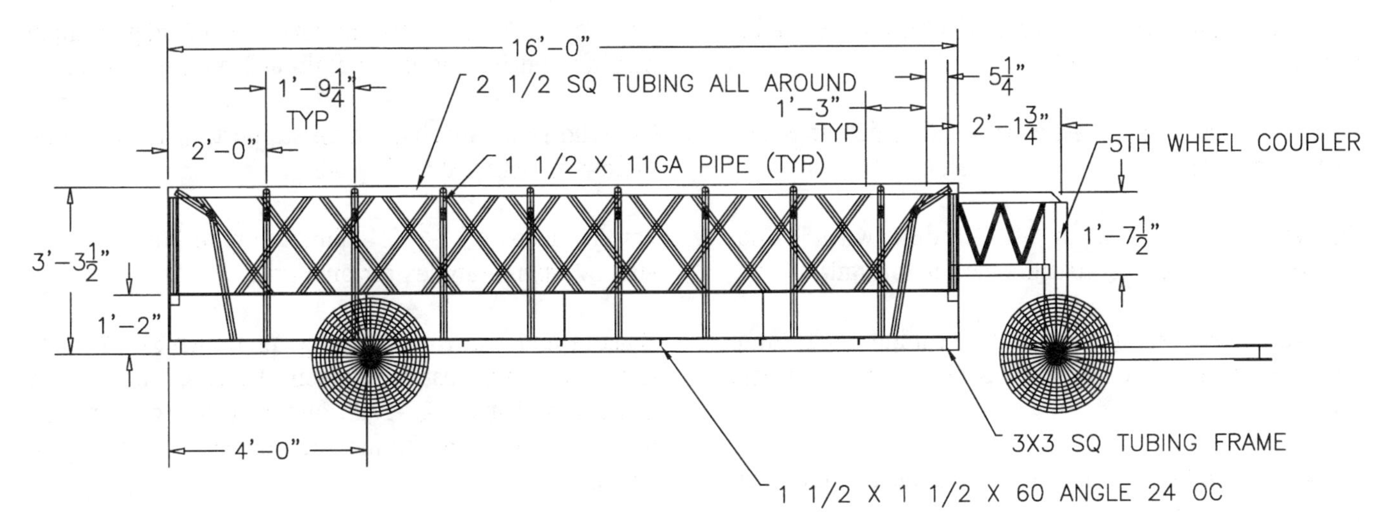

SIDE VIEW

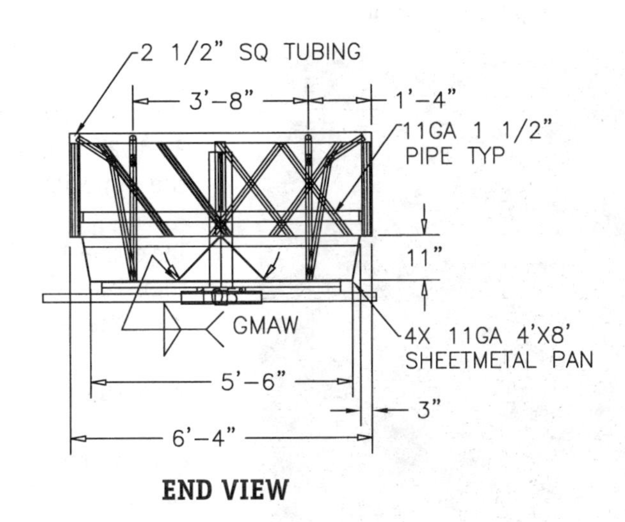

END VIEW

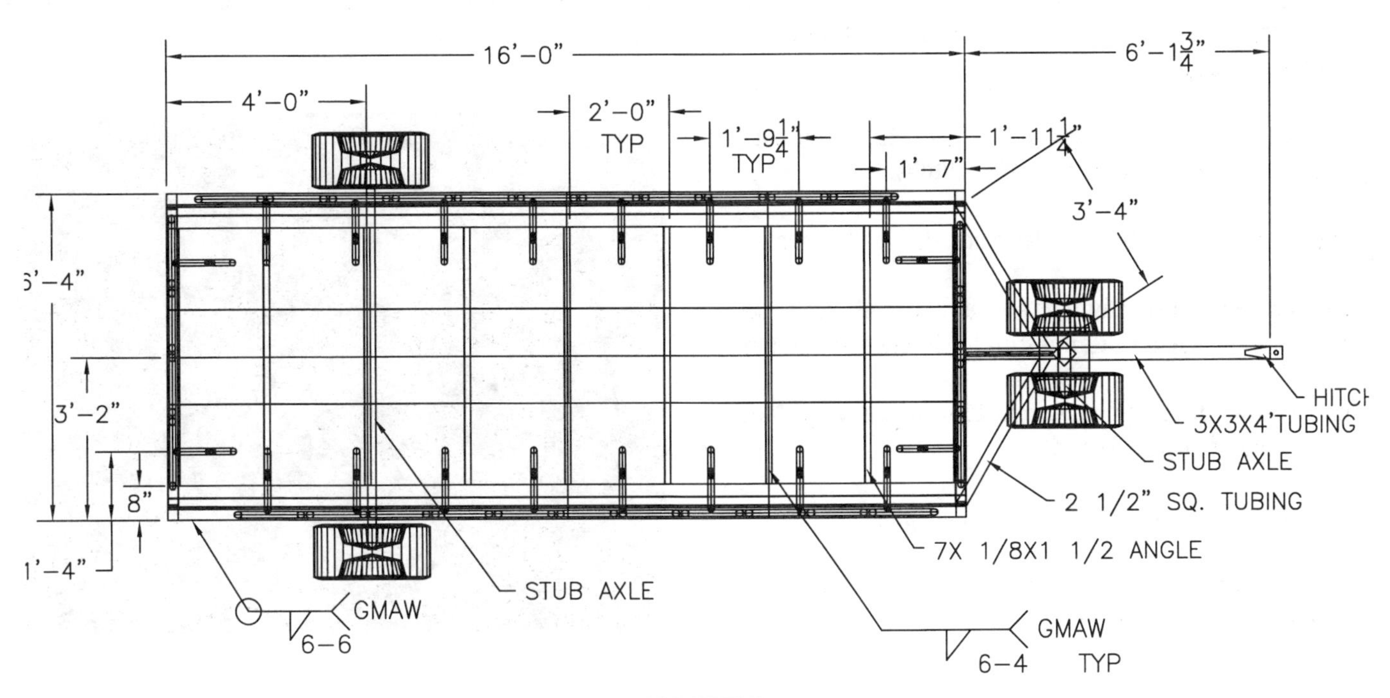

TOP VIEW

Weld the bedpan to frame with 6″ intermittent welds with 6″ spacing all around.

Weld pan to cross members with 4″ intermittent welds with 6″ spacings.

Cut two pieces of 2-1/2″ by 2-1/2″ tubing with 5′ long with 45 degree angles at the ends for top rail.

Cut two pieces of 2-1/2″ by 2-1/2″ tubing 16 ft. long with 45 degree angles on the ends also for the top rail.

Cut 4 pieces of 2-1/2″ by 2-1/2″ tubing 23″ long for the corner posts.

Align and tack corner posts 1″ in on each corner. Check for plum then weld completely around.

Align and tack top rail frame pieces to the four corner posts. Check top rail frame for square then weld completely around.

Lay out for 1″ drain holes in the bed, 12″ from each side, 12″ apart, with a total of 16 holes.

Drill 1″ holes in the bed with magnetic drill press.

Lay out for center divider pieces 11″ tall with 12″ sides, 8′ long with 1″ flanges on both sides of divider.

Bend center divider pieces not 105 degree angle and to the specifications listed.

Align at 37″ in the center of the bed pan.

Weld one continuous bead the full 16′ length of the center divider on both sides.

Cut two pieces of 3″ tubing with 20-1/2 degree angle on each end for the front and back pipe supports.

Weld the front and back pipe supports with 5″ intermittent welds with 10″ spacing.

Cut 22 pieces of 1-1/2″ diameter pipe 30-1/2″ long with 37 degree angles on one end.

Measure and mark on the top rail frame 5-1/4 ″ out from the corner post on opposite sides of trailer, then measure and mark 22-1/2″ out from the corner post of the bottom bed flange on opposite sides of the trailer.

Align the 1-1/2″ pipe on the marks of the top rail and bottom flange and then mark opposing angle.

Cut the 22 pieces to that angle with a template.

Measure and mark locations for the pipe on each side 22-1/2″ at the bottom and 5-1/4″ at the top then did the opposite on other side.

Tack the 11 pieces of the 1-1/2″ pipe on one side, then tack the other 11 pieces of the 1-1/2 ″ pipe on the other side.

Weld the pipe completely around at top and bottom.

Cut one piece of 3″ by 3″ tubing 23″ long for the center post in the front of the trailer.

Align and center post 36-1/2″ from each end, tack and check for square.

Weld center post with a tee joint weld completely around.

Cut 20 pieces of 1-1/2″ diameter pipe 32″ long with a long 30 angle on one side and a 10 degree angle on the other end.

Bend one pipe in hydraulic bender, from the 30 degree angle end of the pipe until it's aligned with the top rail and the inside of the bed.

Using the first pipe bent as a template, bend the other 19 pipes the same.

Measure and mark for the pipes, 24″ from the end and 21-1/4″ between each pipe. Chalk a line on the bed floor for the placement for the bottom of the pipe.

Tack the pipes in, check for plumb, and then weld with 1/4″ fillet welds around on both the top frame and the bed floor.

Align and tack the four pipes for the front and back and then weld completely around with 1/4″ fillet welds.

Cut four pieces of 1-1/2″ tubing 30-1/2″ long with 37 degree angles for the front angled tubing, and then weld completely around with a 1/8″ fillet weld.

Lay out a sheet of 10 ga. for the axle 60″ long and 4″ wide at the bottom and top, then bend the axle.

Cut two pieces of 1-1/2″ by 2-1/2″ tubing 60″ long to set in between frame to weld axle too.

Align and tack the tubing in place for the axle, then weld with 6″ welds with 12″ spacing on both pieces.

Align and tack axle in place, then weld with 6″ welds 12″ apart.

Cut a piece of 2-1/2″ by 2-1/2″ 1/4 walled tubing to 78″ long for the axle.

Drill two 3/16″ holes in 1-1/2″ and 3″ from the end on both sides of axle tubing for plug welds.

Weld 12″ stub axles into axle tubing with 6010 rod with plug welds. Cover root weld with 6018 rod wave bead.

Align hubs on the axle, grease all bearings and secure them. Check for proper operation.

Secure tires on the hubs.

Cut one piece of 1″ steel plate 12″ by 12″ square for the pull plate.

Align in the center of them and tack two stub axles to the pull plate, then weld completely around with 6010 rod.

Cut 2″ hole in pull plate for the 2-1/2″ ball.

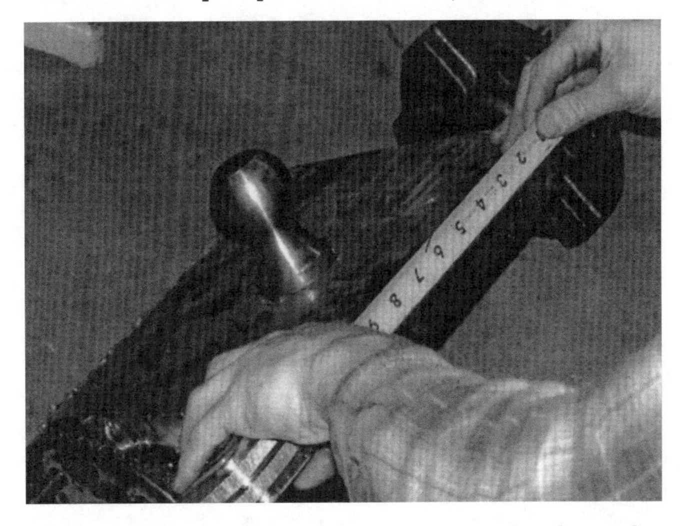

Cover the 6010 welds with 7018 cover bead on the pull plate.

Packed bearings and hubs and attached to the pull plate.

Tighten hub to the 2-1/2″ ball in the pull plate.

Cut three pieces of 1″ inside diameter pipe 2″ long.

Weld 2 pieces of the 2″ pieces of pipe to the pull plate for the hinge.

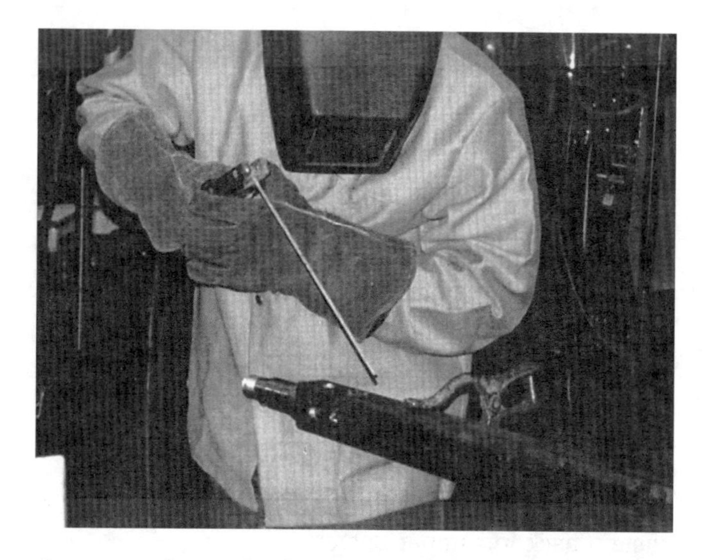

Cut one piece of 3" by 3" tubing 4' long for the tongue.

Cut one piece of 1/4" plate 3" by 9" for the end of the tongue.

Drill two 7/8" holes in the tongue plates to receive a 7/8" hitch pin.

Weld the other 2" piece of 1" inside diameter to the tongue to make the other end of the hinge.

Cut one 8" piece of 1" solid round stock for hinge pin. Cut a 1-1/2" grade 8 washer to the other end.

Put tires on. Tighten lug nuts.

Level pull plate and trailer.

Put fifth wheel coupler assembly on the pivot ball.

Level coupler assembly so it's plumb.

Cut two pieces of 2-1/2" by 2-1/2" tubing for members 40" long to join the trailer to the fifth wheel coupler assembly.

Cut 30 degree angle on one end then mark other angle while aligned on the frame then cut angle.

Align and tack the cross members. Check for level, then weld completely around on both the couple assembly and the frame.

Cut one piece of 3" by 3" tubing 20" long to join the center post in the trailer to the coupler assembly.

Weld the tubing to the frame and coupler assembly completely around.

Weld a 4" by 4" piece of 1/2" plate to the top of the couple assembly.

Cut a piece of 2-1/2" by 2-1/2" tubing 20" long for the top of the trailer to the top of the coupler assembly with a 30 degree angle on one end.

Weld 2-1/2" by 2-1/2" tubing to the frame and the coupler assembly completely around.

Cut four pieces of 1" diameter round stock with 22-1/2 degree angle on both ends.

Weld the round stock at alternating angles between the upper and lower members on the trailer/coupler assembly.

Cut one piece of 1" pipe 3/4" long to hold the hitch pin when not in use, then weld 12" down from the top of the coupler assembly.

Cut two 2" by 4" square holes in the rear of the trailer to allow for drainage.

Quick Attach Bale Spear

Author: Michael Schmidt
Instructor: David H. Murray
School: Ferris State University
City & State: Big Rapids, MI

Bill of Materials

Qty	Description
1	3″ x 3″ x 20′ Square Tubing
1	3″ x 3/8″ x 20′ Flat Stock
1	4″ x 3/8″ x 20′ Flat Stock
1	1-1/4″ x 52″ Round Stock
1	1-1/2″ x 1/4″ x 20″ wall, Round Tubing
1	2-1/2″ x 1/4″ x 39″ wall, Round Tubing
1	48″ x 48″ x 3/8″ Plate

Cut List

The project pieces were cut to size on a band saw and shear. The parts that were cut to size are listed below.

Qty	Description
2	3″ x 3″ x 1/4″ wall x 51-1/4″ Square Tubing
2	3″ x 3/8″ x 19″ Flat Stock
2	4″ x 3/8″ x 19″ Flat Stock
2	1-1/4″ x 8-1/4″ Round Stock
2	1-1/2″ x 1/4″ wall x 5-1/2″ Round Tubing
1	2-1/2″ x 1/4″ wall x 16-1/2″ Round Tubing
2	13-1/4″ x 25″ x 3/8″ Plate
2	3″ x 3/8″ x 25″ Flat Stock
4	21″ x 4″ x 3/8″ Plate
2	3″ x 3/8″ Plate
2	3″ x 3/8″ x 6-3/4″ Flat Stock
2	3″ x 3/8″ x 9″ Flat Stock

Introduction

This front mount bale spear features quick attach and detach brackets that allow it to be installed and removed quickly, making work easier on the farm when changing jobs. The bale spear is used every other day on the farm to feed 40 head of cattle. The bales it moves weigh between 1200 lbs. and 1800 lbs.

Fabrication Procedures

The first step in this project, after figuring out the dimensions and materials needed, was to order the material. Once the material was on hand, the pieces had to be cut to the correct length on the band saw and shear. All the round tubing, square tubing, and round stock were cut on the band saw. All of the 3/8″ flat stock and plate were cut on a 48″ by 3/8″ Mechanical Shear. Some of the material was cut out on a CNC Plasma Burning Table.

After the pieces had all been cut out, the sharp edges were filed off. The main frame of the project was laid out on the floor, so the measurements could be checked. The first piece that was welded up was the center part made out of 3″ x 3/8″ and 4″ x 3/8″. Both pieces had 3/4″ sheared off and were cut to the length of 19″. It was cheaper to buy the 3/8″ flat stock than it was to purchase 3″ x 4″ rectangular tubing. These pieces were welded up in the shape of a rectangle and had a hole drilled in the center. The next pieces were the square tubing that were 3″ x 3″ x 1/4″ and were cut at 51-1/4″ long. The two side pieces were 3″ x 3/8″ x 25″ flat stock.

Once everything was cut out and laid out, it was necessary to make sure that it was squared up prior to welding it. It was tacked and checked again for squareness before being completely welded. The welds that were set up were edge joints, corner joints and butt joints, all welded with flux cored arc welding with 0.035 diameter wire and CO_2 gas. After the welds were finished, it was necessary to chip the flux off with a chipping hammer and clean it with a steel brush. The main frame was completely welded and ready to have the holes drilled.

Three large diameter holes were drilled for the bale spears. Two holes had a diameter of 1-1/2″ and were drilled with a Hole Saw. They were drilled 16-5/8″ from the sides and in the center of the 3″ square tub-

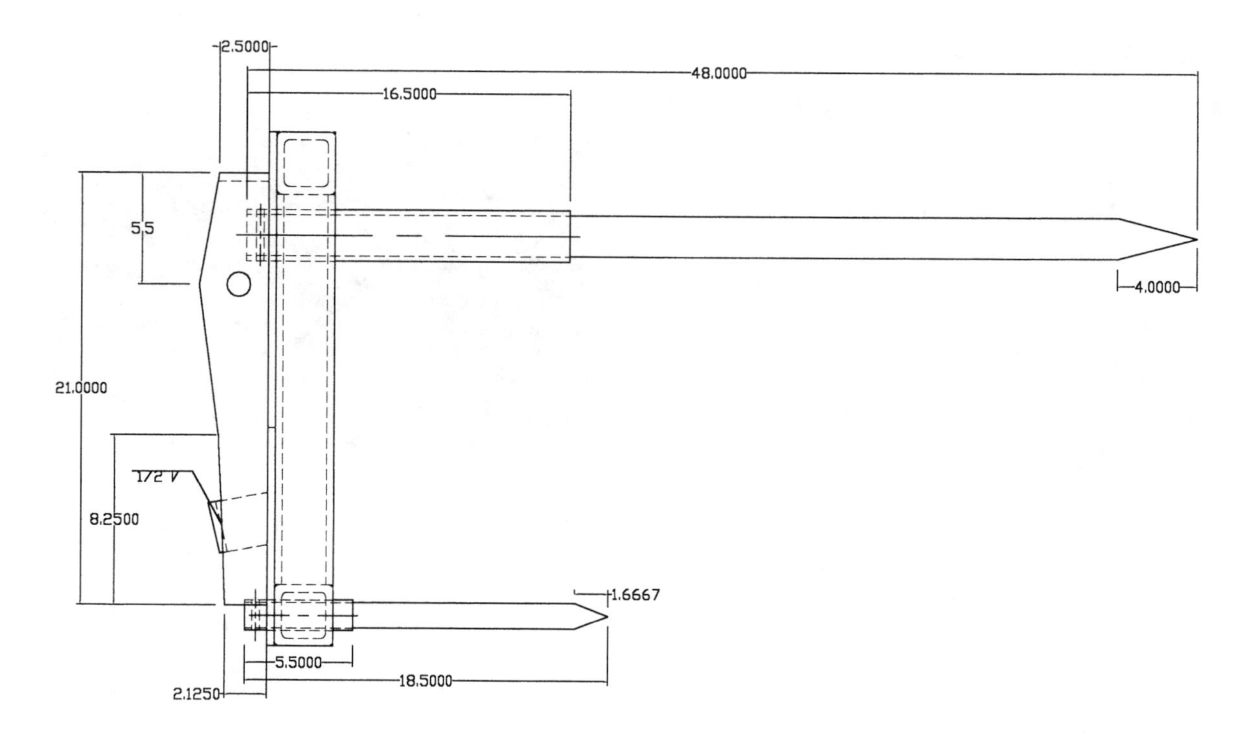

SIDE VIEW

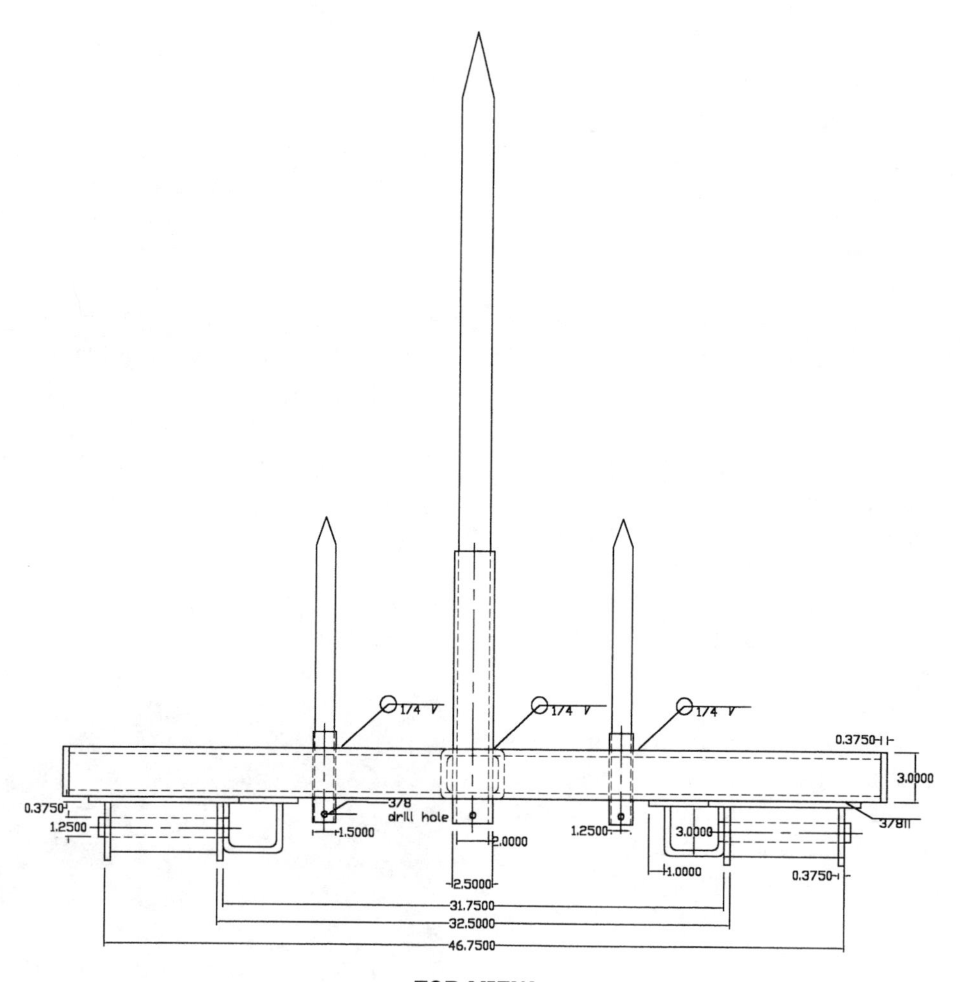

TOP VIEW

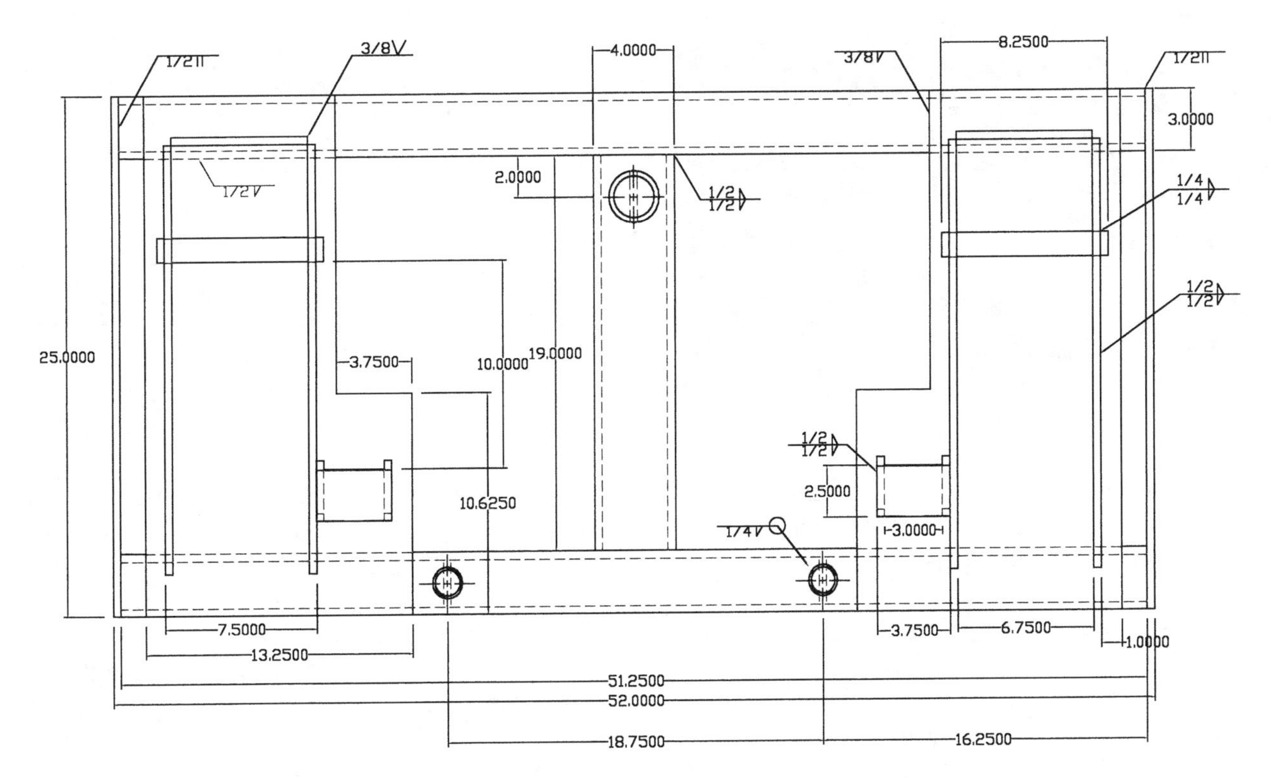

FRONT VIEW

ing. The third hole had a diameter of 2-1/2″ and was also drilled with a Hole Saw. The third one was 5″ from the top and in the center of the 3″ by 4″ tubing. After the holes were drilled and filed to remove the sharp edges, the pipes were then inserted into the holes and welded in Tee joint, completely welded in a 360-degree circumference.

Each piece of round tubing had 3/8″ holes drilled in it prior to being inserted and welded. The holes were for bolts to hold the spears in place. It was important that the round tubing be completely squared with the frame so that the spears were level and straight. The two 1-1/2″ round tubing used as bale spears were cut 5-1/2″ long. The 2-1/2″ round tubing was cut 16-1/2″ long. It was cut longer to give more strength and stability for the large spear. The main frame was now completed with all the round tubing inserts welded in.

The only thing left on the project was to make the quick attach and detach brackets that attach to the loader. The quick attach and detach brackets were cut out with a 400 Amp Plasma and a CNC Burn Table. The brackets were drawn on AutoCAD 2000. It was then transferred to the CNC Burn Table, where the parts were duplicated and auto nested for less scrape. Four pieces had to be cut out for the sides that were 21″ x 4″ x 3/8″ place with a 1″ hole cut out 5-1/2″ from the top and centered in the plate. The two back pieces were 13-1/4″ x 25″ x 3/8″ plate that was in the shape of an "L" and had a cut out of 3-1/4″ x 14 3/8″ x 3/8″ for the back plate.

The next step was to clean up the six cut outs and prepare them for welding. The four sidepieces needed a 1-1/4″ hole drilled where the 1″ hole was cut out. This took out the rough edges and the tapered cut made by the plasma making and made all the holes the same size. To fit this together, two scrap pieces were cut at 6-3/4″ and were placed on the inside of the two sidepieces. The 1-1/4″ pin was placed in the holes to help line it up. Where the scrap pieces were inserted, a C-clamp was placed on each one to hold the sides on and square them up. Then the side brackets were set on the back plate where it was squared up, two inches from the bottom and one inch from the side. This was then clamped down and tacked up. The other side was done the same way. After these pieces had been tacked up, they were welded up with FCAW in a horizontal Tee joint. Two pieces were cut to

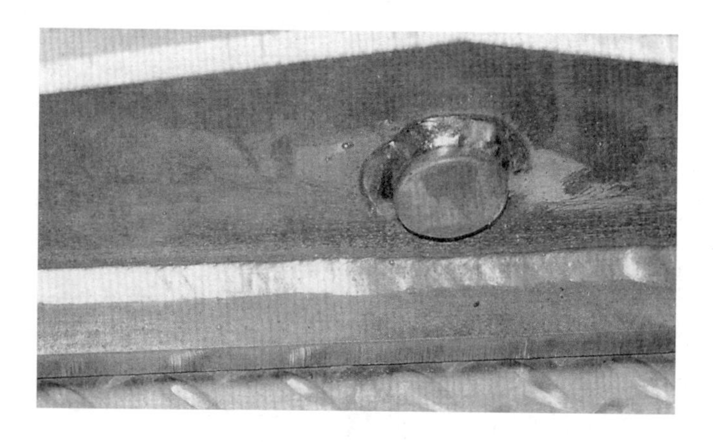

6-3/4″ x 2-3/4″ x 3/8″ to be welded on the top of the sidepieces, above the pin. This was a 2F corner joint. Both pins were welded in 360 degrees in a Tee joint on the outsides of the side plates.

The next step in the project was to cut and bend two brackets; a pin will go into it, holding it to the loader frame. The two pieces were cut 9″ long by 3″ wide out of flat stock. Then they were sheared down to 2-1/2″ by 9″ long. The two pieces were bent on a Press Break. The first one was bent at 3″ on a 2″ die. The second bend was heated up with Oxy-Acetylene Rose Bud. The pressure was set at 7 psi oxygen and 7 psi acetylene, and had the flame set to a neutral flame. The piece was clamped in a vice at 3″ from the second bend and a pair of vice grips was clamped on the top of the piece of metal for bending the piece of metal. Both bent pieces were cut on the band saw on an angle of 10 degrees. The two pieces were then welded to the quick attach frame and were 10-3/8″ from the bottom of the 1-1/4″ solid rod. The quick attach frame was then finished and ready to be welded to the bale spear frame.

The quick attach was laid on the backside of the bale spear frame. Then it was squared up to the bottom of the 3″ x 3″ square tubing and it was 31-3/4″ wide on the inside of the two side brackets. This is one of the most important measurements of the project. If it was measured wrong, it would not fit onto the loader brackets. After it was squared and lined up properly, it was then tacked and welded with FCAW in the Tee and lap joints. This completed the fabrication of the project.

The final step was cleaning up the project with rags, steel brush, chipping hammer and a grinder with a steel brush on it. The main frame was then painted with three coats of green paint. All three spears were painted with three coats of yellow paint. The purpose of the yellow spears was to caution a person to notice the sharp points on the bale spear. The three spears came from a dismantled bale spear off our farm. The spears were inserted into the bale spear frame and bolted on with three 3/8″ bolts and nuts. The project was completed and ready for use on the farm.

Till & Pack

Authors: Cody Searle & Jordan Searle
Instructor: Lex Godfrey
School: Burley High School
City & State: Burley, ID

Bill of Materials

Qty	*Description*
1	4″ x 4″ x 0.250 40′
1	6″ x 4″ x 0.250 40′
1	5-5/8″ x 0.258 sch. 40 BPE
1	72″ 6″ x 6″ x .250
1	4″ x4″ x 0.250 40′
1	2-3/16″ CR Rd. 72″
1	6″ x 4″ x 0.250 40′
1	6-5/8″ pipe 60″
1	2-3/16″ CR RD
1	10″ sch. 40 22′
1	3/4′ 36 x 48 plt
1	36″ 6″ x 6″ x 0.250
1	HBA2516 2-5/16 GN Coupler
1	3/8″ plate
1	Axle Shelby tube
3	Hydraulic pipe fittings
2	1/2″ pipe

Procedure
First, cut all the metal. Everything is cut out of 1/4″ wall. Cut two pieces of 6″ x 4″ tubing, 22-1/2 ft. long, for the main frame. After that, cut the 4″ x 4″ tubing into 28″ lengths. The 6″ x 6″ will be parallel to each other with the 4″ x 4″ sections going between every 42″.

Square everything up and weld the parts together. For the goose neck tongue, use two pieces of 6″ x 4″ and two pieces of 4″ x 4″. Weld the 6″ x 4″ in a "T" formation. Use the 4″ x 4″ for the braces, which will come off the tongue back to the other 6″ x 4″. Now, weld all the parts of the tongue together.

Use four foot long 4″ x 4″ to hook the tongue to the main frame. Weld them to the main frame standing up straight. Next, set the tongue on the 4″ x 4″ that you have welded up, and weld it. Now, weld the ball hitch on.

Now it is time to fabricate the wheelbase. Use 6″ x 6″ tubing 6″ long that came off the main frame. Weld a 6-5/8″ pipe on the end of the 6″ x 6″ tubing. Next, for the axles, use 5-5/8″ pipe that goes in the 6-5/8″ pipe.

Weld the wheel struts on and then weld Shelby tubing into the 6″ x 6″ struts to mount the wheels on. Use 3/4″ flat metal to for the sides and weld them on.

Next, weld the stub shaft in the end of the 10″ pipe. Put the rings and the bearing on the pipe, and then mount the pipe and rollers on the frame. Finally, build a bridge to go on the tongue to give greater bracing and cut back on the bouncing.

Six-Yard Dirt Scraper

Authors: Kyle Klees & Heath Kurrie
Instructor: Dan Froneberger
School: Burlington High School
City & State: Burlington, IA

Bill of Materials

Plate

6′	3/4″ x 5″ Flat bar
240 sq. ft	5/16″
10 sq. ft.	3/8″
8 sq. ft.	1/2″
13 sq. ft.	3/4″
2 sq. ft.	1″ x 4″ Flat bar

Round Stock

6″	3-1/2″
5″	3″
6″	2″
1′	1-1/2″

Angle

8′	2″
66′	3″
12′	3-1/2″

Channel

45′	8″
7′	6″

Pipe

7′	4″ Schedule 80

Tubing

6″	Mechanical Tubing 2-1/2″
12′	2″ x 5″ x 1/4″
5′	6″ x 6″ x 1/2″
6′	4″ x 6″ x 1/4″
12′	2″ x 4″ x 1/4″

5 ea.	Cylinders
91′	Hoses/Fittings
103′	ID Tube
	Bolts/Washers
	Used Truck Axle

Introduction

As awareness of soil stewardship grows, the need to enhance productivity through soil conservation becomes increasingly important. The primary purpose of this project is to provide a cost-effective way to maintain existing conservation practices and build new ones. A secondary use for this form of equipment will be construction site preparation.

Pan Floor/Box Sides

Cut eight pieces of 3″ x 3″ x 3/8″ angle to length for pan frame.

Cut two pieces of 5/16″ x 84″ x 76″ plate steel for pan floor.

Position pan floor plates together. Tack weld.

Lay out 3″ x 3″ x 3/8″ angle for pan floor frame on pan floor. Tack weld. Check squareness. Weld out all joints.

Lay out length and holes from 6″ x 1/2″ flat bar for cutting edge support.

Position in place cutting edge support on pan floor. Tack weld. Check squareness. Weld out.

Lay out and cut steel plates for sides.

Position side plates on edges of pan floor. Tack weld. Check squareness. Weld out. (Use temporary braces to hold side square.)

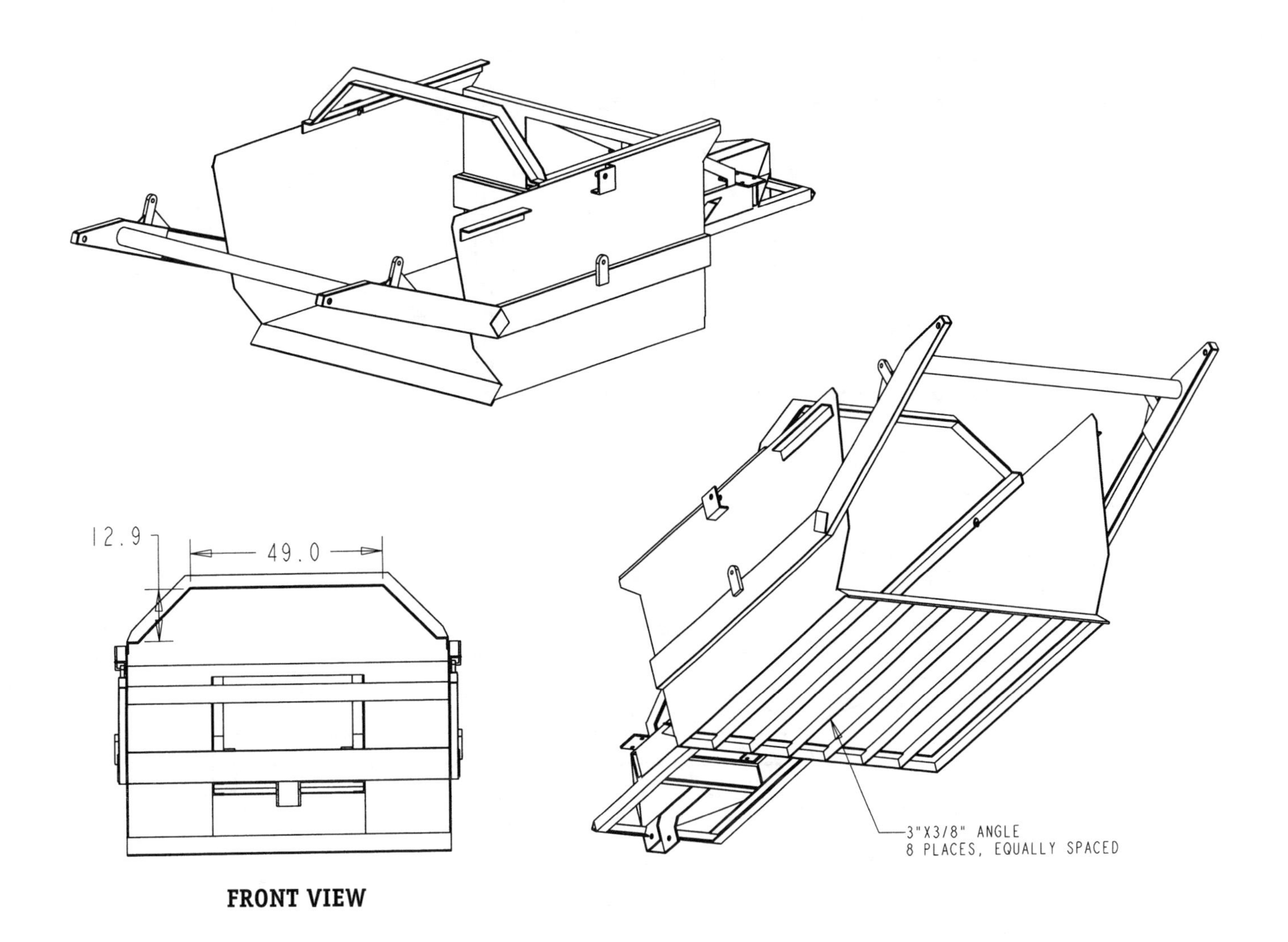

FRONT VIEW

Main Frame/Axle Assembly
Cut five pieces of 8″ standard channel for rear axle supports.

Box channel for rear axle main frame. Tack weld together.

Position channel "tubing" in place. Tack weld. Check squareness. Weld out all joints.

Cut 3″ x 3″ x 3/8″ angle for rear top brace and diagonal axle braces.

Clamp braces in place. Check squareness. Weld out.

Lay out axle support frame assembly.

Tack frame members together. Check squareness. Weld out.

Position axle support assembly in place on axle support.

Put diagonal axle braces in position. Tack weld. Check squareness. Weld out.

Lay out and cut two 1/2″ steel axle fastener plates with four 7/8″ holes in each.

Tack weld axle fasteners in place. Check hole alignment. Weld out.

Lay out, cut gussets for strengthening rear axle frame, axle assembly and fastener plates.

Place gussets in needed locations on rear axle assembly. Tack weld. Check squareness. Weld out.

Cut four pieces of 8″ standard channel for upper and lower main frames.

Cut two pieces of 1/2″ plate steel for boxing of 8″ channel for pulling points.

Clamp plates in place on upper channel ends. Weld out.

Lay out and drill 1-1/2″ holes for pulling point pins.

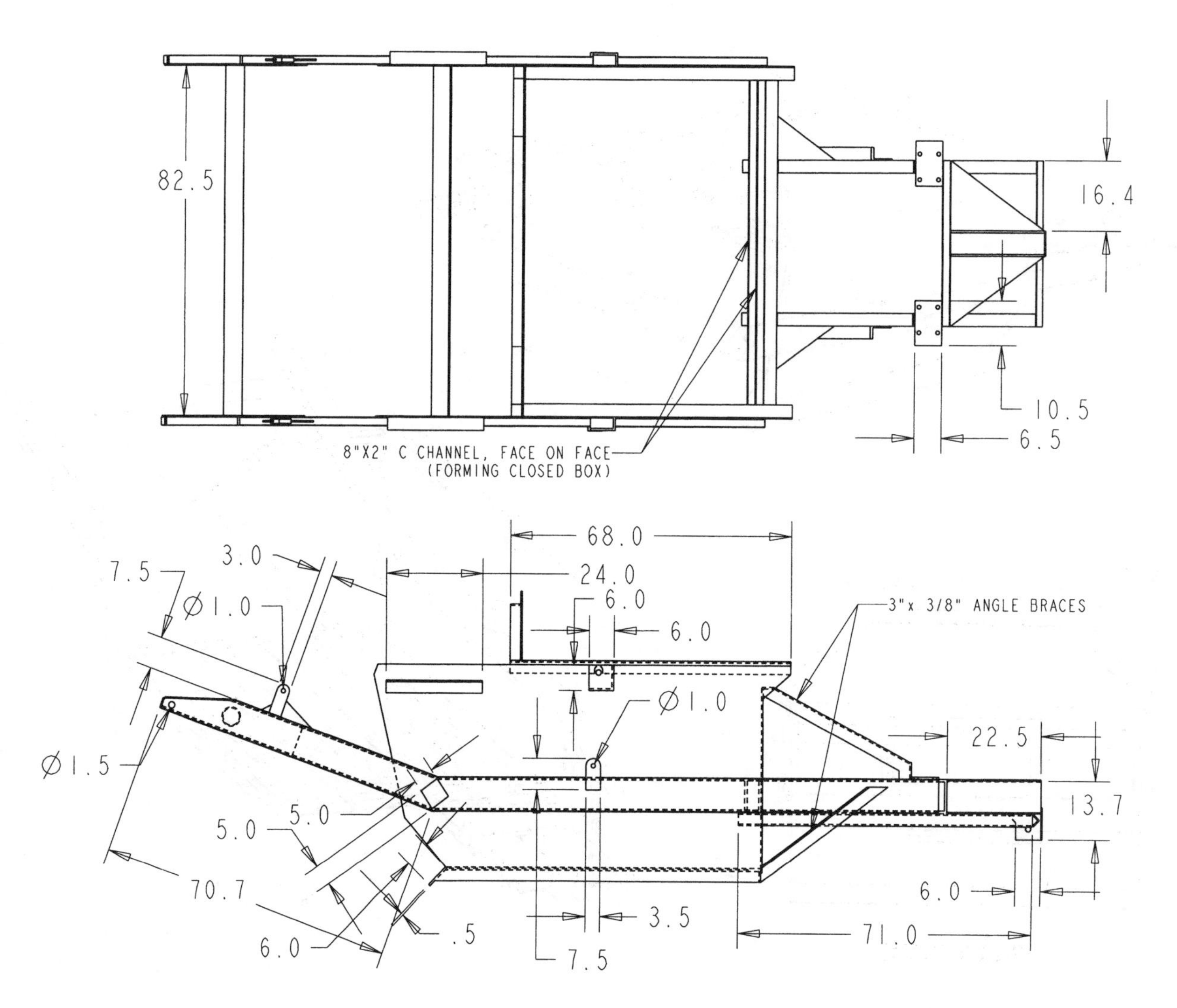

Position lower 8″ channel from pieces on sides of scraper box. Tack weld into place. Check that it is parallel to pan floor frame. Weld out all joints.

Position upper side frame in place. Tack weld together at angled ends. Measure distances for levelness and squareness (very critical). Weld out.

Cut 4″ schedule 80 pipe 82-1/2″ long.

Position behind pivot point on frame ends. Tack in place. Check squareness. Weld out.

Cut two pieces of 3″ x 3″ x 3/8″ angle iron 75″ long for top ejector guides. Notch at 31 ?″ back for clamshell pivot bolt head. Check that it is parallel to pan floor frame. Weld out all joints.

Remove temporary bracing inside scraper box.

Cut three pieces of 3″ x 3″ x 3/8″ angle iron for front box arch brace.

Tack arch together, position on front of ejector guide angles. Tack weld to guide angles. Weld out all joints.

Ejector

Cut two pieces of 5/16″ plate 40-1/2″ x 60″ for ejector faceplate.

Cut 2-1/2″ x 5/16″ flat bar for vertical face stiffeners, notch for 6″ channel.

Cut one piece of 6″ standard channel 81″ long for horizontal stiffener.

Position plates together. Tack weld in place.

Lay out 6″ channel for horizontal stiffener in place on ejector plate. Tack weld in place.

Lay out vertical face stiffeners on ejector plate. Tack weld in place. Weld all joints.

Lay out, cut and drill 7/8″ hole in 5/16″ steel plates for ejector support rollers.

Cut two pieces of 5-1/2?″ x 1-1/2″ CCR center drill 7/8″ hole for ejector support rollers.

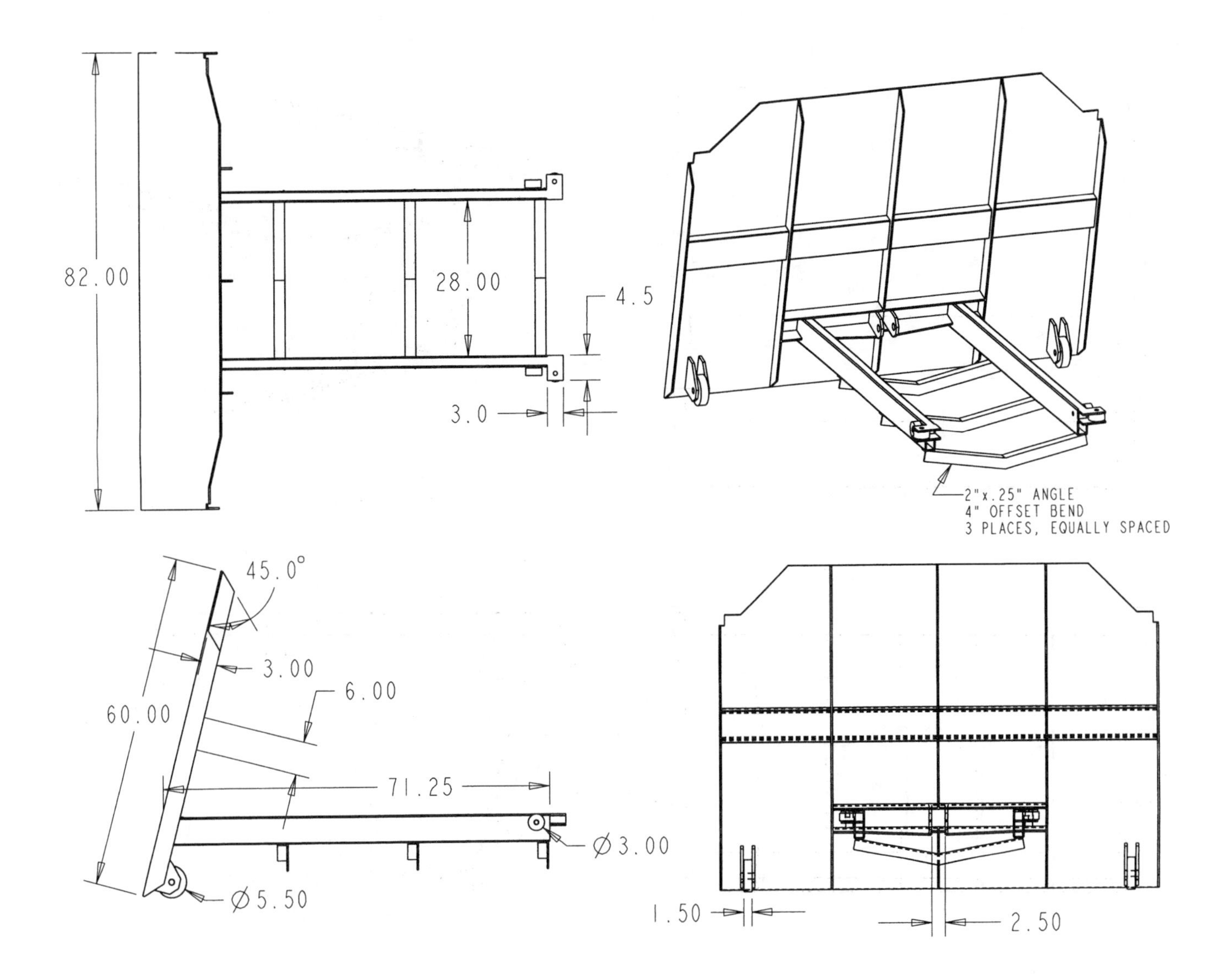

Use two plates on each side of ejector for roller support. Tack weld in plate. Check hole alignment. Weld out.

Cut two pieces of 3-1/2" x 3-1/2" x 3/8" angle, two pieces of 3-1/2" x 1/4" flat bar for ejector track guide. Tack flat bar to angle to make 3-1/2" x 3-1/2" x 6' "C" channel for ejector track guide.

Clamp ejector track guide to axle support frame. Check for edge flushness and proper placement on frame. Weld out.

Lay out, cut and drill 1-1/2?" hole in two pieces of 3/4" plates for rear cylinder ejection attachment brackets.

Position plates on rear of axle support frame. Tack in place. Check hole alignment (critical).

Lay out and cut gussets for rear cylinder attachment plates.

Position gussets in place, tack weld, weld out all joints on gussets on cylinder attachment plates.

Cut two pieces of 3" x 3" x 3/8" angle 14" long to connect ejector track guide to rear cylinder attachment.

Position 3" x 3" x 3/8" angle in place on ejector track guide in rear cylinder attachment plates. Tack weld in place. Check for levelness. Weld out.

Lay out, cut and drill 1-1/4" hole in two pieces for front ejection cylinder attachment plates for attaching cylinder to ejector using 4" x 5" x 1/2" steel plates.

Center cylinder attachment plates vertically on ejector plate. Tack weld in place. Check hole alignment (very critical).

Lay out and cut front cylinder attachment gussets from 1/2" steel plate.

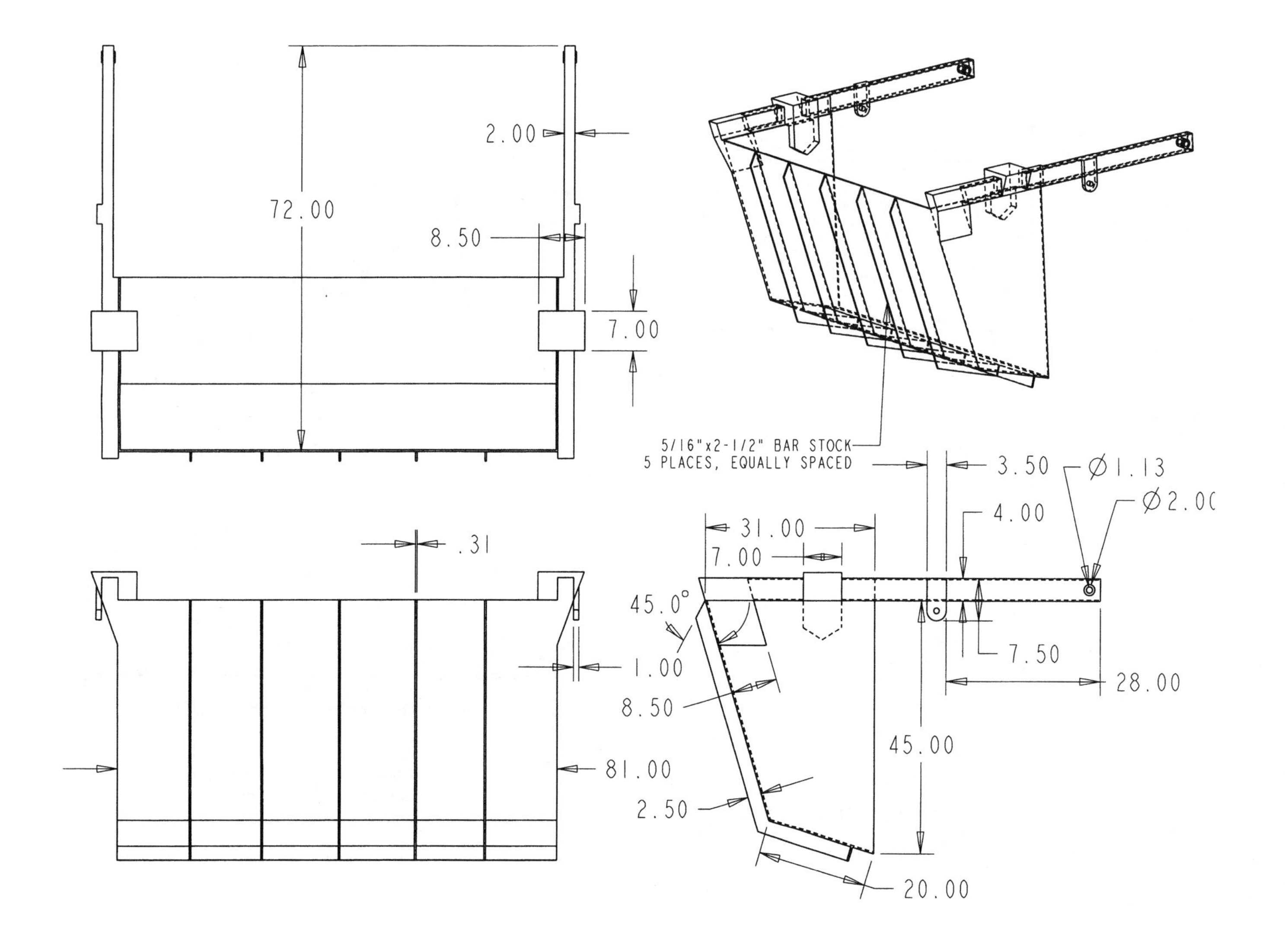

Tack gussets to front cylinder attachment plates. Weld out all joints.

Lay out, cut 6′ long and drill 3/4″ hole in two pieces of 2″ x 5″ x 1/4″ rectangular tubing for ejector guide.

Lay out, cut and drill 3/4″ hole in two pieces of 3″ x 4″ x 5/16″ flat bar.

Position flat bars perpendicular to ends with hole of ejector guide tubing. Tack weld; check squareness, hole alignment and spacing. Weld out.

Cut four pieces of 3″ CRR 1-1/4″. Center drill 3/4″ hole for ejector guide rollers.

Bolt rollers on ejector guide tubing.

Install ejector plate in the most forward position inside scraper box. Temporarily weld in place with scrap metal. (Note: A lifting eye was welded on for moving ejector plate.)

Position both guide tubes with roller in track and ejector ends touching ejector plate. Tack weld in place. Check measurements for tubes to be parallel with pan floor, square with ejector plate.

Cut three pieces of 2″ x 2″ x 1/4″ angle 28″ long and cut one side of angle in middle of each piece. This is for ejector tubing spacers.

Bend angle in middle so there is a 4″ "arch" in middle.

Lay out and cut gussets from 1/4″ plate for center of "arch."

Clamp gussets to ejector tubing spacers. Weld out joints.

Space ejector tubing spacers equally on bottom of ejector tubing. Tack weld in place. Weld out all joints on ejector tubing and ejector plate.

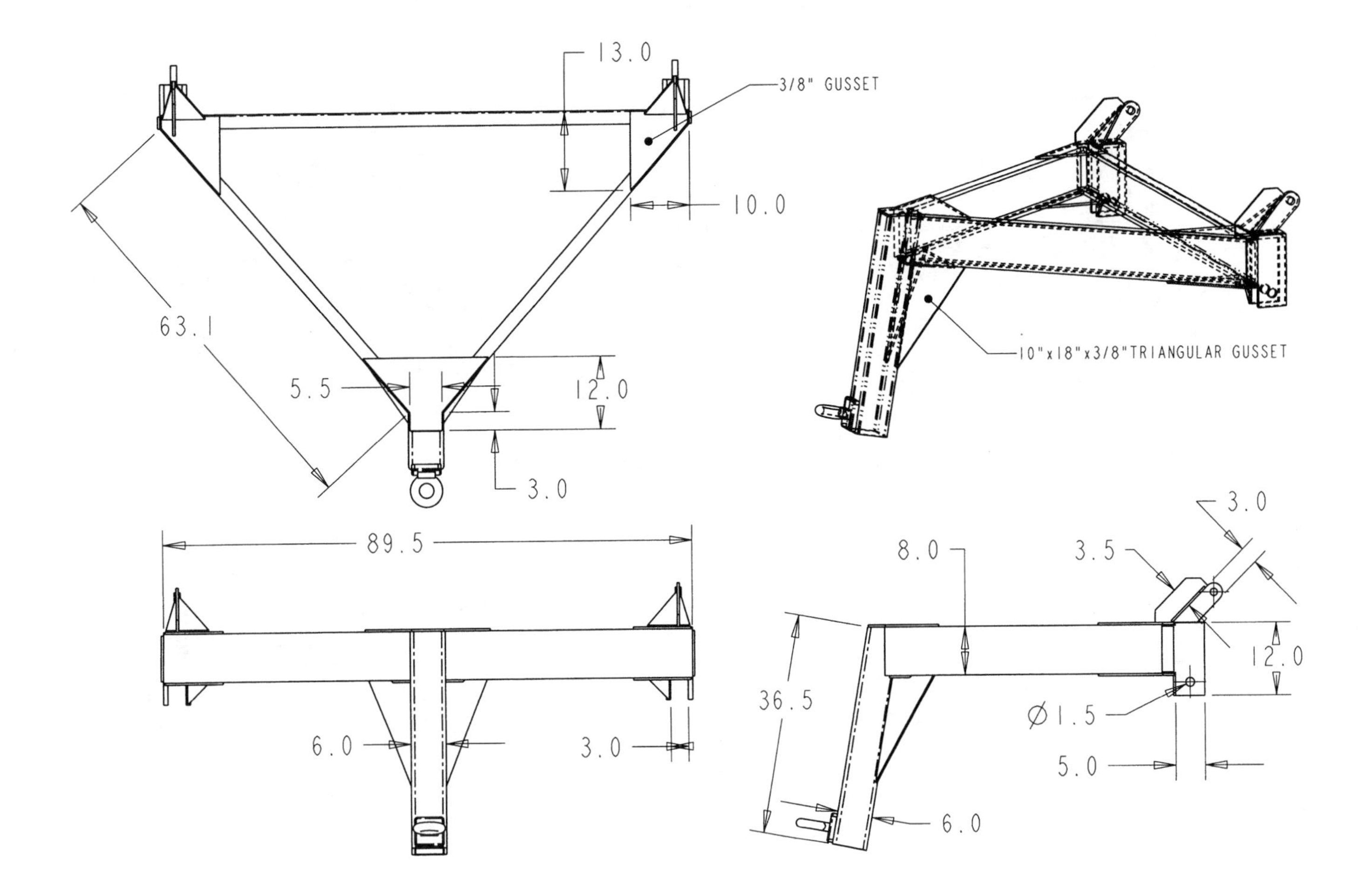

Clamshell

Cut four pieces of 5/16″ plate steel for clamshell.

Position sides of clamshell into place on upper front clam plate. Tack weld.

Position lower front clam plate in place. Tack weld. (Note: use scrap steel to hold sides square.)

Cut 2-1/2″ x 5/16′ flat bar for clamshell stiffeners.

Lay out placement of stiffeners on clamshell. Tack weld in place. Weld out all joints.

Cut two pieces of 2″ CCR. Center drill 1-1/8″ hole for pivot bushings on clamshell raise arms.

Cut two pieces of 2″ x 4″ x 1/4″ rectangular tubing 6′ for clam raise arms.

Lay out 2″ hole for raise arm bushings using cutting torch for making holes.

Install raise arm bushing. Tack weld in place. Check for squareness. Weld out.

Lay out and cut six pieces of 1/2″ plate steel for out pivot point supports for clamshell arms.

Position pieces for each arm in place. Tack weld. Check squareness. Weld out.

Lay out for hole. Drill 1-1/8″ hole through box side and outer pivot supports.

Cut two pieces of 3″ x 3″ x 3/8″ angle 24″ long for clamshell rest.

Clamp angle 28″ forward and level with bottom of pivot point supports. Weld out.

Bolt raise arms in place with 1-1/8″ bolt.

Cut plates for attaching clam raise arms to clamshell.

With arms in place, lift clamshell into position. Space arms evenly apart, clamp attachment plates to raise arms. Securely tack weld to clamshell raise arms. Check operation with "A"-frame hoist. Weld out. (Note: A lifting eye was welded on for moving clamshell.)

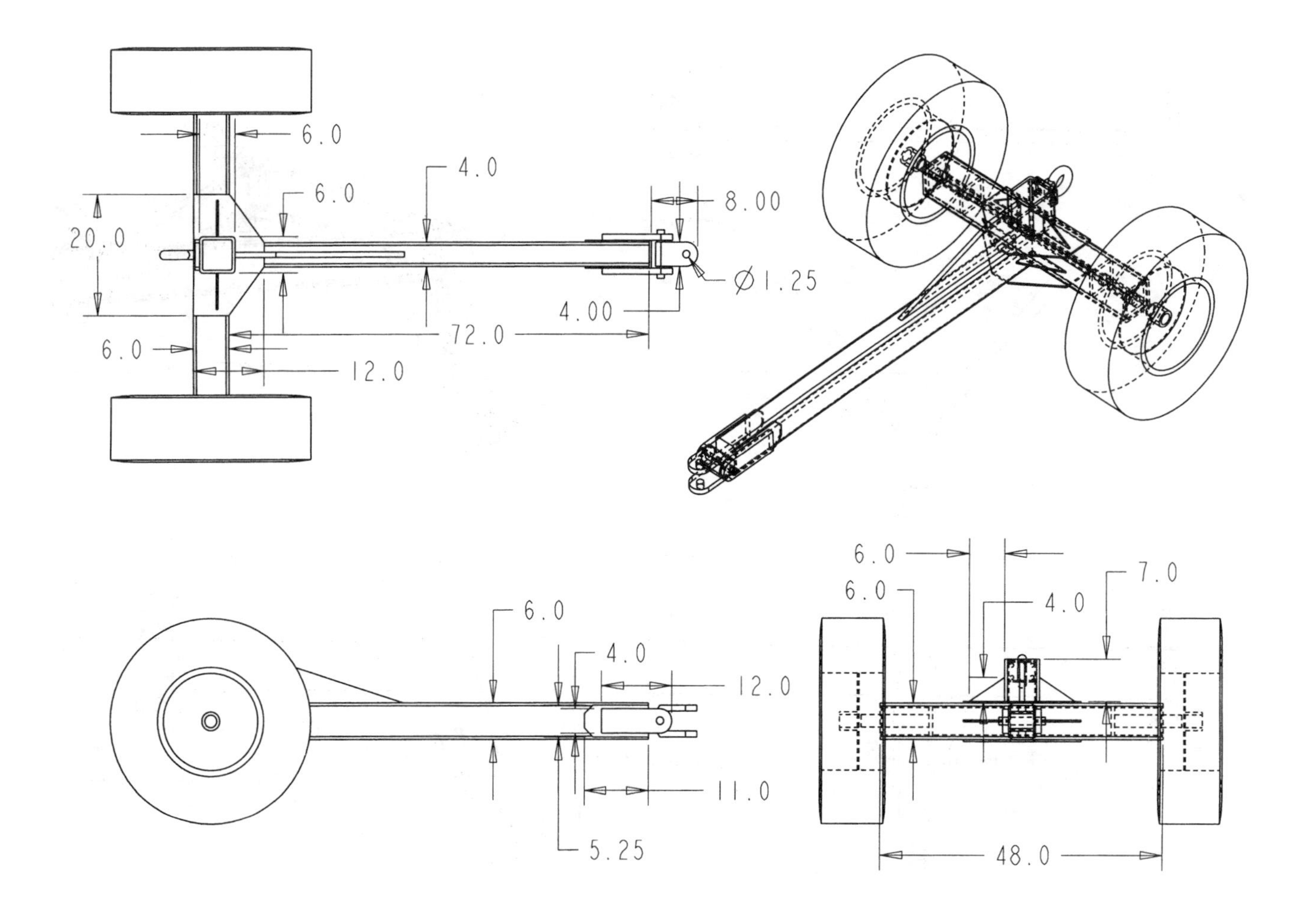

Lay out and cut four 1″ x 4″ flat bars 8″ long for cylinder attachment.

Lay out and drill 1″ hole in each piece.

Position flat bar on 8″ channel frame. Tack weld in place. Check squareness. Weld out both sides.

Clamp flat bar in place on clamshell raise arms. Weld out. (Note: center to center of holes is 28″ in "home" position.)

Tongue

Lay out and cut 6″ x 6″ x 1/2″ rectangular tubing 30″ long for dolly attachment hitch.

Weld "Pintle donut" to end of dolly attachment in hitch tube.

Lay out and cut three pieces of 8″ standard channel for front transition hitch. (Note bevel front ends on channel flanges to accommodate dolly attachment tube.)

Arrange 8″ channel in a triangle. Tack weld. Check bevel angles. Tack weld temporary bracing on front of transition hitch.

Arrange dolly attachment tube to transition hitch frame at 126° forward angle. Tack weld in place. Check squareness. Weld out all joints. Remove temporary bracing.

Lay out, cut and drill 1-1/2″ hole in four pieces of 5″ x 3/4″ flat bar, 12″ long, for pivot point on transition hitch.

Position pivot point flat bar on transition hitch frame. Tack weld. Check alignment and squareness. Weld out.

Lay out, cut and drill 1-1/2″ hole in four pieces of 1″ x 3″ flat bar for lift cylinder attachment brackets.

Position lift cylinder brackets in place. Tack weld. Check squareness, center to center for holes. Weld out. (Note: this is critical for level raising of implement.)

Lay out and cut gussets for transition hitch lift cylinder brackets and dolly attachment tubing from 3/8″ plate steel.

Dolly

Lay out and cut one piece of 4″ x 6″ x 1/4″ rectangular tubing 72″ long for dolly tongue and 6″ x 6″ x 1/2″ square tubing 4′ long.

Position pieces in Tee position in center of axle housing tube. Tack weld. Check squareness. Weld out.

Lay out, cut and drill four pieces of 5" x 5" x 1/2" plate steel for spindle suspension support.

Position spindle suspension supports on spindle. Tack weld. Check squareness.

Weld out. (Note: allow for hub clearance.) Position inside ends of axle housing tube. Weld out all joints.

Lay out, cut and drill four holes in 6" x 6" x 1/2" rectangular tubing 8" long for vertical dolly Pintle hitch support.

Lay out, cut gussets for vertical dolly Pintle hitch support.

Center vertical dolly Pintle hitch tube on axle tube. Tack weld. Check squareness.

Weld out all joints.

Position gussets in place. Tack. Weld out.

Lay out, cut and drill 1-1/4" hole in four pieces of 1" x 4" flat bar 8" long for dolly drawbar pivoting hitch.

Cut one piece of 2-1/2" x 4 5/8" long mechanical tubing with 1-1/4" center hole for dolly drawbar hitch.

Position two pieces of the 1" x 4" flat bar on hitch end of 4" x 6" tongue tube.

Tack weld. Check hole alignment (very critical). Weld out all joints.

Align holes and space 2" apart the remaining two pieces of 1" x 4" flat bar. Tack weld to mechanical tubing. Check squareness. Check hole alignment. Weld out.

Weld Schedule
Remove slag, rush and paint as necessary.

Grind weld joints where needed.

Weld using 1/8" E7018 @ 115 amperes DCRP.

Weld using 1/8" E308-16 @ 115 amperes DCRP (truck axle only).

Weld using 30 CFH 25/75 mixed gas.

Finish Schedule
Weld 3/4" flat washers for hydraulic bulkhead fittings as needed.

Cap all open ends of tubing and channel ends on project.

Align truck axle, weld spindles rigid.

Repack wheel bearings on truck axle. Pack dolly wheel bearings with grease.

Chip weld spatter off all welds with chisel or wire brush.

Sandblast project.

Wipe down project. Prime and paint.

Bend hydraulic tubing to fit as needed. Install hydraulic cylinders and hoses.

Assemble project.

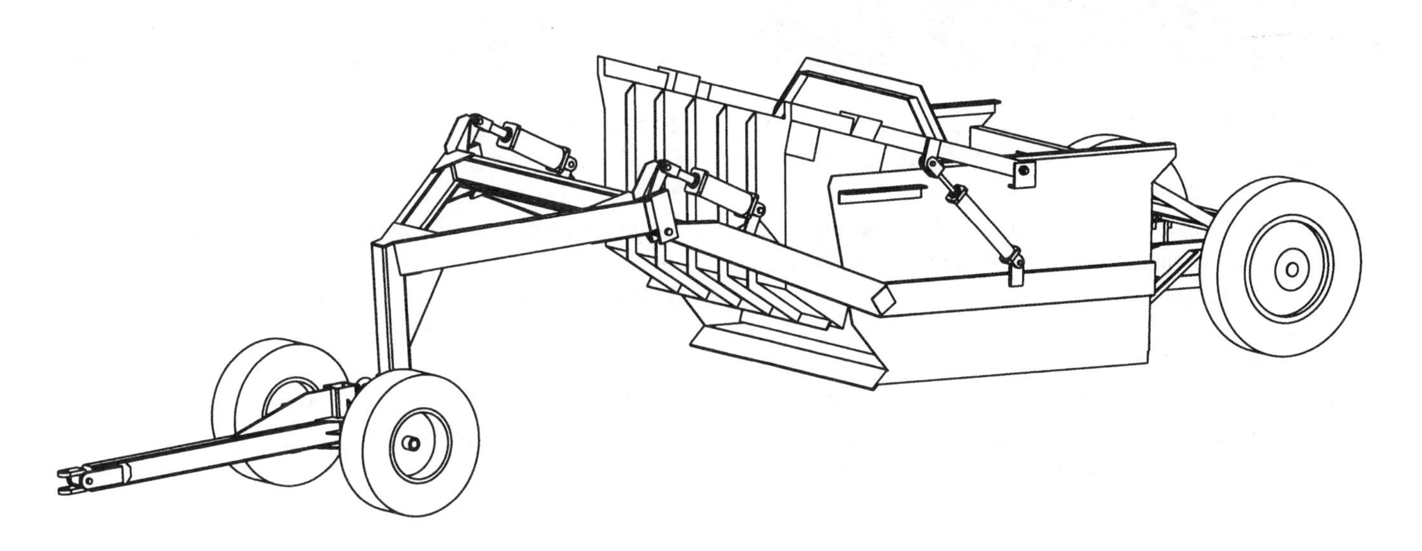

10-Foot Triple Axis Blade

Author: Kyle Bowers
Instructor: Lex Godfrey
School: Burley High School
City & State: Burley, ID

Bill of Materials

Qty	*Description*
1	6″ x 6″ x .250 20′
0.5	6″ x 3″ x .250 20′
1	3/8″ plt 4 x 10
1	Shear and bend blade
1	4″ od 3″ id x 70″ Shelby
1	C1045 2-15/16″ rd
1	1″ 4 x 4 Plate
0.2	1-1/8 CR rd
1	1″ sq
1	4 x 8 cylinder
1	Hoses and fittings
5	Grease zirks
1	Paint

Three-Point Hitch

For the 3-point hitch to the tractor, first cut the 6′ x 6″ and the 3″ x 6″ tubing. The distance between the bottom of the 3″ x 6″ and 6″ x 6″ has to be correct for the stacked 6″ x 6″ to fit correctly, so the angles must be precise.

After fabricating the main frame of the 3-point attachment, cut out the dog ear attachments for the 3-point and drill all the holes. When the holes are finished, put a piece of 1-1/8″ CR round through the holes to ensure proper alignment and weld them in their proper locations.

6 x 6 Extension Frame

After finishing the 3-point frame, cut the main frame that extends from the 3-point back to the blade. This is two pieces of 6″ x 6″ stacked. Clamp the pieces together and stitch-weld them together on one side, then the other, to prevent warping. This assembly can be stitch-welded—it is not necessary to weld it solid, but it is vital that the pieces do not twist. You can also see where the Shelby tube was welded into the cut out circles of the 6″ x 6″ in both the 3-point and the stacked pieces. Note that the outside 3-point attachment is used to cap the 6″ x 6″.

Major expenses in building the blade will be the Shelby tubing and the 2-15/16″ 1045 CR round. The next step is to cut the Shelby tubing and the cold rolled round. Check and double-check your math and measurements before cutting. Next, fit everything up and tack in place before welding to ensure proper and correct fit up. When completed, weld the Shelby tube to the tubing. (I used a wire feed dual shield process.)

Pressure Plates

The large 1″ pressure plates on the back of the blade will transfer the stress of the large blade over an even larger area to prevent bending and breaking.

The top plate is an 18″ circle; the bottom plate is an 18″ circle with two dog-ears with holes for the hydraulic rams. When welding the top plate to the stacked 6″ x 6″ extension frame, stop often to let everything cool to prevent distortion and ensure a proper fit between the two plates.

As shown, the Shelby tube extends through the 6″ x 6″ and also the top 1″ plate. Before welding the Shelby tubing in, drill a hole for a grease zirk.

Blade Attachment

Before starting on the blade attachment, send the 3/8″ 4 x 10 plate to a local fabrication shop to have it sheared and bent. While this is being done, begin work on the box, which attaches to the blade. This is

an intricate box, a wear plate that sits against the back of the blade with a Shelby tube collar for the pivot point. It also has two side plates that have to match the curve of the blade and attach to the back plate, which goes behind the 6″ x 6″ upright.

The channel in the center of the back plate holds the 2-15/16″ pin that allows the blade to pivot. A lot of pressure will be applied to these points, so it is built out of 1″ plate.

Next, cut one piece of 1″ plate that will fit inside the 6″ x 6″ tubing and then cut a circle in the middle of it large enough for the Shelby tubing. Then put the tubing into the 6″ x 6″ tubing and ensure a proper fit-up before welding.

Now, weld the 6″ x 6″ upright to the 18″ circle with the dog-ears. This will be the upright from the blade box to the 6″ x 6″ stacked frame. On the bottom of the 6″ x 6″ upright, cut a hole for the Shelby tube pivot point and weld the Shelby tube in after making sure you have proper fit.

At this point, the blade is ready to weld to the box. Measure and check to make sure the box is in the proper location on the blade. Next, encase the blade out to within 6″ of the end of the blade. Angle it to conserve material and make it look better.

Now fabricate the dog-ears for the hydraulic ram attachment. Form one of the dog-ears into the gusset, which is under the bottom 1″ plate. Make the other four from 3/4″ plate – two for the bottom of the stacked 6″ x 6″, one on the front of the 3-point frame and the last one for the box on the back of the scraper.

Assembly & Finishing

Now it is time to put all the pieces together. Fit the pieces together and weld a cap on the 2-15/16″ CR round to prevent the shaft from falling out. Grind the rough edges. To make sure the blade is completely clean, wipe it down with paint thinner, then paint.

Add the hydraulic cylinders, hoses and necessary fittings. With the hoses on, the blade is ready for the tractor.

Hay Bale Squeezer

Author: Ryan Cessna
Instructor: Dan Saunders
School: Monroe High School
City & State: Monroe, WI

Bill of Materials

Length	*Description*
15′	3/16″ x 3″ square tubing
43″	3/8″ x 3″ square tubing
3′ 8″	3/16″ x 2″ x 3″ rectangular tube
15′ 2″	3/16″ x″ 2-1/2″ square tubing
15′	3″ schedule 80 pipe
2′ 6″	2″ x 1/4″ flat bar
	Quick-attach plate
	Log chain & hook
	2″ x 24″ cylinder
	1/2″ hydraulic hose & fittings
	Case skid loader hydraulic couplers
	Paint and primer

A bale squeezer is an attachment for a skid loader that lifts big square bales or round bales by squeezing them from the ends or sides of the bale and then lifting them. The square bales are 3″ x 3″ and 6″ long; round bales range in size from 5′ to 7′ in diameter. The bales weigh between 900 and 1,200 pounds. This bale squeezer is designed so bales that are wrapped in plastic can be moved without puncturing the airtight bale.

For this project, I used a MIG machine with 0.045 metal core wire to weld the bale squeezer together. The metal core wire creates larger and hotter welds, which increase the strength of the weld. I used a spray transfer method because of the deep penetration that it provides and its ability to weld in many different positions. Metal core wire and spray transfer combination works well with thicker metals such as 1/8″ or greater and when multipass welds are required.

Procedure

Use seamless 3″ schedule 80 pipe for the "U's" of the squeezer and weld them to pieces of 3/8″ x 3″ square tubing 21-1/2″ long. Put a triple pass all the way around where the "U's" meet the 3″ tubing for added strength, since the greatest amount of pressure is exerted on this portion of the squeezer.

To create nice rounded 90-degree bends, cut the ends of the pipe at 22.5 degrees. Then put a 22.5-degree angle wedge in between the two pipes. By making the 90-degree bend with multiple angles in it, the bend is less sharp and it creates more of a rounded look. Space the joints a sixteenth of an inch apart to get better penetration and fusion with the metal core wire.

Cap the ends of the square tubing with pieces of 1/2″ x 3″ x 3″ flat bar. After welding the joints, smooth out all the sharp edges by grinding and air sanding. This ensures there will not be any sharp edges for the plastic-wrapped bales to catch on while they are being moved.

Make the back of the squeezer next. Use 3/16″ x 3″ square tubing and 3/16′ x 2″ x 3″ rectangular tube to hold the 3′ tubing together. The 3/16″ square tubing provides a 1/16″ clearance between the back of the squeezer and the sliding portion of the "U's."

Cut the four 3″ square tubes 45″ long and line two of them up and weld them together with the welded seam of the tubes up so later the inside tubes will not bind on them. Do the same thing for the remaining two. Cut 45-degree angles on the band saw and capped the ends of the 2″ x 3″ square tubing with a piece of 1/4″ flat stock to give it more of a smooth, finished look. Also, cap the ends so that water cannot get inside the tubing and rust it out. The 45-degree angles make the edges less sharp and so they provide more of a finished look.

For the stationary side of the hydraulic cylinder, weld a short piece of 3/16″ x 3″ square tube 10″ long behind one of the 2″ x 3″ rectangular tubes. For added safety, put raised expanded metal on the top of the back part of the squeezer so you do not slip on it when getting in and out of the skid loader.

Assemble the portion of the "U's" that slide in and out of the back of the squeezer out of 3/16′ x 2-1/2″ square tubing. Cut the 2-1/2″ tubing 45-1/2″ long and slide the two tubes into the back of the squeezer.

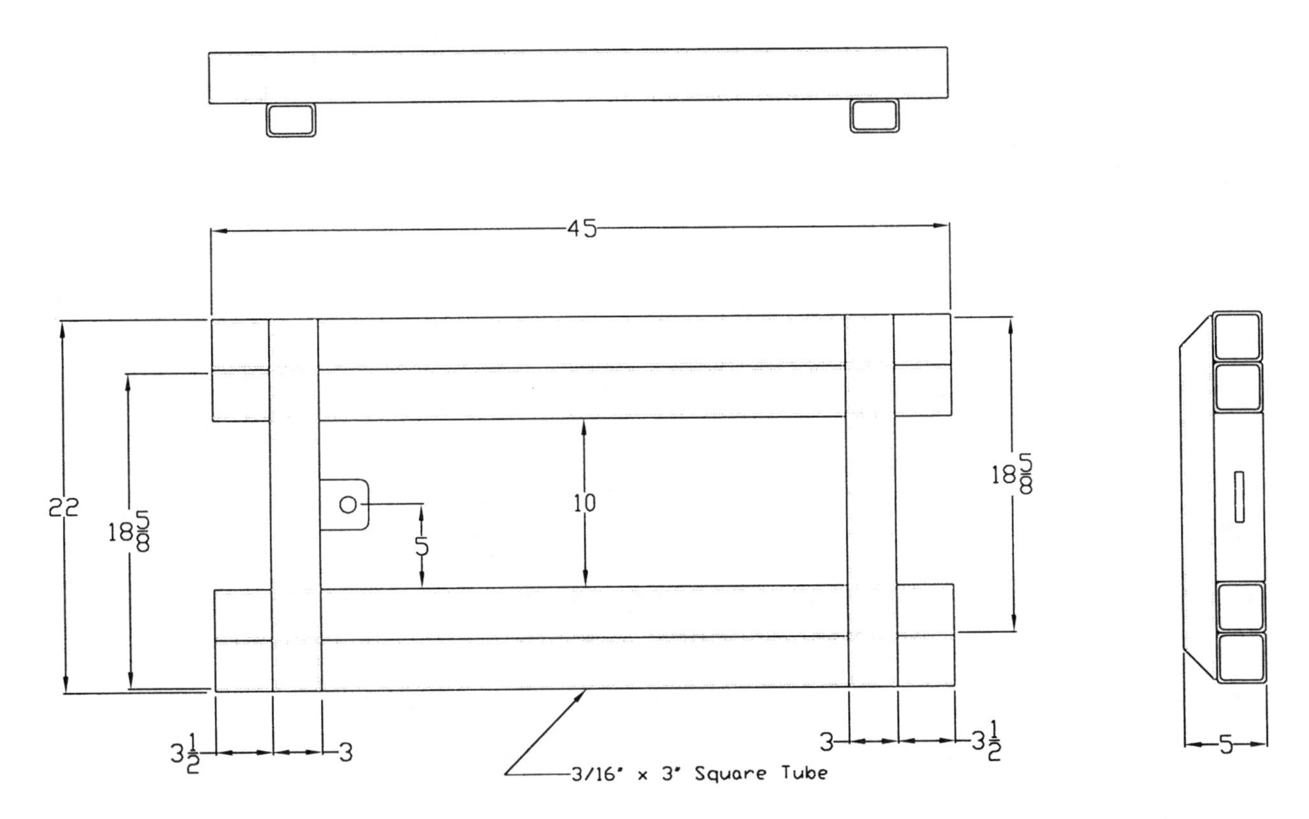

BACK FABRICATION

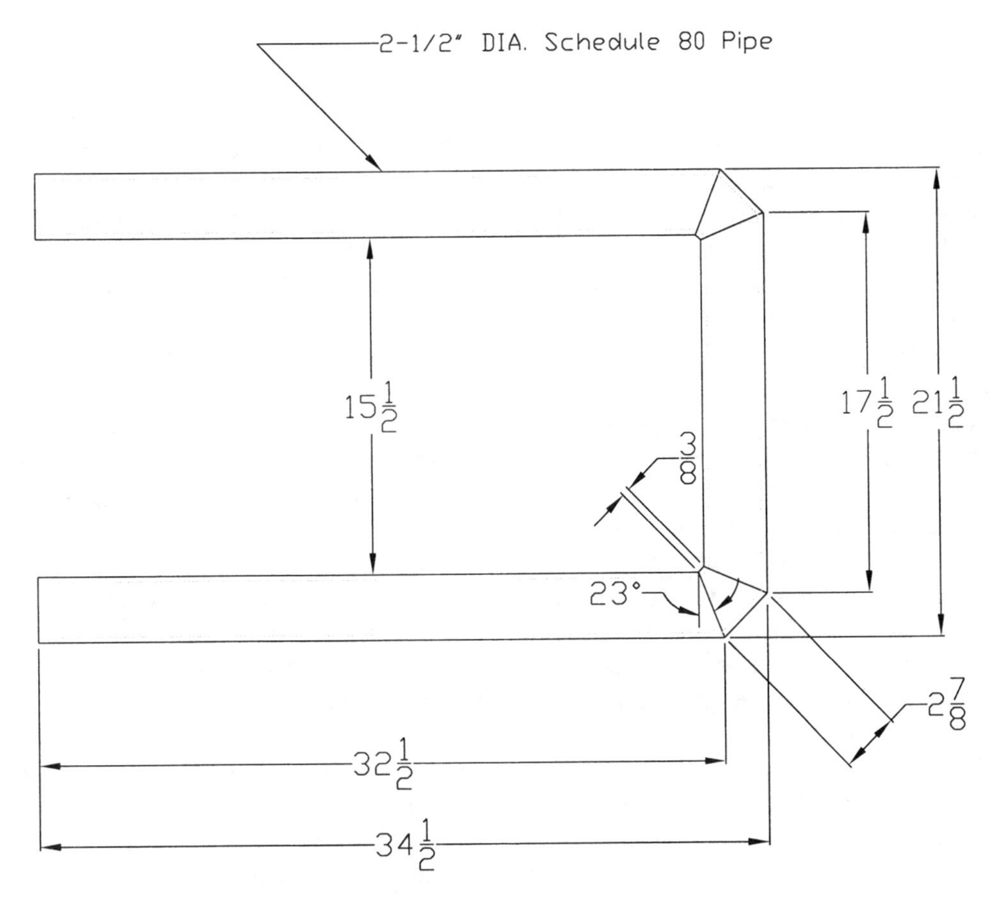

"U" FABRICATION

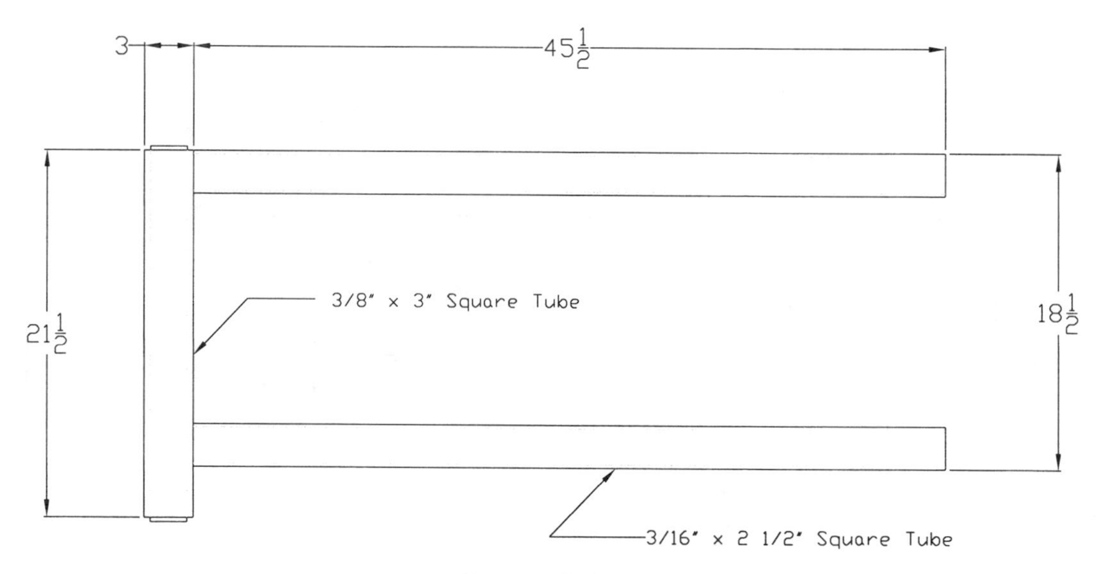

COMPLETE "U" FABRICATION

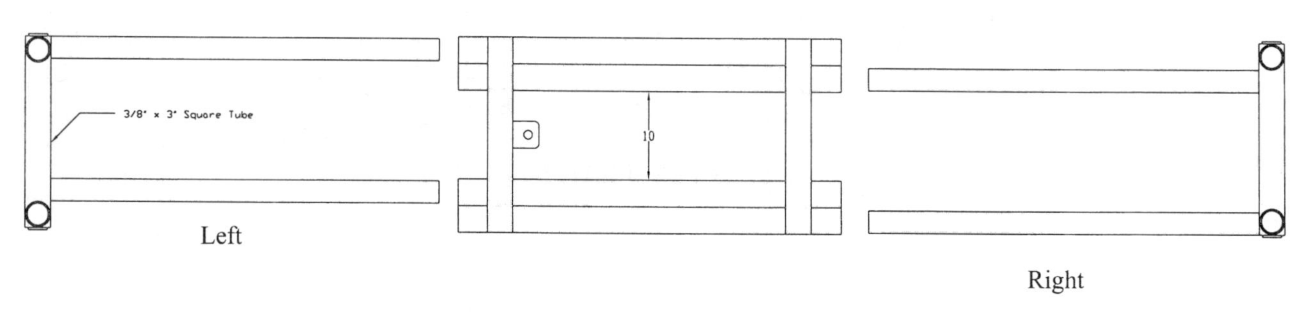

ASSEMBLY

Wedge both of them to the same side of the tubes so when you weld them to the "U's" they will line up correctly. After tacking the "U's" in place, remove them from the back of the squeezer and tack a piece of angle iron on the free end of the tubes to help prevent them from moving while they are being finish welded together. Do the same thing for the remaining tubes.

Weld the quick-attach plate on the backside of the squeezer. The quick-attach plate allows you to switch skid loader attachments easily and efficiently. The plate is also compatible with most skid loaders. It can be purchased from a local welding shop to save time. Place the plate in the center and as far to the bottom of the squeezer as possible to take advantage of the maximum lift capacity of the skid loader. Put 2" stitch welds along the edges of the plate where it contacts the back portion of the squeezer.

Make the hydraulic cylinder brackets out of a piece of 1/2" plate. Then weld 1/4" spacers on both sides to bring the thickness up to 1". The brackets should be 1" thick due to the tremendous amount of pressure that the cylinder puts on the brackets and also the hitch on the cylinder is an inch apart. This provides a tight fit between the cylinder hitch and the brackets to prevent the brackets from bending while under pressure. Also, weld in 1/4" triangle supports on both sides perpendicular to the brackets to help prevent them from bending under pressure.

When installing the hydraulic hoses, route them so they will be out of the way of the bale and other moving parts.

To prevent the side that is not attached to the cylinder from sliding out while under pressure, build in 2' of 3/8" log chain and matching chain hook. Weld the chain to the back of the squeezer and the hook to the "U" so the chain side of the squeezer can still move. The chain limits how far it is allowed to extend out. When you squeeze a bale, it pushes the bale against the chain side until the chain is tight.

After everything is welded, go over the whole project with a grinder to remove all sharp edges that would puncture the plastic on the bales. Use an air sander to smooth the surfaces even more to prepare for paint. Apply a coat of primer with rust inhibitors and three coats of finish paint.

Irrigation Diker

Author: Brent Bean
Instructor: Lex Godfrey
School: Burley High School
City & State: Burley, ID

Bill of Materials

Qty.	*Material Used*
1	1″ x 4′ x 8′ Plate
4	2″ x 2″ x 0.250 Square Tubing
3	0.50″ x 8′ Flat
1	0.50″ x 6′ Flat
2	2″ x 20′ Cold Rolled Shaft
1	0.750″ x 4′ x 8′ Plate
3	4″ x 4″ x 0.250″ x 40′ Square Tubing
1	300″ x 7″ x 5″ x 0.250 Square Tubing
1	240″ x 4″ x 3″ x 0.250 Angle Iron
1	Advanced Key Way
0.5	4″ x 4″ x 0.50 Angle Iron
2	Bolts and Nuts
4	Bearings
9	Hub and Inserts
1	12′ x 0.50″ Key Stock
4	Turnbuckles

Procedure

First, shear the three 8′ flat plates into lengths 12″ long, making 54 paddles. Next, shear each side to create a 4″ wide point centered on the opposite end. Cut a 45 degree angle on the small end of the plate to make an edge for the Diker paddles. Punch three 1/2-in. holes in the center of the plate so that they can be adjusted as they wear down with time.

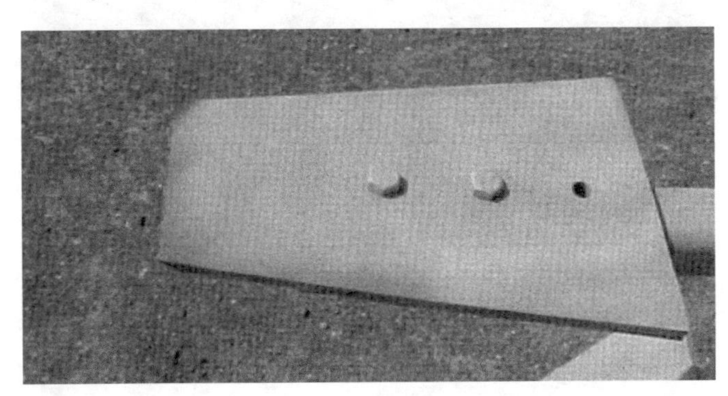

Take the 2″ x 2″ x 0.250″ square tubing and cut 54 arms at 18″ lengths. On one side, cut a 45 degree angle and drill two 1/2-in. holes to bolt the paddles to.

Plasma cut the 1″ inch plate into nine 16″ circles with a 5″ center hole and six 2″ x 4″ wide by 4″ deep, cuts off center at a 45 degree angle. In these holes, weld in the 2″ tubing for the paddle arms on the diker. When welding in the center hole, offset each key 15 degrees so that the paddles do not hit at the same time.

Key the 2″ shaft so that the diker wheels can be held in place; cut it into two sections, one 14′ long holding five diker wheels; and the other 11′ long holding four diker wheels.

Using the 7″ x 5″ x 0.250 rectangular tubing, find the center point, and measure 18″ off center, and then every 36″ after that, down each side. This tubing is where the rippers are going to be mounted. Offset the frame bars so they are not where the rippers and spiders need to bolt onto the frame. Using the 4″ x 4″ x 0.250″ tubing, cut six 11″ lengths and weld them onto the 5″ x 7″ tubing for the top bar on the frame, then weld the 4″ x 4″ tubing on top of it. Angle down both sides of the tubing to support the weight of the frame better.

Cut eight 6″ lengths of 4″ x 4″ x 0.250″ tubing and weld them onto the 5″ length of 5″ x 7″ tubing. This is the part of the frame that you will mount the spiders onto. Cut eight 20″ pieces and weld them onto the 4″ x 4″ tube in line with the other ones. Cut another piece of 4″ x 4″ tubing 25′ long and weld it to the back end of the 20″ cuts.

Cut four pieces of 4″ x 4″ tubing at 60 degree angles and weld them down form the top 4″ x 4″ frame, and run them to the farthest 4″ x 4″ tubing to give additional support to the frame.

Next, using the 4″ x 4″ x 0.250″ angle iron, cut eight 18″ lengths and drill two 1-in. holes on each piece of angle iron. Weld them onto the 4″ x 4″ tubing with a 4″ space between each set for mounting the Diker wheels.

Using cardboard to make a pattern, cut 1″ plate into four copies of an ear needed to make the three-point mount system. Drill the two 1″ holes in them while they are tacked together to make sure they are in line and will line up once mounted.

Repeat this process for the top two ears by making another pattern for them. Also, weld these together for drilling to make sure they line up and will mount nicely on the frame. After mounting them, cut 1″ thick support pieces of flat metal to support the ears and prevent them from twisting or warping from the rigors of working the field.

Finally, create the bracket needed to connect the diker wheels to the frame using 4″ x 4″ x 0.250″ lengths of tubing. Drill two holes on one end to bolt the bearings to. Then drill clear through the other side and weld in a pipe so that the pin will ride more smoothly in it and support the weight more evenly throughout the connection bracket. Also, weld a 1″ thick piece of scrap metal with a 1″ hole in it to connect the turnbuckle to the frame and the diker. The turnbuckle allows the depth of the paddles to be adjusted and controlled better from field to field.

15-Foot Enclosed Loading Chute

Author: Christopher Wieneke
Instructor: Marlin Buus
School: Adrian High School
City & State: Adrian, MN

Bill of Materials

Qty.	*Description*	*Length*
Metal for Frame		
2	4 in. I-Beam	15 ft.
7	4 in. Channel Iron	5 ft.
14	3 in. Channel Iron	83 in.
7	3 in. Channel Iron	81 in.
2	3 in. Channel Iron	8 in.
24	2 in. Flat Iron	12 in.
14	1-1/2 in. Square Tubing	3 ft.
Door		
2	3 in. Channel	51-1/2 in.
2	3 in. Channel	32-1/2 in.
Tongue		
1	2 in. Shaft	6 ft.
1	2-1/2 in. Round Pipe	4 ft.
1	2-1/2 in. Square Tubing	6 ft.
2	2 in. Flat Iron	6 in.
2	3 in. Channel	1 ft.
1	3 in. Square Tubing	1 ft.
2	2 in. Flat Iron	10 in.
Wood for Sides		
6	2 x 12	144 in.
2	2 x 10	144 in.
2	2 x 8	144 in.
2	2 x 12	36-1/2 in.
1	2 x 10	36-1/2 in.
1	2 x 8	36-1/2 in.
2	2 x 12	32-1/2 in.
1	2 x 10	32-1/2 in.
1	2 x 8	32-1/2 in.
Wood for Floor		
2	2 x 12	15 ft.
3	1 ft. wide timbers	15 ft.
Tin		
7	small front, side, top	35-1/4 in.
7	big back, side, top	144-1/2 in.
8	side channel for tin	83 in.
4	side channel for tin	60 in.
2	side channel for tin	53-1/2 in.
2	side channel for tin	144-1/2 in.
Additional Items		
2	4,000 lb. Shelby Jacks	
1	used wagon axle	

Fabrication

Depending upon the used axle you have to work with, setting the frame on it may require modifying the axle. For a 5 ft. wide wagon axle like the one used for this project, weld two pieces of 8-in. long, 3 in. channel, four inches from the wheel. On top of the channel iron, weld a piece of flat iron so you will have more to weld the frame to. Next, take four pieces of 1-1/2 in. square tubing and cut a 45 degree angle on each end. Weld them to each end and each side of the axle to give more support.

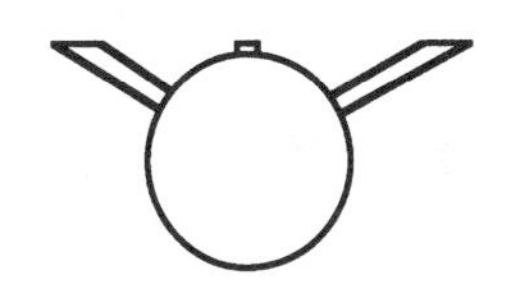

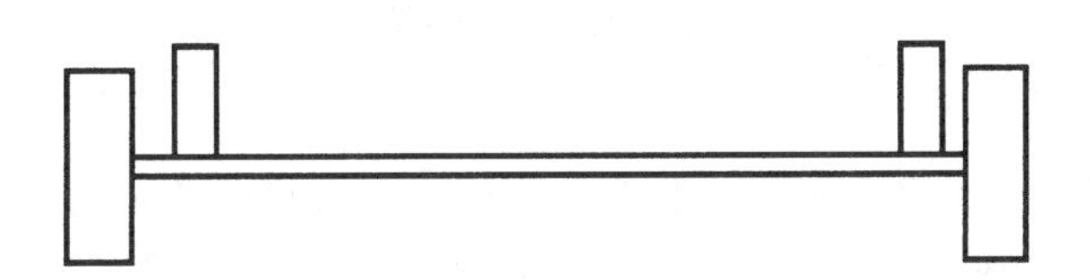

To assemble the frame, take two 15-ft. long, 4-in. I-beams and space them four feet apart. To keep the frame square, weld two of the supports on right away. Put the 3-in. channel flush with the end of the I-beams. Measure to make sure they are square and weld them into place.

Next, attach the frame to the axle, making sure the front is 36 in. off the ground so it will line up with the back of a trailer. The last 3 ft. have to be flat, so measure back 3 ft. and set that mark 36 in. above the floor. Then slide the axle back up under the frame until the two are touching. When the axle is lined up to the frame, weld the supports to the frame.

The next task is to flatten the one end so it will sit flat on the ground. Make a cut right underneath the top of the flat part of the I-beam. Cut up about 15 in. Next, make a cut in a straight line back to the end of the I-beam, but cut towards the bottom of the I-beam. Heat it up where the bend will start. When it is red hot, pound it down flat, then weld the gap shut. Repeat this process on the other side.

To make sure the 3 ft. length on the other end stays flat, make a sawhorse 36 in. tall. Set the edge of the sawhorse on the 3 ft. mark. With the torch, cut out a triangle shape on each side. Bend down the end until it is perfectly level, and to hold it there, weld the gap shut.

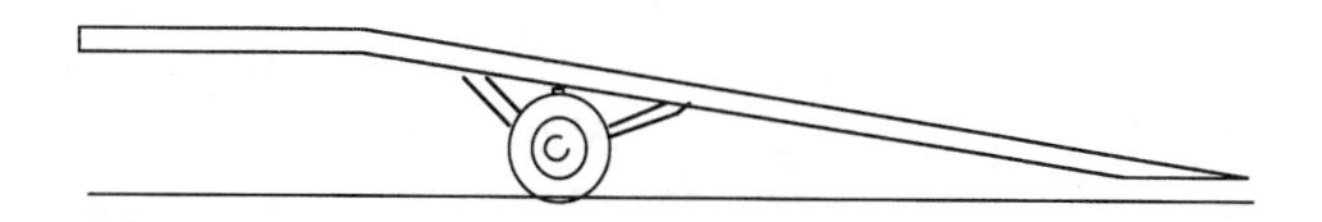

After the frame is assembled to the axle, weld the rest of the supports on, three feet apart. At the bend, put two right next to each other for extra support.

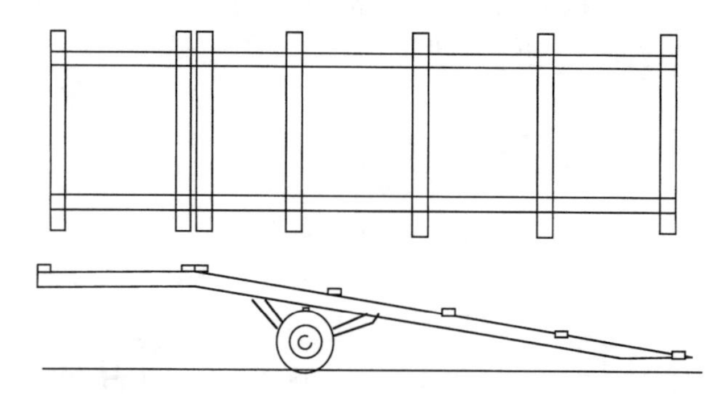

Now, start attaching the uprights, using the 3 in. channel. On the top end of the frame, the channels will be 83 in. tall. When you get to the bend, you will have to make a 15 degree cut on the bottom and top of each channel iron. Weld the uprights to each of the

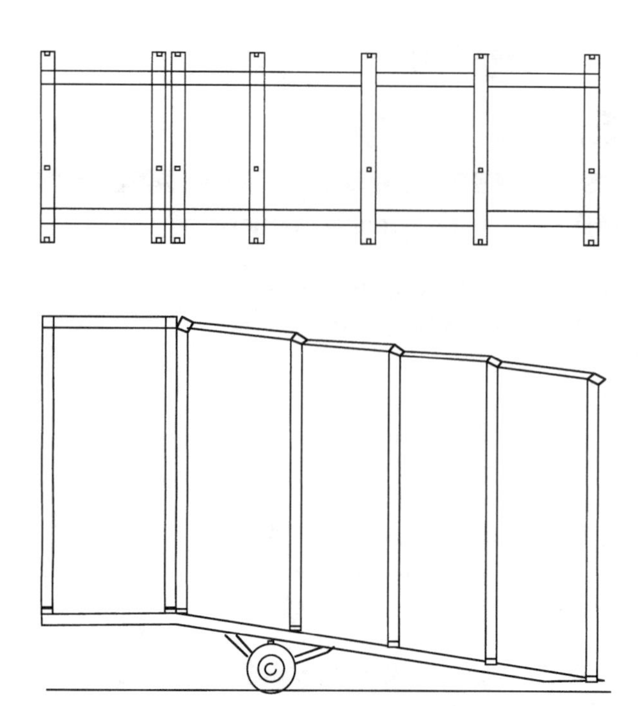

4 in. cross pieces on the frame. Center and move them to the edge of the cross pieces, then weld them into place. On the bend, put two right next to each other for strength.

After attaching the uprights, weld the 3 in. channel across the top of the uprights and weld into place. Next, weld the 1-1/2 in. square tubing in between the uprights to keep them from moving. Using 3 in. channel, put on the dividing row, which will divide the animal side from the human side.

After welding all the uprights on, weld cross sections of 2 in. flat iron from the uprights to the cross piece of 3 in. channel. Put these on the back side of every section.

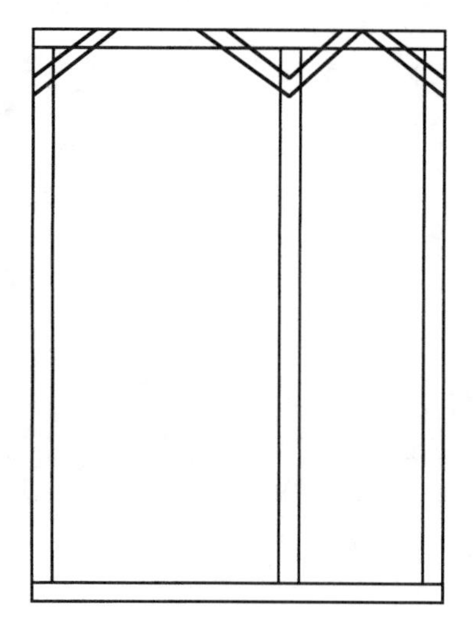

The door is a basic rectangle made from 3 in. channel iron. The two long sides are 51-1/2 in. long and the top and bottom are 32-1/2 in. long. Weld them together to make the door.

To make the tongue, use a piece of 2 in. solid shaft 6 ft. long. Drill two holes on each end of your jacks. After the holes are drilled, put a piece of 2-1/2 in. round pipe, 4 ft. long, over the solid shaft. This will pivot over the shaft. Weld a piece of 2-1/2 in. square tubing to the center of the round pipe. Weld two pieces of 2 in. flat iron, 10 in. long, on the top and bottom of the tubing, letting them stick out 6 in. over the end so you can put the pin through it. After they are welded on, cut holes in them with a plasma cutter, for the pin to go through.

Next, put braces on the tongue so it can be dropped out of the way when the trailer is backed up to the chute. To do this, weld two 6 in. long pieces of 2 in. flat iron to the bottom of the first 4 in. channel. They should have two holes in them to put a pin through. Next, weld on two 1 ft. long pieces of 3 in. channel. Cut a half circle in one end so it fits into the round piping. Angle them for support and weld into place.

To make the tongue, cut the 2-1/2 in. tubing in half a foot from the frame. Take a 1 ft. piece of 3 in. square tubing and slide it over the piece of the tongue not attached to the frame. Slide it over 6 in. and weld it into place. Drill holes for the pin and put the pin through. After the tongue is finished, put on the two 4,000 lb. Shelby jacks.

Prime and paint the metal. Put the wood on, only on the two sides where the animals go. Cut the timbers for the floor 15 ft. long. At the bend, cut an angle on them so they fit close together. Drill a holes through the wood and the frame, then use bolts to hold them down. For cleats, screw 2 x 4's every 2 ft. For the floor on the other side of the divide, bolt 2 x 12's to the frame. For the walls on the animals' side, start with a 2 x 12, 144 in. long. Bolt them to the divider down the middle. Next, put on another 2 x 12, then two more on top of them. Then add one 2 x 10 and one 2 x 8. Raise the 2 x 8 up 2 in. from the 2 x 10.

Next, put the wood on the door. Use the same pattern, except these are 36-1/2 in. long instead of 144 in. Repeat the same process on the other side.

Now put the tin channel on all edges of the chute. Also, put two right next to each other on the bend because the tin will have to be cut at that point. The channels hold the tin in place and make it look nicer because the edges will show. For the big side, cut them 144-1/2 in. long. Also, cut an angle on them so they fit nicely into the channel. On the short piece, cut the tin 35-1/4 in. long. Use the same process on the other side. The top is easy to put the tin on because you don't have to cut an angle on them. To attach the tin, drill a pilot hole through the tin and the channel iron.

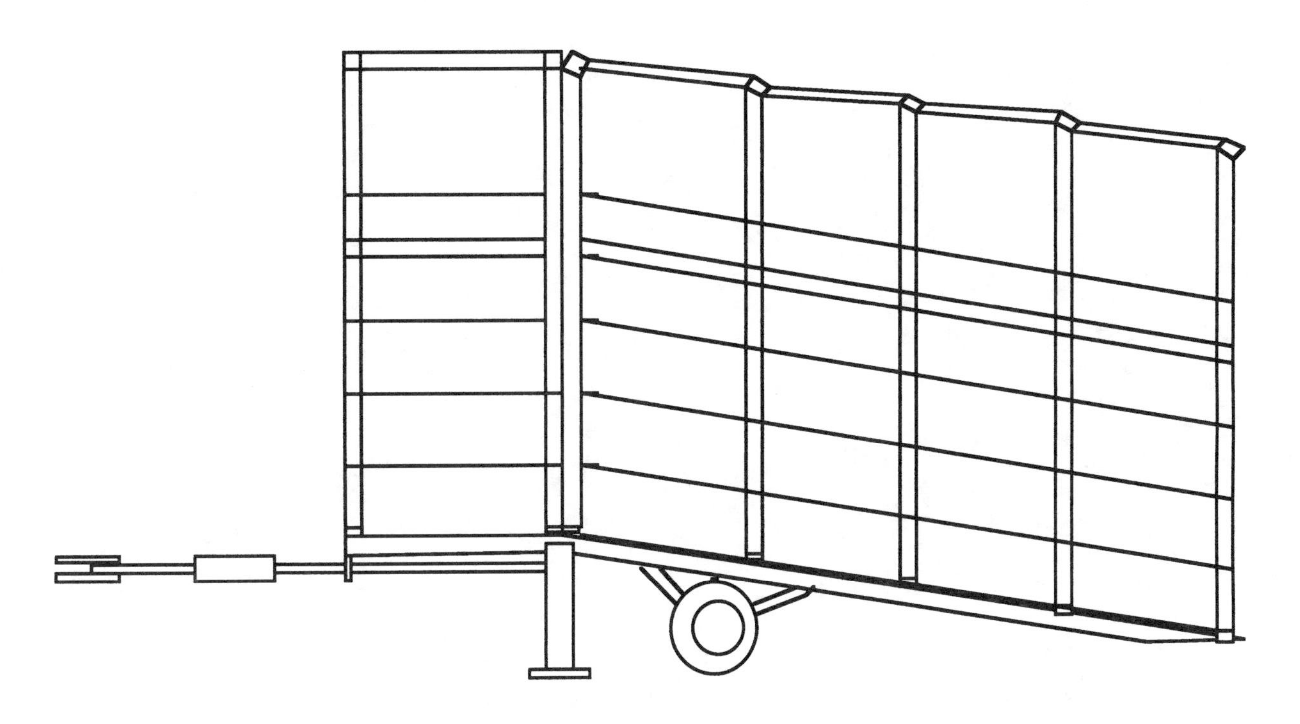

Trailers

Table of Contents

E-Z Dumper Tandem Trailer

Authors: Austin R. Diewold and Joe R. Lindquist
Instructor: James L. King
School: Burlington High School
City & State: Burlington, IA

Bill of Materials

96′	2" x 6" x 3/16" Tube steel
168′	2" x 4" x 1/4" Tube steel
7′	2-1/2" x 2-1/2" x 1/4" Tube steel
48′	2" x 2" x 1/4" Tube steel
80′	1" Round bar stock
2 sheets	4" x 8" x 1/8" Deck plate
168 sq. ft.	6" x 12" x 3/16" Flat plate (and labor to bend)
1	2" Bulldog ball coupler
1	Bulldog screw jack
1	HL-20001-00A Twin cylinder kit (with hydraulic pump)
2	13 815 865 Torque flex axles
4	16" Tires
4	1" Locking collars
10′	5/16" Steel chain
6	Weld-on D-Rings
2	Clevis sling hooks
1	6600-8 Xtra heavy tee
6	318-8 Nut
6	314-8 Coupling
1	26" Hydraulic hose
2	3/8" Hydraulic hose
1	3/8" Flow control valve
1-1/2 Gal.	Hydraulic fluid
2	3/8" Xtra heavy black pipe 90
3	3/8" Xtra heavy black pipe 45
1	3/8" Xtra heavy black pipe Tee
1	3/8" Xtra heavy black pipe Plug
2	3/8" Xtra heavy black pipe Coupling
6	3/8" x 4-1/2" Black pipe nipple
6	1/2" Hydraulic vibration clamp
12′	1/2" Steel tubing
20′	1/2" EMT Conduit
2	1/2" Rigid T-body
1	1/2" EMT Set screw
1	12-volt Marine auto craft battery
1	31" 4-gauge Wire
1	24" 4-gauge Wire
1	Battery tie down
6	Amber reflectors
2	Red reflectors
1	Wiring harness
2	Lite-mate tail lights
4 gal.	Jet black paint
1	19"W x 20"H x 31"L Box

Step-by-Step Process

1. Cut two pieces of 2" x 6" x 3/16" tube steel and miter at a 45-degree angle at each end for the sides of the main trailer frame.

2. Cut two 6′6" pieces of 2" x 6" x 3/16" tube steel and miter at a 45-degree angle at each end for the ends of the main trailer frame.

3. Grind face of all 45-degree angles to make sure they're all true.

4. Take framing square and check the frame.

5. Double check for squareness by measuring corner to corner.

6. Weld all corners.

7. Cut and tack into place two 2" x 6" x 3/16" x 30" long braces to strengthen the side frames.

8. Cut and tack into place two 2" x 6" x 3/16" x 74" long cross braces for back end of trailer.

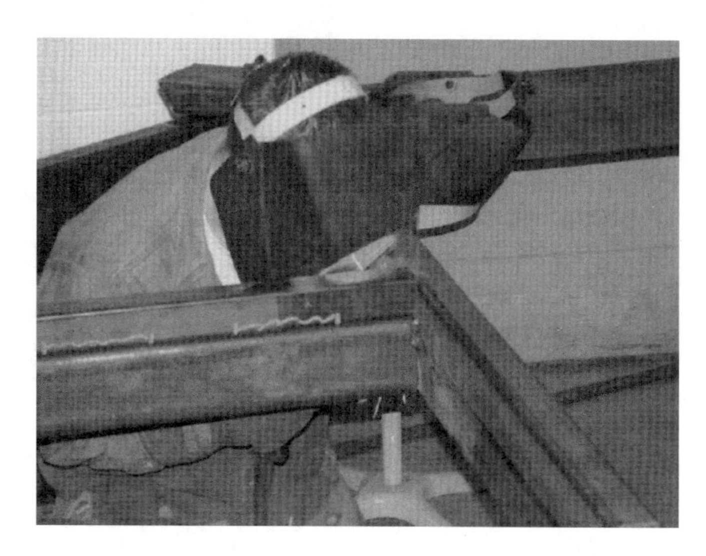

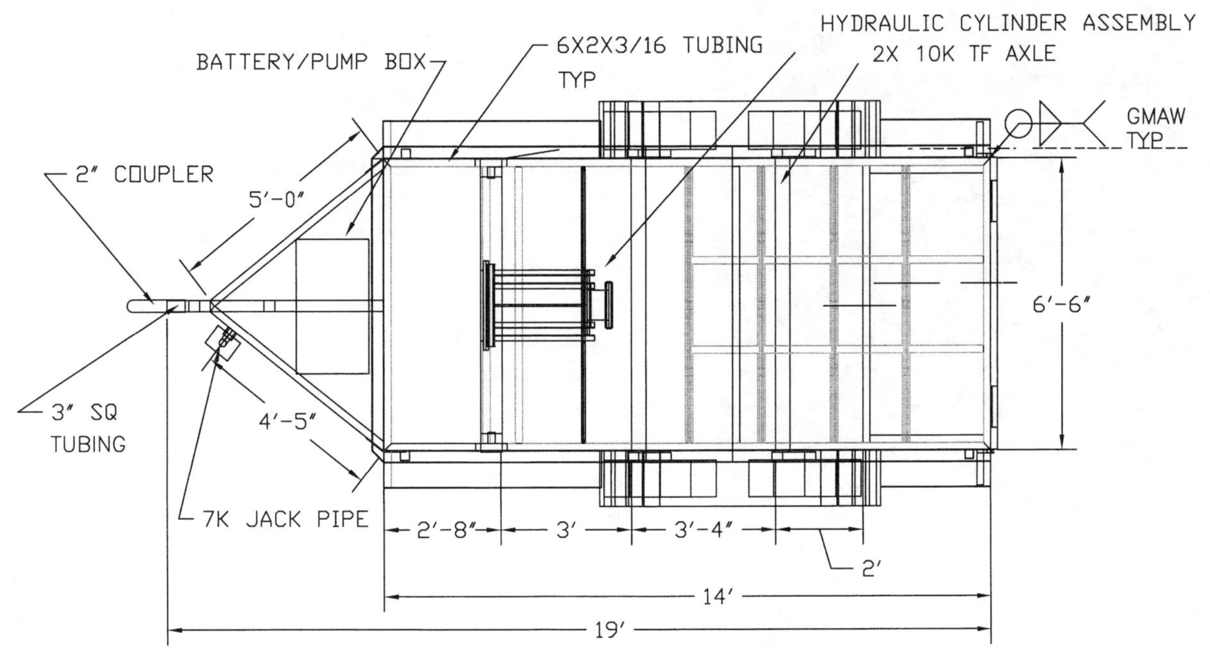

TOP VIEW-FRAME

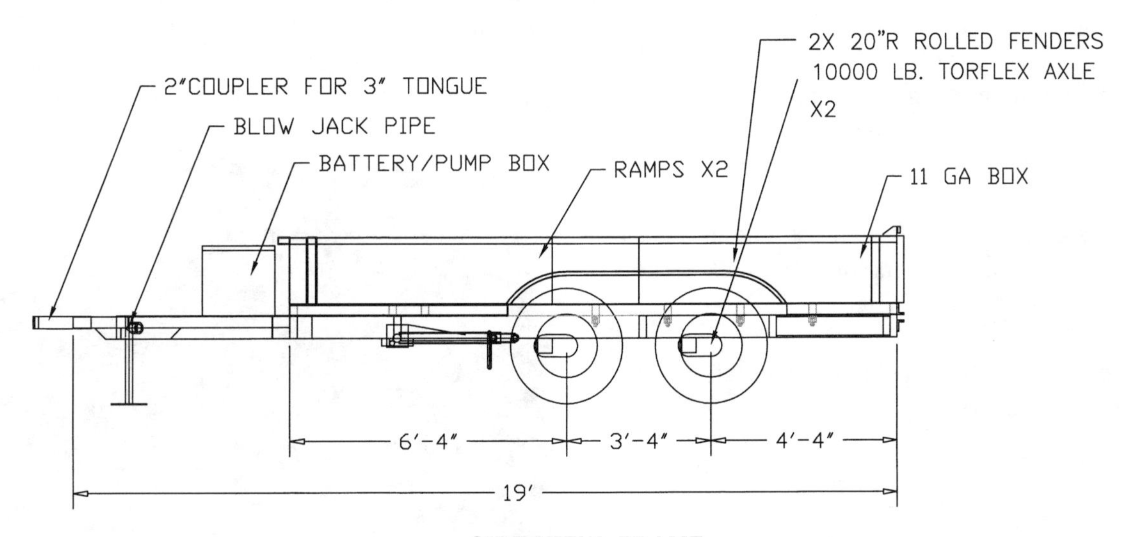

SIDE VIEW-FRAME

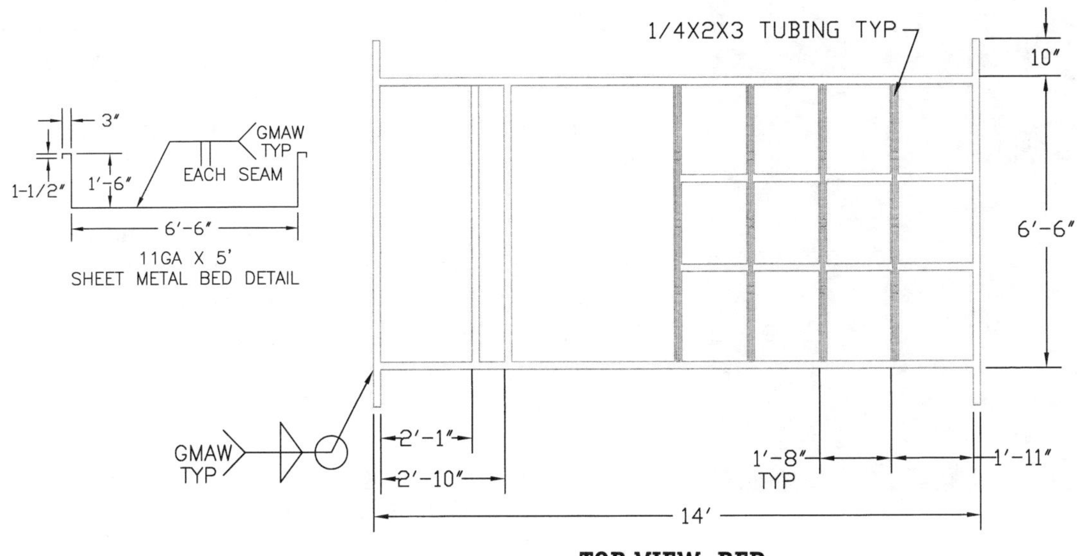

TOP VIEW-BED

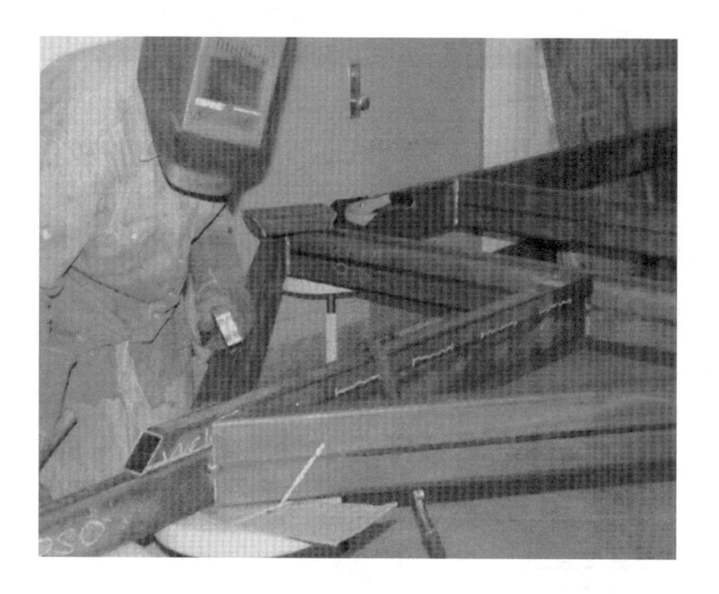

9. Cut and tack into place two 2" x 6" x 3/16" x 74" long middle cross braces.

10. Check all braces for being square and weld out all braces.

11. Cut trailer tongue out of 3" x 3" x 3/8" steel tubing. Tongue is 5′ long and welded to trailer and angle pieces from corners of the trailer. Install 2" Bulldog coupler on end.

12. Cut and weld two 2" x 6" x 3/16" x 5′ long pieces to support from front corners of the mail trailer angled to the trailer tongue.

13. Cut, bend, and weld 6" x 12" x 3/8" gussets to add to the corners of the two pieces in step 12.

14. Cut two pieces of 2" x 4" x 1/4" steel tubing 14′ long and miter each end. These pieces are to be the sides for the dump box.

15. Cut two pieces of 2" x 4" x 1/4" x 6′6" long steel tubing and miter both ends. These two pieces are the ends of the dump box frame.

16. Square frame with framing square and tack at corners. Double check for squareness by measuring corner to corner. After double checking, weld out each corner.

17. Cut four cross pieces 74" long for the dump box. Tack the first on in 2′ back and the other three on 3′ centers.

18. Check cross braces for squareness and then weld into place.

19. Use A-frame and move the dump box frame out of the way.

20. Cut 1/8" deck plate to fit front of the trailer. This is where the hydraulic pump box will sit. Plate is 74-1/2" wide in the back, 2" in the center in front and 59" from the center 2" in front to the back of the plate on each side. Weld intermittently every 6" around the plate.

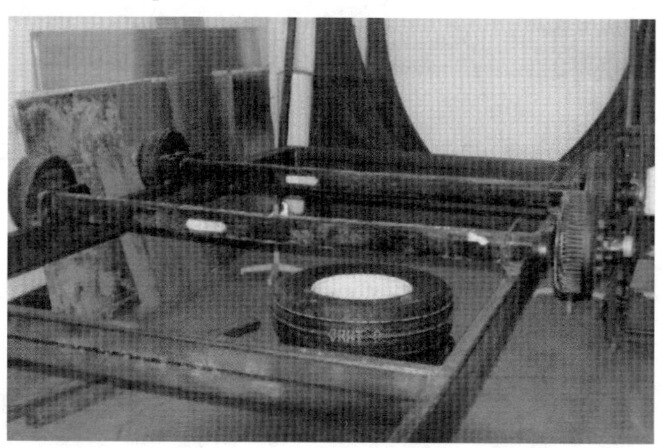

21. Unload the axles and tires. Set axles in place and check for squareness. Then check from the center of the tongue to each side of each axle to ensure that the trailer will pull straight. After this, weld the axles into place.

22. Cut 1" square tubing two pieces 1" long and one piece 2" long and space them on the bottom of the tongue with the 2" piece in the center. Go from the tongue over the spacers and back to the tongue with 1" round stock for bridge timbering to add strength to the tongue.

23. This method of bridge timbering should also be used on the last three cross braces of the dump box frame.

24. Weld the Bulldog jack on the front of the trailer and use the A-frame to flip the trailer frame over on the tires.

25. Block up under the center of the trailer with 4" x 4" pieces of wood and set the cylinders on, then figure out the best way of mounting them.

26. Cut one piece of 2" x 6" x 3/16" tube steel and weld to the bottom of the trailer for the bottom cylinder hinges to mount on.

27. Make two L-shaped supports out of 6" x 2" x 3/8" and weld to the outside of the trailer frame and under the 2" x 6" x 3/16" hinge brace.

28. Make two triangular gussets for each side. These gussets should be welded to the inside of the trailer frame and to the top side of the 2" x 6" x 3/16" hinge brace.

29. Set the dump box frame previously built on top of the trailer frame.

30. Take two scrap pieces of 3" x 3" x 3/8" angle iron 10" long and weld two pieces of 1" pipe at each end, bend 3/8" x 2" flat strap over the top of each for support. These make the lower rear hinges.

31. Take two scrap pieces of 3" x 3" x 3/8" angle iron 6" long. Weld 3/8" x 2" flat strap over the top of each end for support. These make the upper part of the rear hinges.

32. Run 1" bar stock through all hinge parts. Set each hinge in 3" from sides and weld top and bottom hinges in place. The 1" bar stock will remain across the rear of the trailer for ramps to set on.

33. Cut two pieces of 2" x 4" x 1/4" tube steel to weld between the dump box cross braces for the top cylinder hinges to weld to. Weld hinges in place.

34. Cut and bend 1" round bar stock to weld under the trailer for the back of the two cylinders to rest on when in the down position.

35. Cut 2" x 4" x 1/4" braces to go between each cross brace 2′ from each side of the entire length of the trailer. After they are tacked in place and checked, weld all braces and grind the top of the dump box frame.

36. After the dump box is bent to 6′6" wide x 1′6" tall with a 3" top, square and tack in place three sections of the box; two pieces 6′ long and one piece 2" long. Weld all seams and weld to box frame and all braces.

37. Cut one piece of 3" x 3′ x 1/4" angle iron from scrap 8" long. Cut two pieces of 3/8" plate. Let overhang the angle iron by 2". Leave 1/2" space between 3" x 7" pieces and weld to angle iron in center. On the end that overhangs the angle iron measure 1-1/2" up and 1" back from edge and drill a 9/16" hole through both pieces. This hole will allow a 1/2" pin to pass through. Round off back edge on top lip and weld in place. Repeat process for the other side of the box.

38. Build a rectangular frame from 1-1/2" x 1-1/2" x 1/8" angle iron, frame to be 6′6" long x 6-1/2" tall with one cross brace in the middle. Cut two pieces of 6′6" x 1′6-1/2" x 1/8" flat plate and weld each side of the frame for the tailgate.

39. On ends of the tailgate, install 1/2" x 2" flat strap and extend above tailgate by 3". Come down from top 1-1/2" and drill a 9/16" hole in the center. This will allow a 1/2" pin to pass through and the gate to swing.

40. Install fenders on each side that were rolled to match the radius of the tires. Fenders are 86" long and 15" wide.

41. Bend two pieces of 1/8" deck plate 10" wide with a 4" lip and 53" long. Install in front of each fender a runningboard.

42. Bend two pieces of 1/8" deck plate 10" wide with a 4" lip and 28" long. Install behind each fender for a rear runningboard.

43. Cut and cap one end of a 2" x 6" x 3/16" x 10" long and weld to frame below rear runningboard. Take cutting torch and cut a 3" hole to install combo brake, turn signal and taillight. Repeat process on other side.

44. At front of the trailer, install a 19"W x 20"H x 31"L box for the hydraulic pump. Set box forward enough to all the top to open fully.

45. Install hydraulic pump and 12-volt marine battery in the box from step 44.

46. Wire 12-volt marine battery to the hydraulic pump.

47. Run 1/2" steel tubing from pump to the cylinders, make final connection to the cylinder with hoses to help with vibration.

48. Run 1/2" EMT conduit from the front of the trailer to the back two taillights. Pull wires and make final connections.

49. Install four sleeves 2-1/2" x 2-1/2" x 3/16" x 1'6" tall. Install all sleeves 10" in from each corner up through lip of box. Weld at lip and down sides. These sleeves are for the pipe racks.

50. Build two pipe racks from 2" x 2" x 1/2" tube steel. Each riser is 48" tall and cross braces are 6'6 3/8" long. Come down 12" from cross brace on each riser and weld in a 2" x 2" x 1/4" piece of tube steel. Run on a 45-degree angle as a gusset. These racks will set into the sleeves in step 49.

51. Fill hydraulic tank with fluid and run to work the air back to the tank.

52. Follow finish procedures.

Gas-Fired Cooker Trailer

Authors: Brett Lambert and Lucas Lawrence
Instructor: James L. King
School: Burlington High School
City & State: Burlington, IA

Bill of Materials

120′	1-1/2″ x 1-1/2″ x 1/8″ Angle iron
60′	2-1/2″ x 2-1/2″ x 3/16″ Angle iron
72′	2-1/2″ Square 3/16″ tubing
6-4 x 8 11 gauge Steel 960 lbs.	
3-4 x 8-3/4 #9 Flat 50.63-20%	
40′	1/8 x 1-1/2
20′	1/8 x 3
42′	sch. 40 Untested pipe

Editor's safety note: *Be sure to meet state regulations on the transportation of fuel gas cylinders.*

Step by Step

Box

Lay out one piece of 11 ga. metal 42″ x 84″ for bottom of cooker box cut by plasma cutter.

Lay out two pieces of 11 ga. metal 10′ x 84″ for sides using a plasma cutter.

Cut one piece 11 ga. metal 84″ x 26″ for top using a plasma cutter.

Cut both ends to match bottom to top and sides using a plasma cutter.

Bend 90 degree angles on the sides and end; bend 45 degree angles on top so box will fit easily.

Fit pieces together and tack.

Check for squareness. Weld all places including inside of box.

Cut eight pieces of 3″ x 26″ strap for support on top and sides.

Weld strap iron on box.

Place box on frame and weld all the way around.

Weld inside grill rack holders in place.

Doors

Lay out six 12-1/2″ x 26-3/4″ 11 ga. steel and six 11″ x 26-3/4″ 11 ga. steel for the doors.

Cut out doors with the plasma cutter.

Stainless steel arc weld hinges onto the doors.

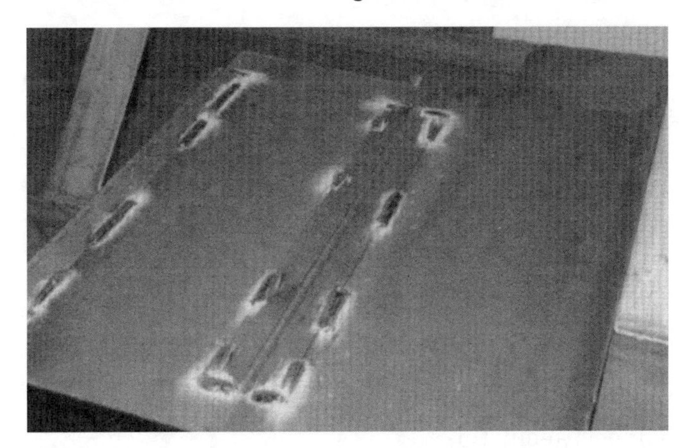

Stainless steel arc weld hinges to cooker box.

Cut out six door handles and weld them onto doors.

Manifold

Cut 7′ piece of 2″ x 2″ square tubing

Cut six 1″ wide holes on square tubing.

Connect all pipe fittings onto square tubing.

Place manifold on side of the cooker box and drill six holes into side of cooker box.

Drill out marks for gas pipes to fit through holes.

Cut six round pipe pieces 42″ long and thread them so they fit pipe fittings.

Cut slits 3/4″ apart all the way down the piping.

Cut six 2″ angle iron for gas pipe holders and weld them to cooker box.

Gas Tank Holder

Cut one 44″ and two 4″ pieces of 1-1/2″ x 3″ square tubing.

Cut one 1/4″ hole 7″ down on square tubing.

Cut one 1/4″ hole 3″ down on square tubing.

Connect regulator.

Drill 1/2″ hole 25″ down on square tubing.

Weld two 9″ pieces of square tubing onto 44″ piece of square tubing to form a T.

Weld T onto front of cooker.

Weld a 29″ triangle of expanded metal to fit front of cooker box.

Weld two 1″ strap metal circles onto expanded metal to hold gas tanks.

Weld two safety chains under the regulator.

Cut two 12″ half-circles out of 2″ strap metal with a connector piece in the middle, forming a W to hold the gas tanks to cooker.

Grill Trays

Cut eight pieces of 1-1/2″ angle iron 40-1/2″ long.

Cut eight pieces of 1-1/2″ angle iron 20-1/2″ long.

Cut 45 degree angles at both ends of all angle iron.

Square up the angle iron, forming rectangles, and weld them together.

Cut four sheets of 40-1/2″ x 20-1/2″ pieces of expanded metal for grills.

Weld expanded metal to the angle iron rectangles.

Fenders

Cut two 12″ pieces of 2″ x 2″ square tubing.

Cut one 22-1/2 degree angle on each side of those pieces.

Cut one 10″ piece of 2″ x 2″ square tubing.

Cut two 22-1/2″ degree angles on each side of this piece.

Weld these pieces together to form a flat bottom V.

Weld V to cooker.

Weld fender to mount on cooker box.

Due to manifold, fender on other side was put on differently.

Cut one 26″ piece of 1″ square tubing.

Weld square tubing to cooker box.

Weld fender to square tubing.

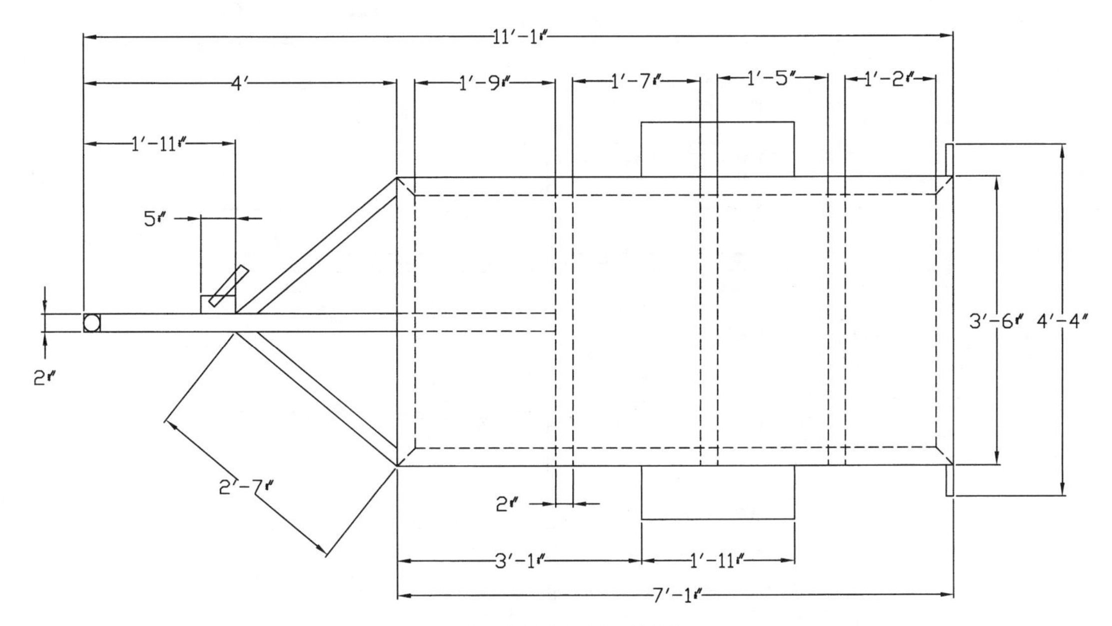

TOP VIEW–FRAME

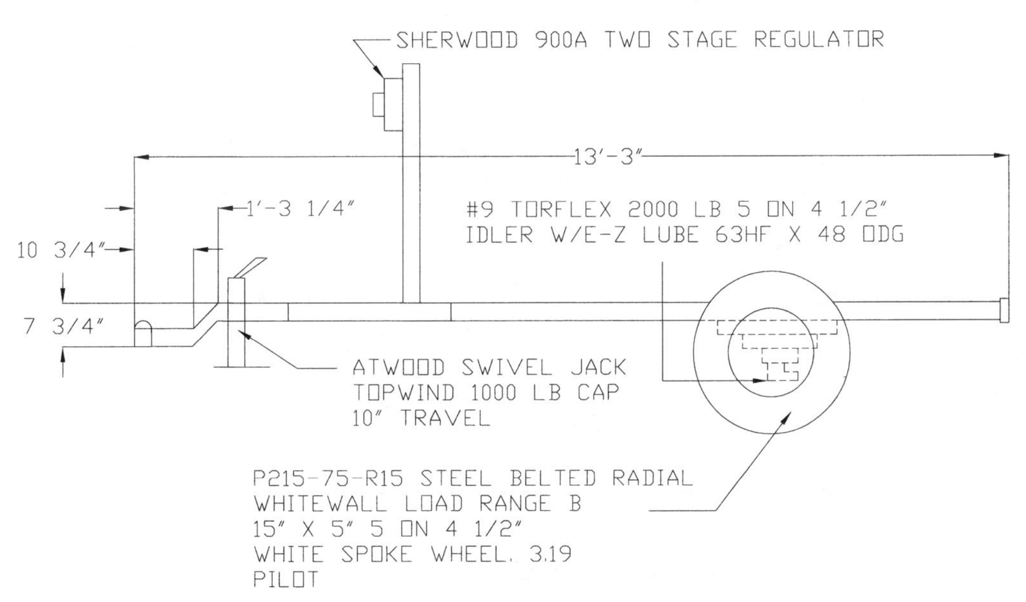

FRONT VIEW–FRAME

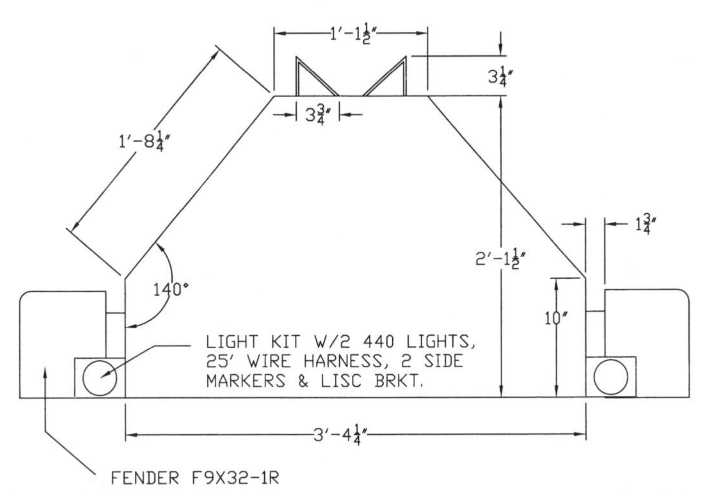

SIDE VIEW–FRAME

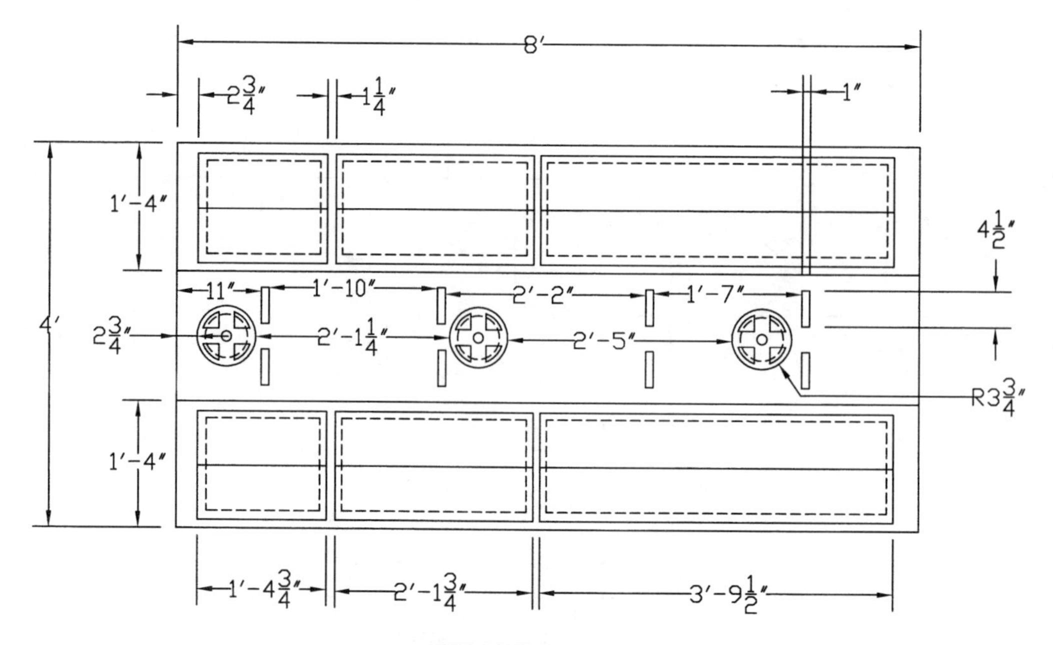

TOP VIEW–BOX

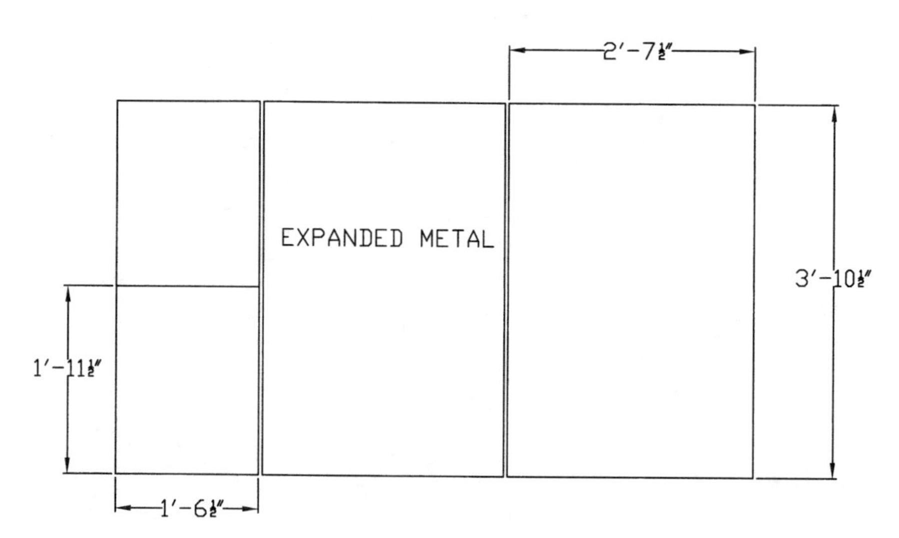

TOP VIEW–GRILL

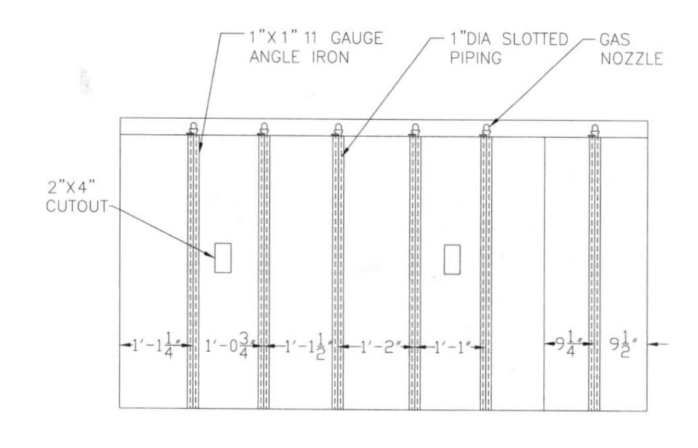

TOP VIEW–GAS LINES

8'
1'-6 3/4"
1'-6 1/2"
2'-6 1/2"
1'-8"
1'-0 3/4"
10"
2"
2" DIA HOLES
GAS MANIFOLD

SIDE VIEW–GAS LINES

Frame

Cut two 84″ pieces of 2-1/2″ x 2-1/2″ square tubing.

Cut two 42″ pieces of 2-1/2″ x 2-1/2″ square tubing.

Cut 45 degree angles on each end of all square tubing.

Cut two 37-3/16″ pieces of square tubing for frame support.

Square up the square tubing to form the frame.

Weld the two shortened pieces inside of frame 20″ apart.

Grind all welds.

Tongue

Cut one 5′ piece of 2-1/2″ x 2-1/2″ square tubing.

Cut two 3′ pieces of 2-1/2″ x 2-1/2″ square tubing.

Cut 45 degree angles in all of the pieces.

Plasma cut angles for the tongue support.

Test fit supports and then weld to the frame.

Grind down all the welds.

Cut one 17″ piece of 2-1/2″ x 2-1/2″ square tubing for hitch axle.

Axle

Cut two 21″ pieces of 2-1/2″ x 2-1/2″ square tubing with 45 degree angles on both sides.

Cut two 16″ pieces of 2-1/2″ x 2-1/2″ square tubing with 45 degree angles on both sides.

Make eight end caps for step 1 and 2.

Weld 16″ piece to 21″ pieces to form the axle lift.

Weld both axles lifts to the frame.

Put axles on axle lifts and tack to frame.

Weld all places.

Put tires on.

24-Foot Gooseneck Trailer

Author: Clayton K. Jensen
Instructor: Kirk Scheidt
School: Buena Vista High School
City & State: Buena Vista, CO

Bill of Materials

Qty.	Width	Length	Thickness	Description
2	5″	20′	0.1885	5″ Channel
2	5″	4′	0.1885	5″ Channel
1	4″	8′	0.1721	4″ Channel
1	4″	2′	0.1721	4″ Channel
15	2.5″	91″	14 ga.	1.5″ Rectangular Tubing
2	2″	20′	1/4″	Band
2	2″	4′	1/4″	Band
26	2″	3″	11 ga.	3″ Rectangular Tubing
4	1.5″	61.75″	14 ga.	2.5″ Rectangular Tubing
1	20″	8′	3/16″	Plate
2	4.005″	24′	0.235	12″ H-Beam
2	4.005″	94″	0.235	12″ H-Beam
2	4.005″	50″	0.235	12″ H-Beam
1	4.005″	66″	0.235	12″ H-Beam
9	9.5″	24′	1.5″	Boards

Shop Materials

2.5 gal	Flat Black Paint
2	Pots of Flat Black
4	24′ Electrical Wire (green, yellow, blue, brown)
18′	6-Wire (14 ga.) Insulated 146C500
1	Tiger Flapper disc
1	Grinding Disc
39′	Reflective Tape

Extra Metal for Ramps and Storage

8	1.5 x 3 x 20.5 Square Tubing
3	1.5 x 3 x 8′ 6″ Square Tubing
24	2.5 x 2.5 x 17.75 Angle Iron
4	3″ x 8′3″ Heavy Channel Iron

Hardware

2	12TF70-865E-EZ Hub face to hub face - 81″ Outside to outside of H-beam 65 5/8″ Outside to outside of brackets - 66 5/8″ Starting angle - 45 degrees down
1	GG2516ADJ Gooseneck Coupler
1	DUALPIN-SL Binkley Gooseneck Jack

Hardware

6	142-18K Mounting Kit
4	142A Amber 2.5″ Round Lights
2	142R Red 2.5″ Round Lights
1	150-3R I.D. Bar
1	1026 Breakaway Kit
4	421KR Red Turn-signal Lights
4	17-309/326080 16 x 6 865 White mod 17-281-7
4	TF01-225 Torx floor screws 1/4″ x 2-1/4″ (QTY 40)
2	TX430-3 Torx screw driver bit (pkg of 3)

Procedure

Starting with the two big 24′ H-beam pieces, square them up and weld some 1.5″ x 2.5″ square tubing between them to keep them from twisting.

Place 15 pieces of the same square tubing across the top of the H-beam for all of the cross members. Weld them using a wire-feed welder.

Next, get help to flip the trailer over. Then, weld two 7,000-pound axles to the bottom of the H-beam, burning the rod hot and multipassing.

Then flip the trailer back over. Build the tongue out of the same kind of H-beam as the main frame. Draw the whole tongue out on the floor and to find the my angles and where the coupler will fit into the H-beam.

Next, put the tongue on the ball of a flat bed truck and use hydraulic jacks to adjust the tongue to get it level. Then back up to the trailer and weld it heavy.

Now, weld 5″ channel iron all the way down the side of the bed to the cross members. Then, weld stake pockets and a rub rail on top of that.

Make the storage area at the back of the trailer out of 1.5″ x 2.5″ square tubing and rock screen. Also, make the door that opens to a storage area out of 3/16″ plate.

Cut holes for the taillights using a plasma cutter.

Clean the dirt, grease and oil from the trailer to prepare it for painting.

Run 14 ga. electric wiring for all of the lighting.

Screw 1.5″ x 9.5″ boards on for the decking.

For the final touches, place reflective tape down the sides of the tongue, bed, and across the back of the trailer. Make two ramps 8′ long out of 3″ heavy wall channel iron and 5/16″ x 2.5″ x 2.5″ angle iron.

Also, make some stake pockets and legs that connect to the ramps and go to the ground for additional support.

Note: Tools to use include: a chop saw for cutting all the metal to length; an oxy-acetylene cutting torch and a rose bud; a wire feed welder; an arc welder using 7018 1/8″ rod for all of the arc welding; grinders and flapper discs for beveling; wire cutters and splicers for wiring; and a drill press and a sliding T-bevel for angles.

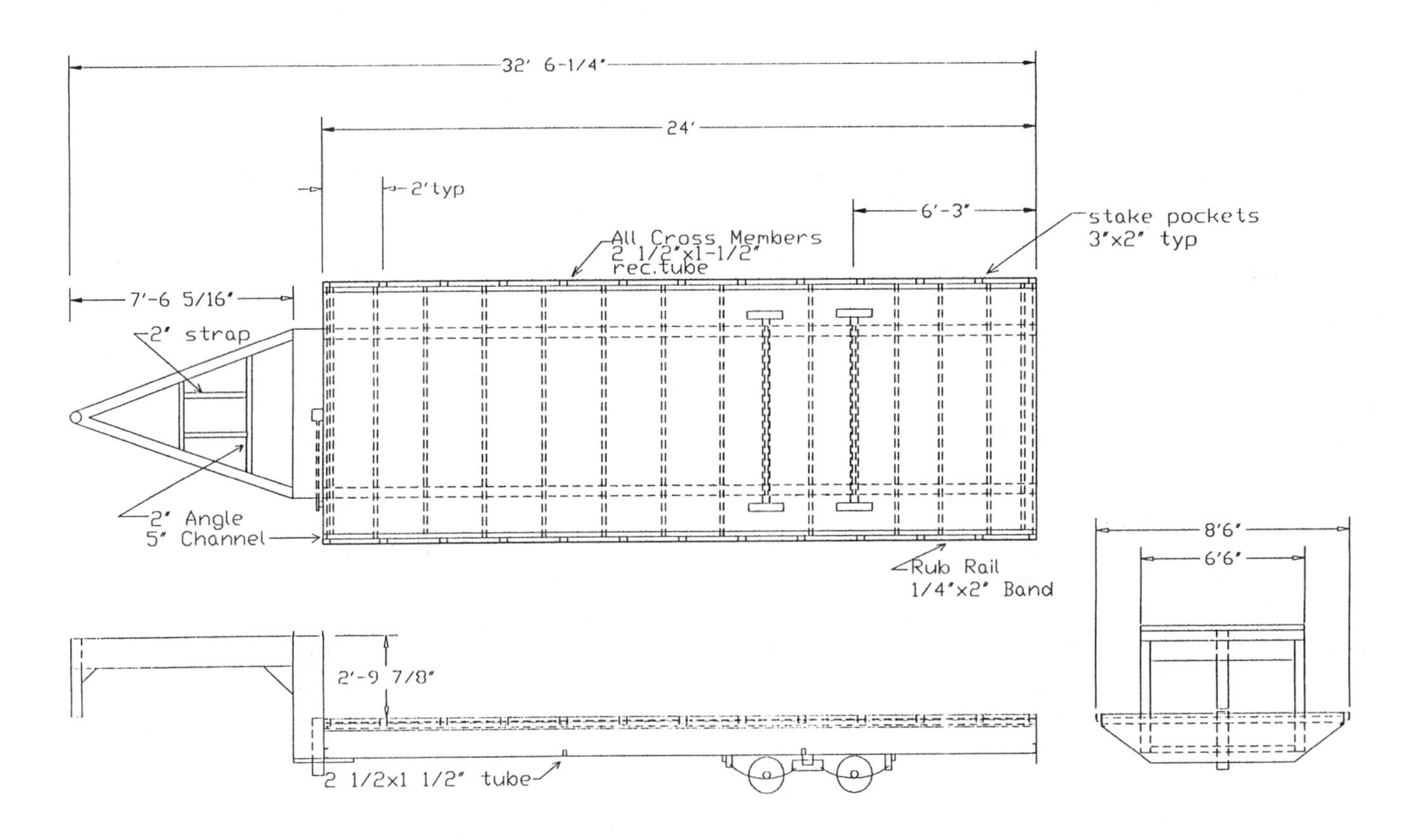

Single Axle Van Trailer

Authors: Thomas Peterson and Kyler Ross
Instructor: James L. King
School: Burlington High School
City & State: Burlington, IA

Parts List

2	fenders
1	axle
2	wheels and tires
2	hub covers
1	frame coupler
1	jack
1	jack foot
6	top bows
1	flat roof cap
1	gravel guard
1	weather strip
4	side lights
1	3a light bar
1	3r light bar
2	Leo lights
1	license plate bar
4	reflectors
4	hinges
1	cam door lock
2	safety cables

Introduction
This trailer is designed to store and transport construction materials, such as large sheets of plywood, power tools and other valuables. It is enclosed to protect the contents from the weather. The trailer has a single axle, with two swinging doors in the back and a sturdy diamond plate floor. Including the 2″ hitch, the trailer is 120″ long and 82″ tall.

Step by Step Process
1. Cut two 118″ lengths of 2x4, 11 ga. tube steel with 45 degree angles on the ends.

2. Cut one 72″ length of 2x4, 11 ga. tube steel and one 60″ length with 45 degree ends.

3. Tack and square the four pieces of steel. Cut four 56-1/2″ lengths angle iron for reinforcement bars and tack them. Check for square measurements to make sure everything comes together evenly.

4. Cut two pieces of metal 48-1/2″ with a 45 degree angle. Cut two 44″ long rails with 45 degree angles and two 48″ rails with 45 degree angles. Square up the tongue and tack it into place.

5. Cut two 12 x 5 squares of 1/4″ steel to reinforce the tongue. Weld everything into place.

6. Now, cut the following out of 1x1 square tubing: four 65″ pieces; four 66″ pieces; two 44″ pieces; two 38″ pieces, and two 54″ pieces for side-walls. Make sure everything is square and weld into place.

7. Cut the following out of 1x1 square tubing at 45 degree angles: two 66″ pieces and two 60″ pieces. Cut two 64″ pieces of 1x1 square tubing for the front wall. Tack and finish.

8. Cut four 60″ 1x1 square tubing and three 114″ 1x1 square tubing for the sidewall reinforcements and weld into place.

9. Weld the axle into place after checking for squareness.

10. Place the tongue hitch; check for squareness and weld into place.

11. Flip the trailer over and put up the side and front wall, check for levelness and weld into place.

12. Cut the tail beams at 77″ and put notches in the side to fit in the sidewall support beams; check squareness and weld together.

13. Put on the roof bows 24″ apart; check squareness and weld in place.

14. Cut eight 22-3/4″ pieces of 1x1 square tubing for roof bow supports; tack and weld in.

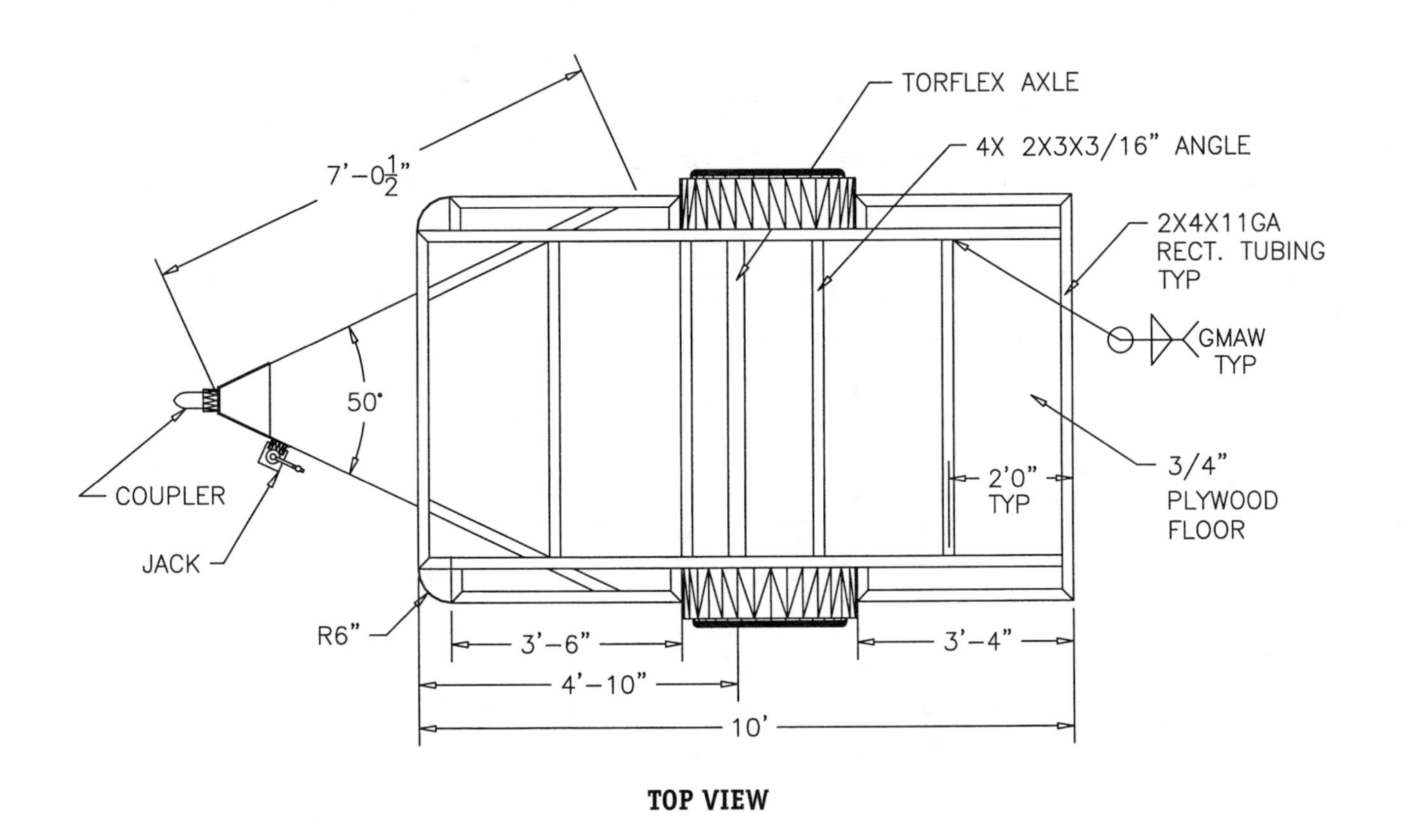

TOP VIEW

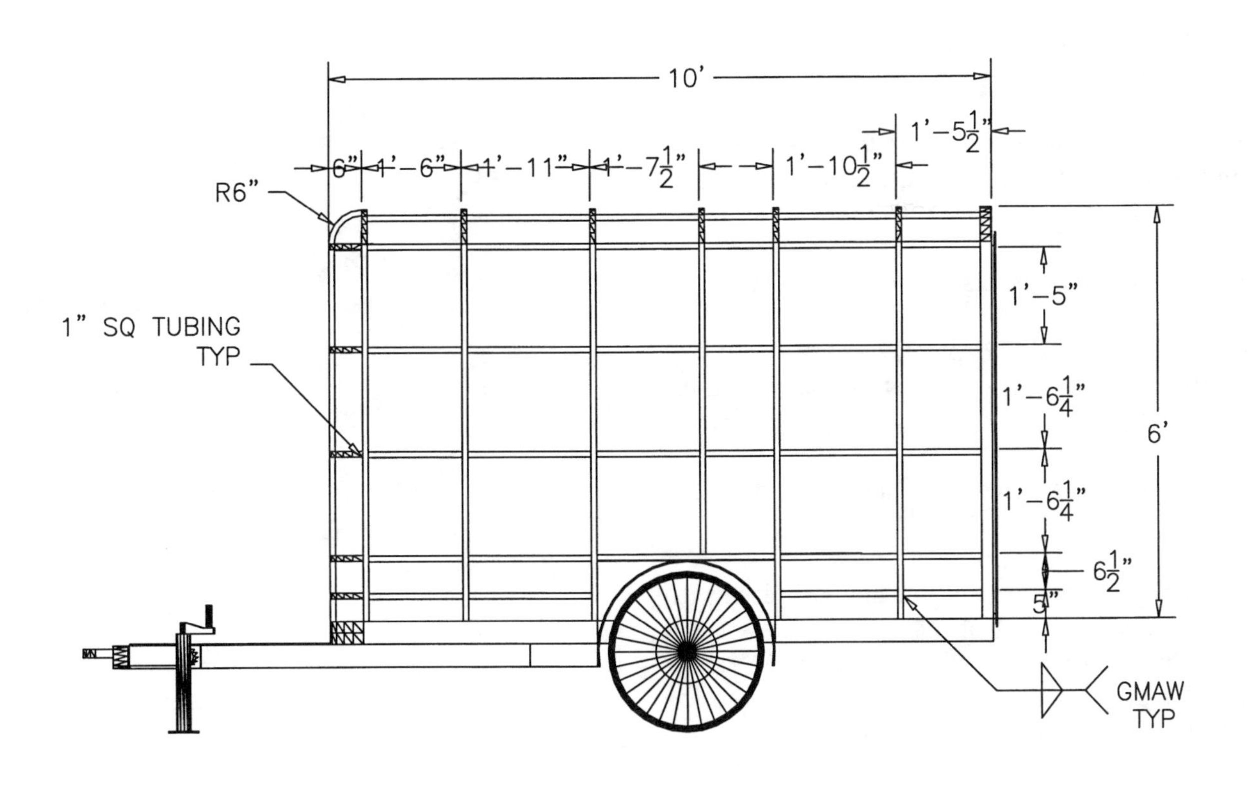

SIDE VIEW

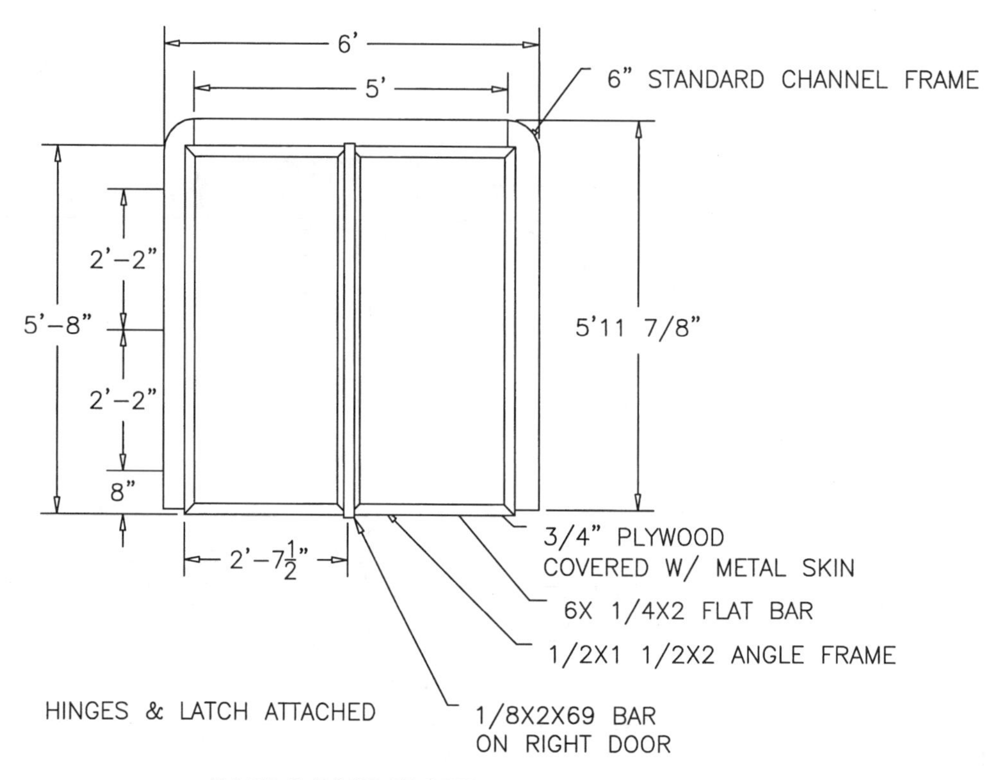

DOOR & DOOR FRAME

15. Put up sheet metal on the roof using come-alongs and rivet in for temporary support.

16. Finish putting all metal on roof and start putting on the rivets.

17. Cut four 36″ pieces and four 65″ pieces of 1x2 square tubing. Check for squareness, weld into place, and grind welds down.

18. Put wheel wells into place; tack and weld.

19. Cut five 71″ by 47-1/2″ pieces of sheet metal. Put them into place, drill holes for rivets and rivet them in place.

20. Finish putting all rivets on the roof and on the walls of the trailer

21. Use a plasma torch to cut out one 52″ piece with a 5″ x 16″ square cut out of the diamond plate metal to fit around the wheel well; then cut a 60″ sheet with a 5″ x 16″ cut out of one end to fit around the other half of the wheel well; then trace the front corners of the trailer to make the floor completely enclosed.

22. Weld in the whole floor, top and under the trailer; then cut two 38″ by 67″ sheets of metal for the doors, and put in rivets to hold the sheet metal to the door frame. Mount the doors on the back of the trailer.

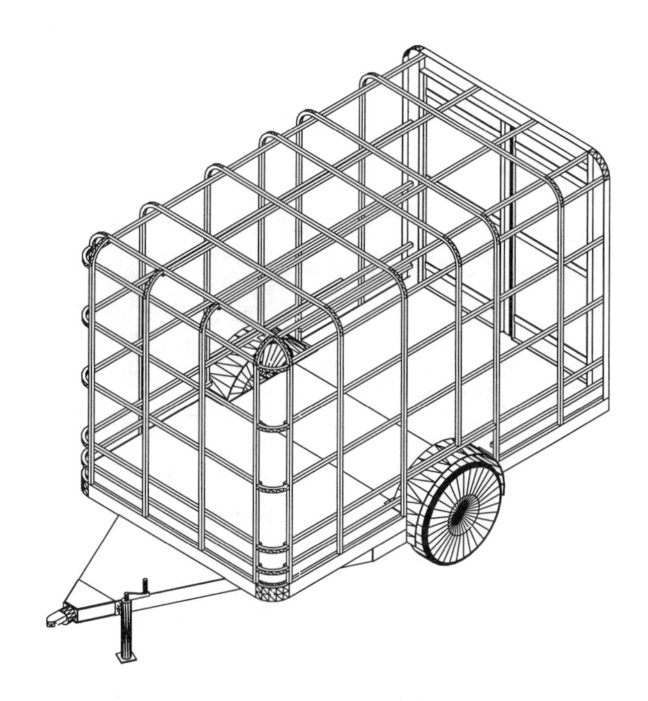

23. Install the cam door locking system.

24. Cut out the holes in the back of the trailer for the lights; grind the holes so they are smooth; install lights, reflectors and license plate holder.

Tilt Bed Utility Trailer

Author: Joel D. Diller
Instructor: David H. Murray
School: Ferris State University
City & State: Big Rapids, MI

Bill of Materials

2 x 24′ Lengths	2″ x 3″ x 11 ga. Steel tubing
1 x 24′ Length	2″ x 2″ x 1/8″ Steel tubing
1	3′ x 3′ x 3/8″ Steel plate
1 x 24′ Length	2″ x 2″ x 1/8″ Steel angle
1 x 24′ Length	1″ x 1″ x 1/8″ Steel tubing
1	4′ x 4′ x 1/8″ Steel sheet
1	2″ x 1/4″ x 1′ Steel bar
1 x 24′ Length	1″ x 2″ x 14 ga. Steel tubing
1 x 24′ Length	1-1/4″ x 1-1/4″ x 1/8″ Steel angle
1	4′ x 8″ x 12 ga Expanded metal
1 x 24′ Length	1″ x 1/4″ Steel bar
1	3500 lb Axle
1	Axle hanger kit
2	13″ Tire
1	2″ Coupler
1	Light Kit
11	10′ Deck Planks

Basic Framework

The first step of this project is to cut the 2″ x 3″ and the 2″ x 2″ steel tubing to length in order to build the basic frame of the trailer. All the pieces of tubing are cut to length on the horizontal band saw while mitered ends are cut utilizing the abrasive cut-off saw. Cut all the pieces indicated on the Basic Frame Cut List out of the corresponding lengths of tubing. Use the Basic Frame drawing to identify and cut mitered ends as needed.

Once all of the pieces have been cut to their final lengths, it is time to weld them together. Use the Welding Parameters table to help set up your particular gas metal arc welder. These parameters are just a guide. Since every welder is different, you may need to adjust some of the parameters to ensure high-quality welds. Once your machine is set up, assemble all of the pieces according to the Basic Frame drawing. Use a tape measure to check the diagonals of the frame before and after tacking pieces into position to ensure that the frame remains square. Once all the pieces have been securely tacked into position, all the final welds can be made.

Tongue Tilt Mechanism

Once the basic frame has been welded together, the tilt mechanism needs to be built to attach the trailer's tongue (part I) to the center supports (parts E and F) of the basic frame. Cut the pieces indicated in the Tilt Mechanism Cut List to length from the appropriate material. Use a metal shear to cut the 3/8″ plate to length. Use the Tilt Mechanism Detail to lay out and cut the 3/8″ plate to its final shape using an oxyacetylene torch.

Once both pieces are cut to their final shape, use a drill press to drill the 3/4″ holes in each as shown in the Tilt Mechanism Detail. You will also want to lay out and drill the 3/4″ hole in the tongue (part I). While you have the tongue at the drill press, you may also want to drill the 1/2″ hole that will be used for the latch pin as indicated on the Latch Detail. Once the holes are drilled, bolt all three pieces together with a 4″ x 3/4″ bolt. This will keep them square

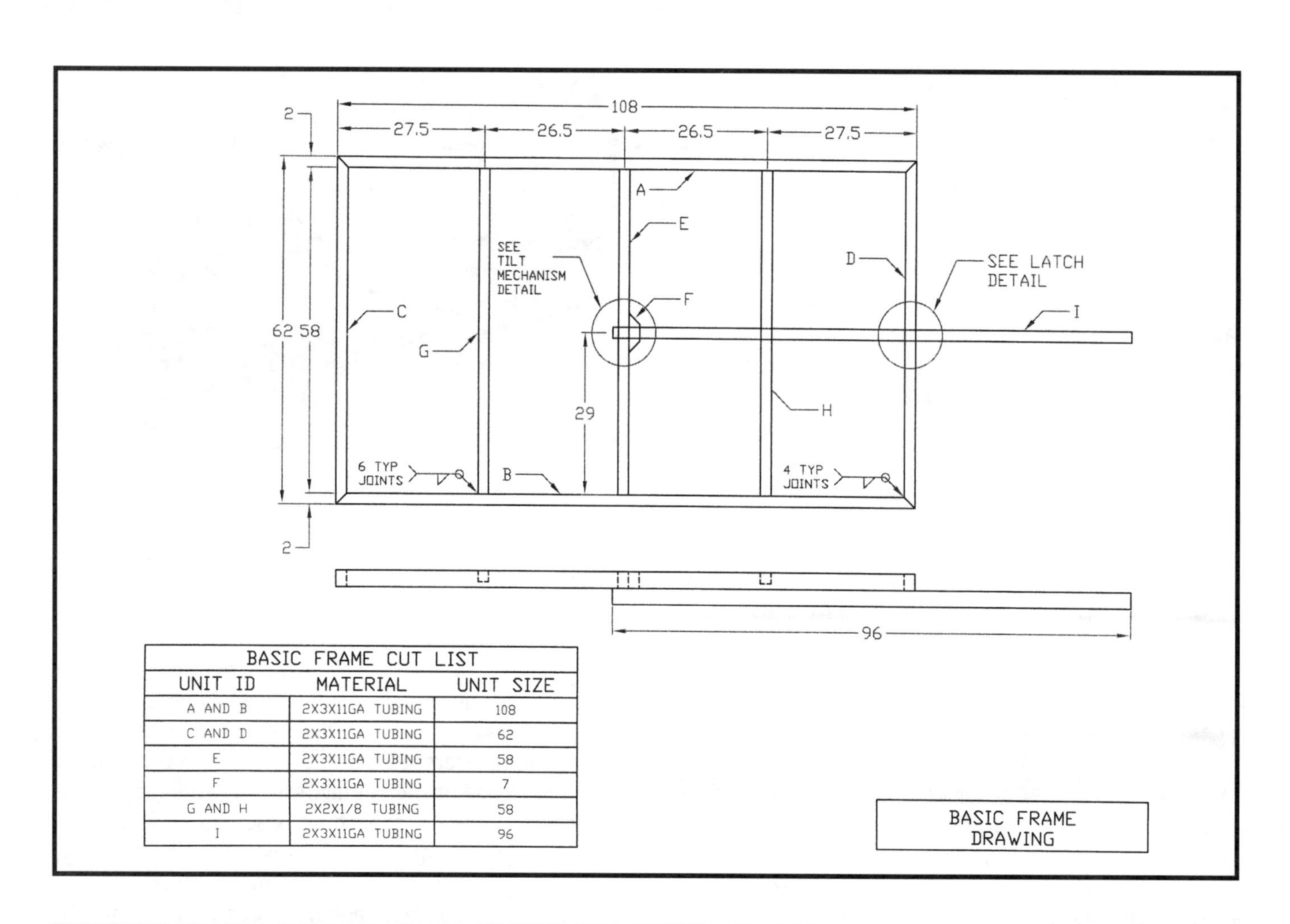

BASIC FRAME CUT LIST		
UNIT ID	MATERIAL	UNIT SIZE
A AND B	2X3X11GA TUBING	108
C AND D	2X3X11GA TUBING	62
E	2X3X11GA TUBING	58
F	2X3X11GA TUBING	7
G AND H	2X2X1/8 TUBING	58
I	2X3X11GA TUBING	96

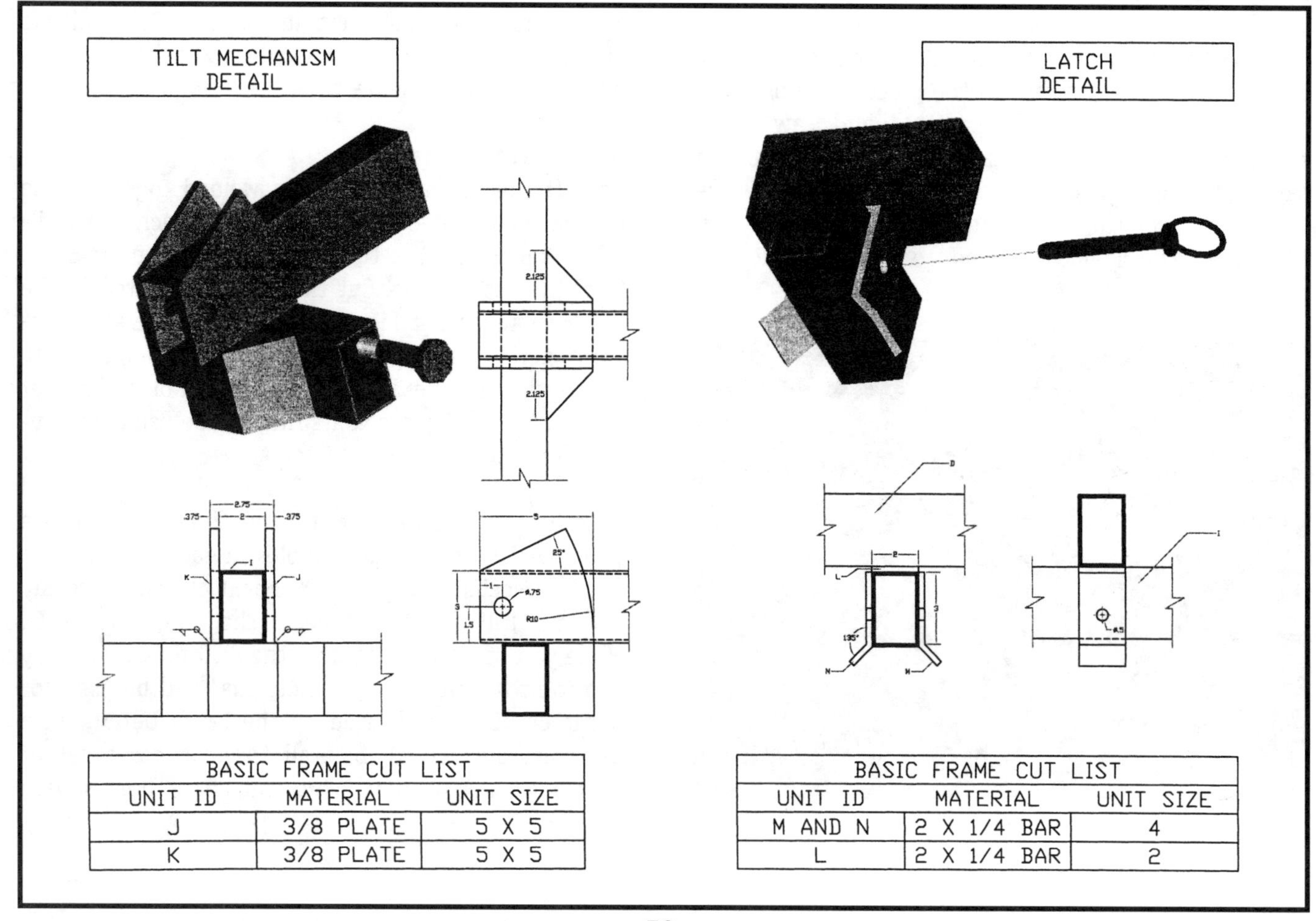

BASIC FRAME CUT LIST		
UNIT ID	MATERIAL	UNIT SIZE
J	3/8 PLATE	5 X 5
K	3/8 PLATE	5 X 5

BASIC FRAME CUT LIST		
UNIT ID	MATERIAL	UNIT SIZE
M AND N	2 X 1/4 BAR	4
L	2 X 1/4 BAR	2

while you tack them into position. Once tacked in place, the final welds can be performed as indicated on the Latch Detail Drawing.

Tongue Latch and Axle

Cut the three pieces shown in the Latch Cut List out of the 2″ x 1/4″ bar stock. Use the press break to bend the ends of pieces M and N, as shown in the Latch Detail. This ensures that the tongue passes easily into the latch without binding. Next, drill the 1/2″ hole in both pieces and attach them to the 1/2″ hole in the tongue with the 3″ x 1/2″ latch pin. Now slide piece L into position between the tongue and frame and tack all the pieces into place. Once all the pieces are securely tacked, perform the final welds.

Since the trailer frame is already upside down, it is a good time to hang the axle and put on the tires. Locate the assembly instructions that came with the axle hanger kit and tack spring mounts into position on the trailer's frame. Make sure that you keep the axle at least 3″ behind the center of the frame to ensure that the trailer has enough tongue weight.

Now that the spring mounts are tacked into place, follow the assembly procedures to attach the springs and axle. Once you are positive that everything is correct,

perform the final welds around the spring mounts. Next, pack the axle bearings and assemble the hubs onto the axle spindle according to the assembly procedures provided with the axle. You can now put the tires on the trailer and flip everything over so the trailer is sitting on the ground.

Trailer Sides

Begin by cutting all of the uprights out of the 1″ x 1″ x 1/8″ square tubing according to the Trailer Sides Cut List. All the pieces are cut from the full length of stock using the horizontal band saw. Once they are all cut to length, miter one end using the vertical band saw. A cap can be welded over the mitered ends to improve the overall appearance. The uprights are now ready to be tacked into place around the basic frame.

Now that the uprights are tacked into position, cut the 2″ x 2″ x 1/8″ angle to length using the horizontal band saw; miter the appropriate corners on the vertical band saw. Next, tack the angle on top of the vertical uprights as indicated on the drawing. Once everything is square, the sides can be welded solid. The two short uprights seen on the fender detail in the Trailer Sides Drawing will need to be welded on once the fenders are in position.

Fenders and Light Protectors

Now that the sides are in place, the fenders and light protectors can be built. Cut the basic pieces out that are indicated on the Fender and the Light Protector Template Cut List using the metal shear. Once the basic pieces have been cut, laid out, and sheared, a press break will need to be used to bend the 90 degree angles indicated as dashed lines. The fenders can now be welded together as shown on the Fender Template drawing. A belt sander was used to sand all of the welds smooth before welding the fenders to the trailer as indicated on the Trailer Sides Drawing.

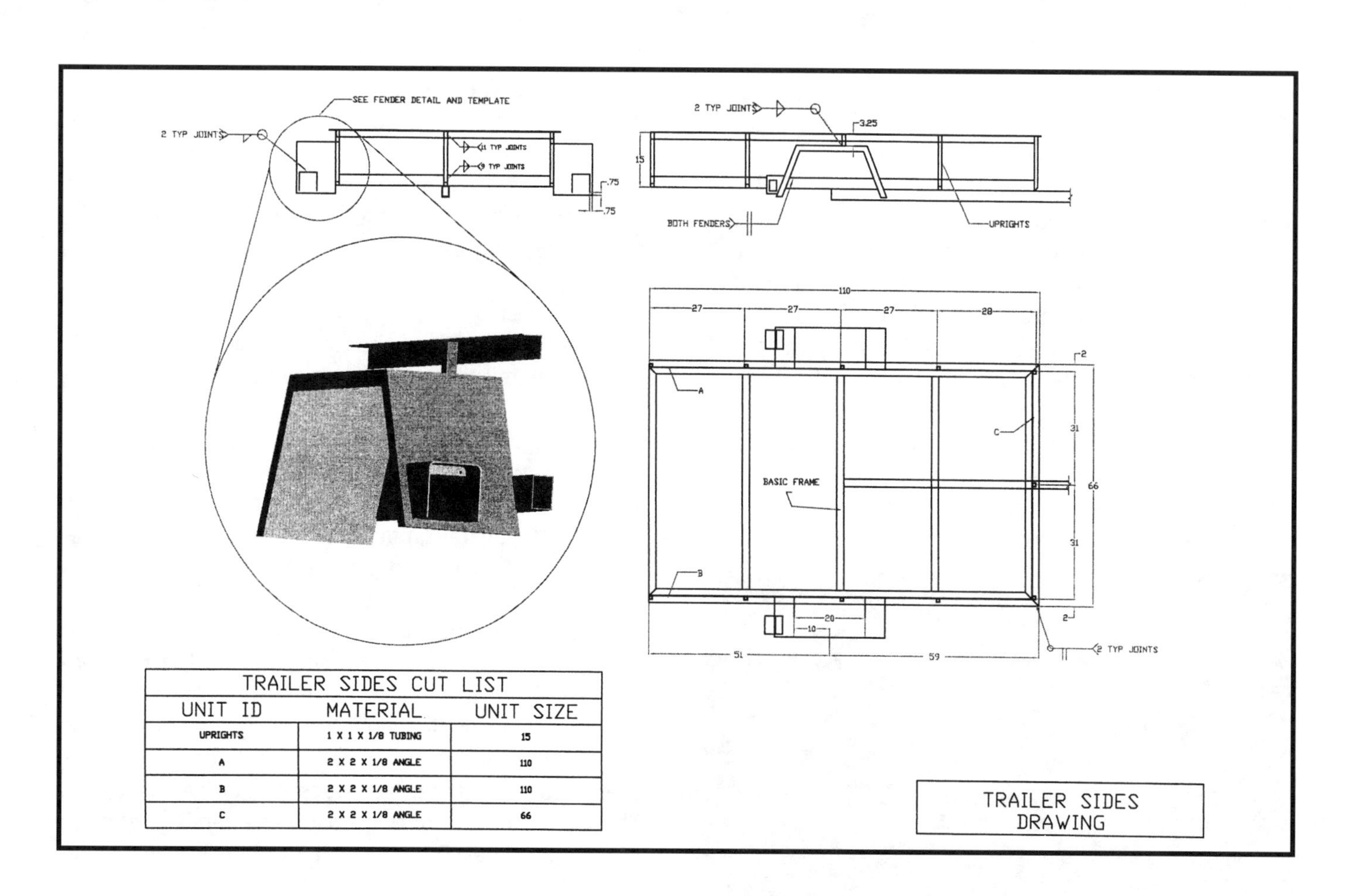

TRAILER SIDES CUT LIST		
UNIT ID	MATERIAL	UNIT SIZE
UPRIGHTS	1 X 1 X 1/8 TUBING	15
A	2 X 2 X 1/8 ANGLE	110
B	2 X 2 X 1/8 ANGLE	110
C	2 X 2 X 1/8 ANGLE	66

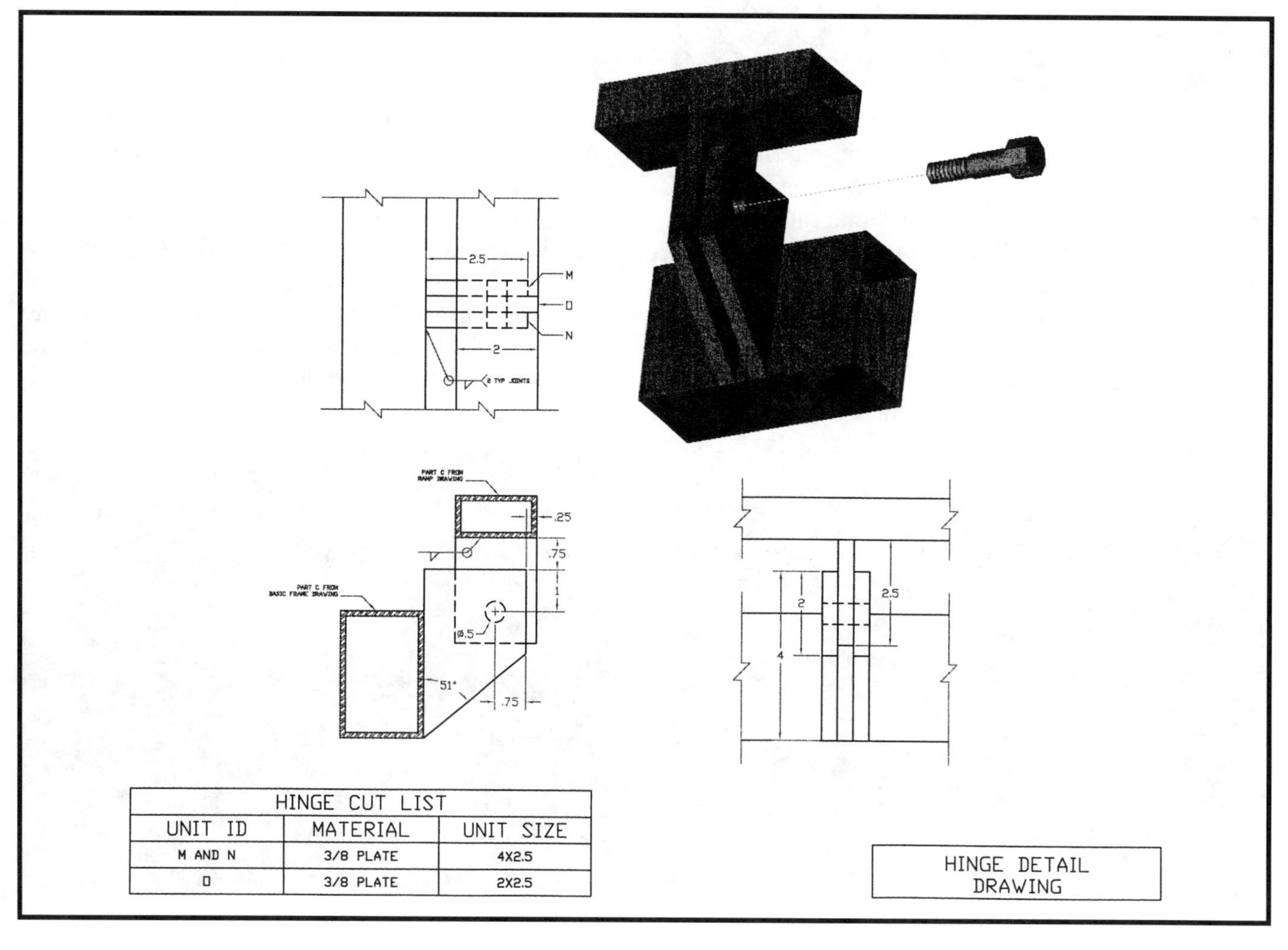

HINGE CUT LIST		
UNIT ID	MATERIAL	UNIT SIZE
M AND N	3/8 PLATE	4X2.5
O	3/8 PLATE	2X2.5

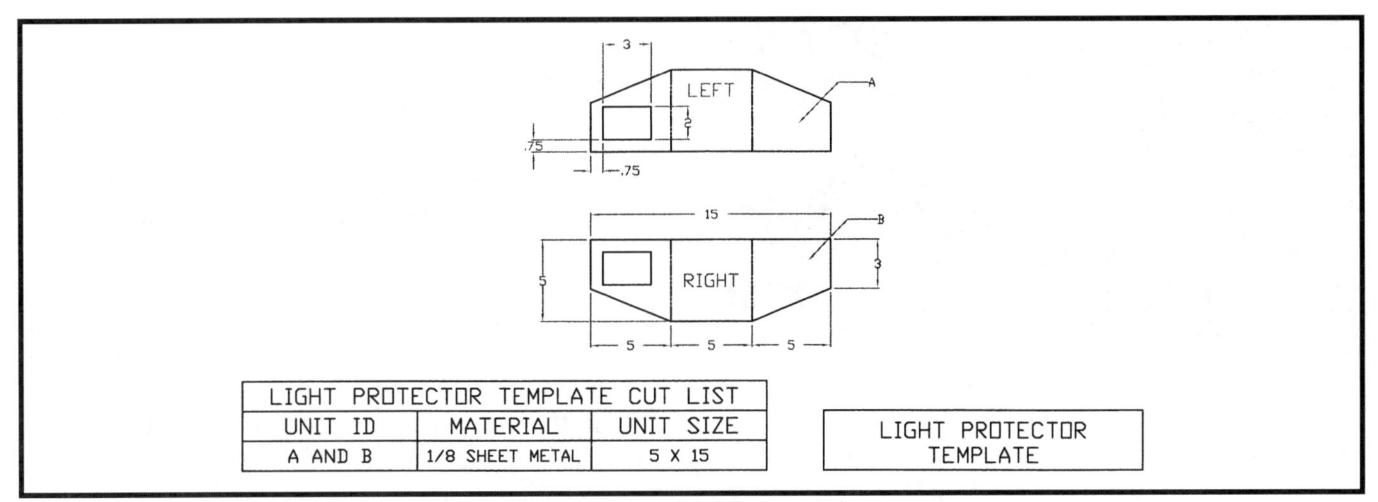

LIGHT PROTECTOR TEMPLATE CUT LIST		
UNIT ID	MATERIAL	UNIT SIZE
A AND B	1/8 SHEET METAL	5 X 15

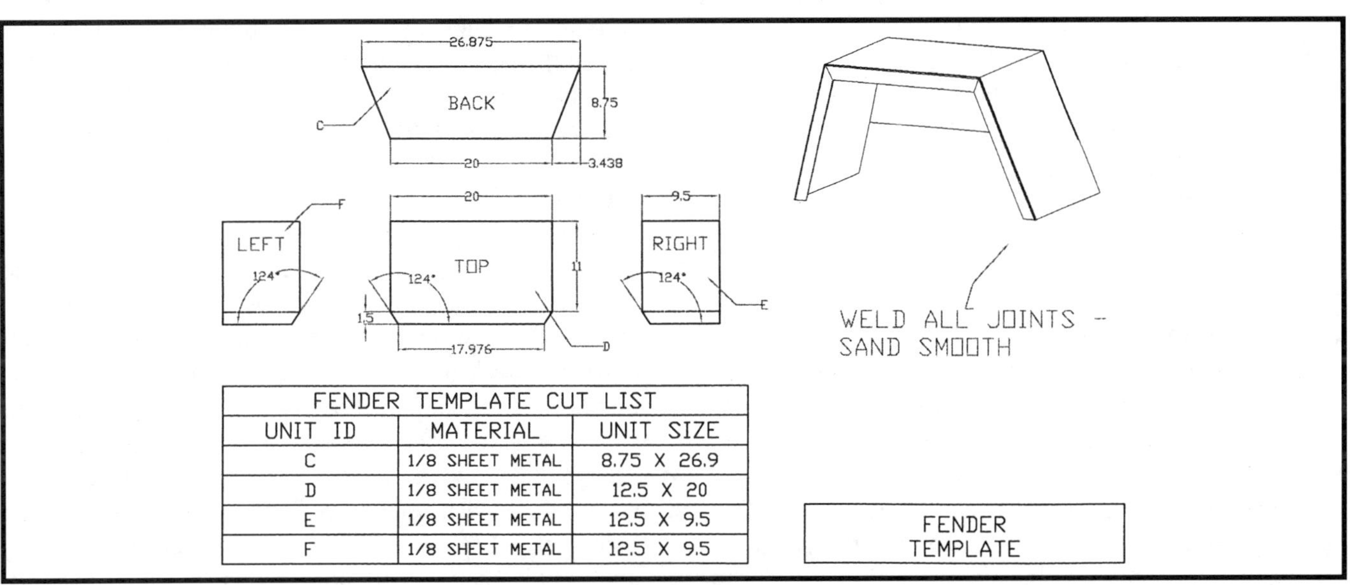

FENDER TEMPLATE CUT LIST		
UNIT ID	MATERIAL	UNIT SIZE
C	1/8 SHEET METAL	8.75 X 26.9
D	1/8 SHEET METAL	12.5 X 20
E	1/8 SHEET METAL	12.5 X 9.5
F	1/8 SHEET METAL	12.5 X 9.5

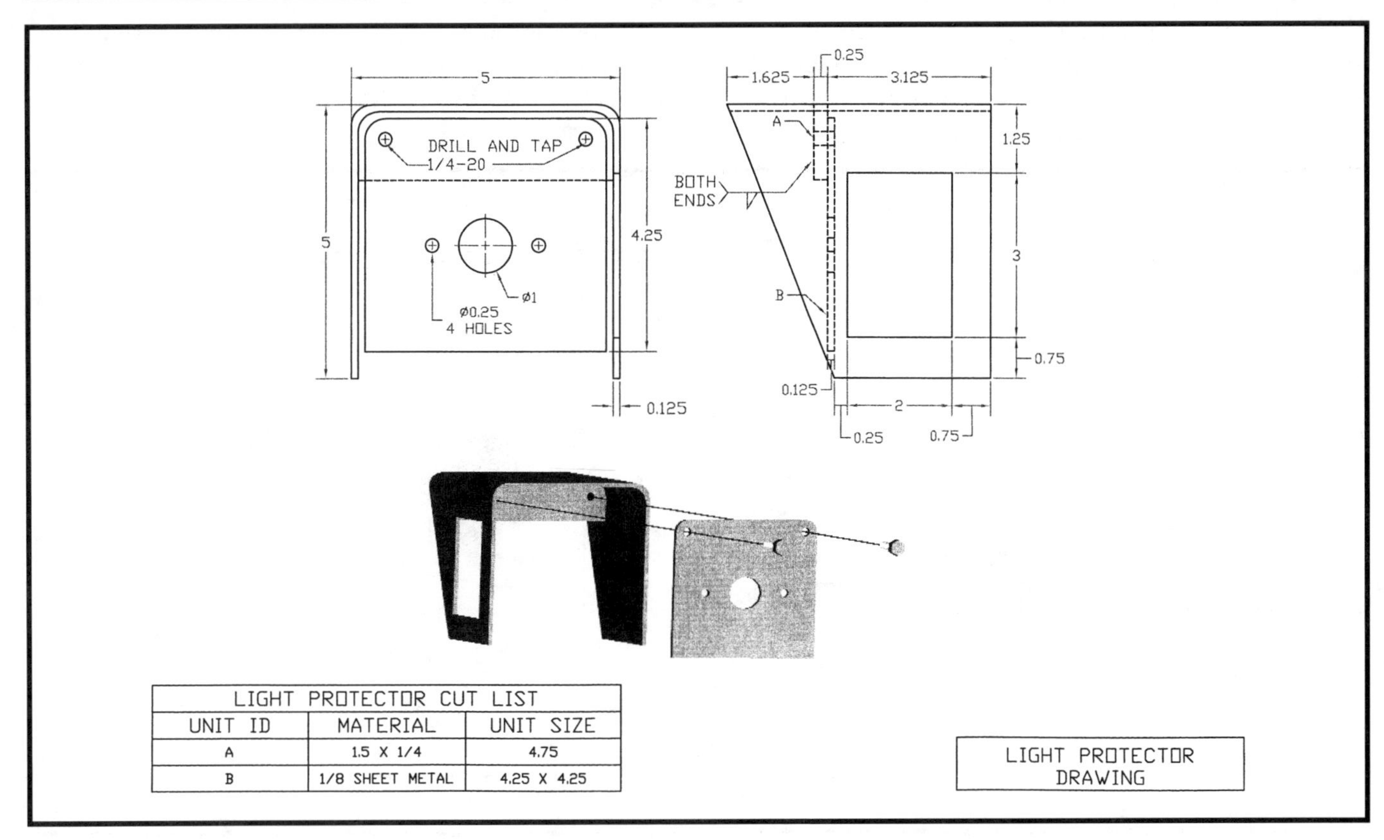

LIGHT PROTECTOR CUT LIST		
UNIT ID	MATERIAL	UNIT SIZE
A	1.5 X 1/4	4.75
B	1/8 SHEET METAL	4.25 X 4.25

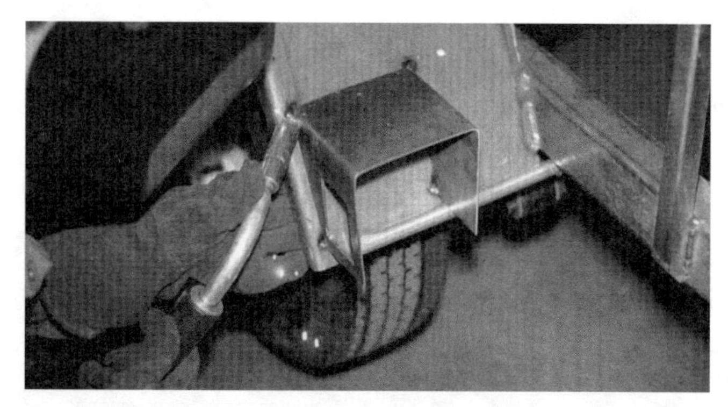

On the light protectors, a reciprocating saw is used to cut the holes for the running lights. Cut the two pieces indicated on the Light Protector Cut List for both the left and right light cases. Drill, tap, and weld the pieces to the cases as indicated on the Light Protector Drawing. The tail lights will now be attached to part B instead of bolted through the fenders. This should help the lights and fenders last longer. Now that the light protectors are complete, they can be attached to the fenders. Finally, weld the last two vertical supports to the fenders and angle iron to complete the sides.

Fold Up Ramp

You will need to cut all of the pieces indicated on the Ramp Cut List. Again, cut the lengths of the steel tubing using the horizontal band saw and the mitered cuts using the vertical band saw. Once the tubing is cut to length, tack it together according to the Ramp Drawing. After checking the diagonals to make sure everything is square, weld the outer frame and the three center supports into position. Remember to keep the center supports down 1/8" so when the expanded metal is set on top it will be flush with the outer frame. Now that you have the basic frame of the ramp built, shear the expanded metal to length. The U-Edging will slide over the edges of the expanded metal and form the inner frame on the Ramp Drawing. A vertical band saw was used to cut the edging to length and miter the ends. Once the expanded metal is inside the U-Edging, it can be welded to the outer frame as indicated on the Ramp Drawing.

Now cut out the hinges and assemble them from the 3/8" plate. Cut the basic pieces indicated on the Hinge Detail Cut List and then use the Hinge Detail Drawing to cut them to their final shape. Drill all the 1/2" holes as indicated on the drawing. Assemble all three hinges with a 2" x 1/2" bolt and tack them into position on the back of the trailer frame according to the drawing. Tack the ramp to the other end of the hinges and fold it up to ensure that nothing is binding. Once everything is moving smoothly, weld the hinges.

Finish Work

Attach conduit to the basic frame with self-tapping screws to run the wires to the tail lights. Once the conduit is in place, weld strips of 1" x 1/4" steel bar to the inner supports on the basic frame. This will allow the deck planks to be bolted down to the strips instead of bolting through the trailer's supports, ensuring that water doesn't get inside the frame and rust it out from the inside. The deck planks can now be cut to length and bolted into place. Cut the two pieces of 1-1/4" x 1-1/4" x 1/8" angle iron 62" long and weld them over both ends of the decking. Capping the ends of the boards ensures that they won't pull up or get damaged when items are dragged on and off the trailer. The only thing left to do is to paint the trailer

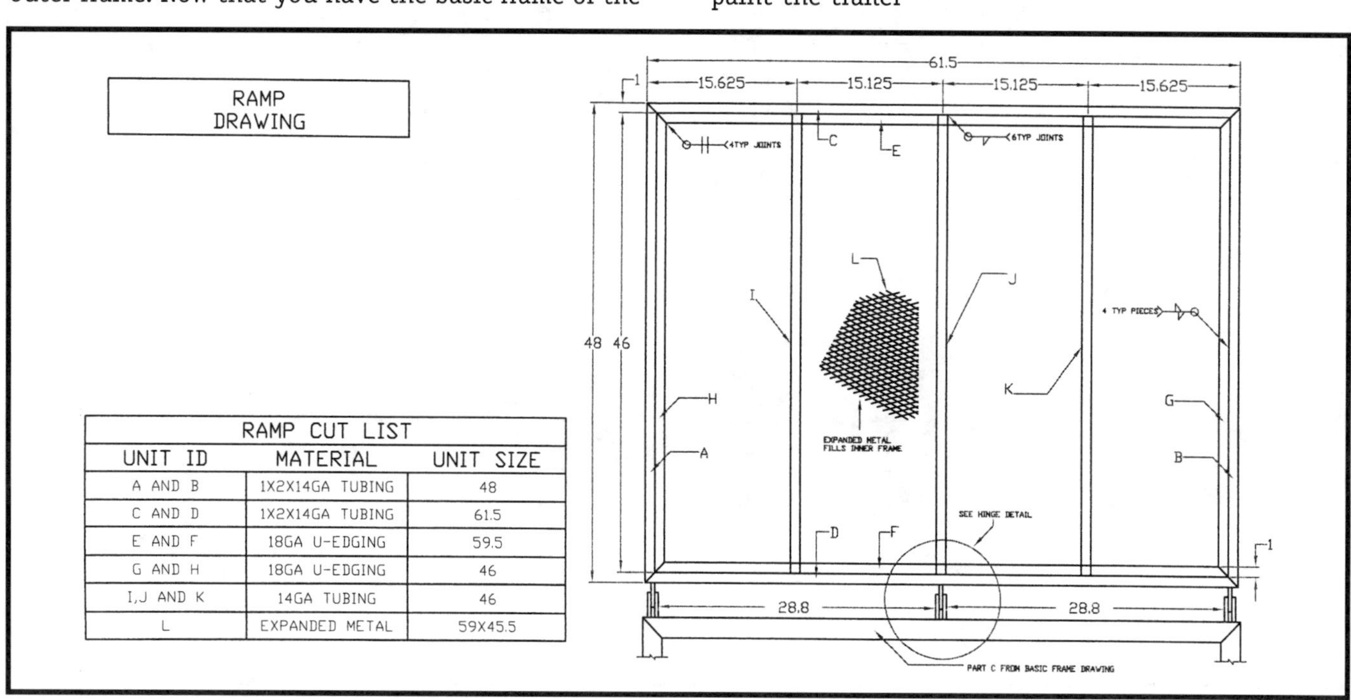

RAMP CUT LIST		
UNIT ID	MATERIAL	UNIT SIZE
A AND B	1X2X14GA TUBING	48
C AND D	1X2X14GA TUBING	61.5
E AND F	18GA U-EDGING	59.5
G AND H	18GA U-EDGING	46
I,J AND K	14GA TUBING	46
L	EXPANDED METAL	59X45.5

Welding Trailer

Author: Colby Gardner
Instructor: Garry Wilfone
School: Tulia High School
City & State: Tulia, TX

Tool List

Hand grinder
Drill
Torch
File
Hammer
Chisel
Stick welder
Wire welder
Plasma torch
Pliers
Paint sprayer
Chop saw
Pedestal grinder
Come-along
Welding rod
Drill-bit
Chipping hammer
Wrench
Tape measure
Tri-square
Square
Bar-clamp
C-clamp
Vice
Vise grip C-clamp

Bill of Materials

2	15-inch Deck Plate Fenders
2	Fender Skirts
1	Bull Dog Jack
1	Bull Dog Wheel for Jack
1	Axle Assembly
2	15-inch Wheels
1	Bull Dog Ball Hitch
2	Tires
1	Floor Plate, 1/8" x 72" x 120"
1	Floor Plate, 4' x 6'
1	Paint and Paint Thinner
1	Light, Light hanger, Wire
20'	4 inch Channel Iron
2	Hasp
4	Weld On Hinges
14'	Rectangular Tubing
1	10 inch Pipe Casing

Frame

The main frame of the trailer is made of 20 wall 2 x 4 inch rectangular tubing. For the tongue, use a chop saw to cut two pieces, 7 foot long. Bend them at a 60-degree angle using a cutting torch. These pieces will extend 3 foot into the trailer.

For the side runners, cut two 8 foot long pieces of the 4-inch rectangular tubing. Then, cut four pieces of 4-inch channel iron to be equally spaced for the cross-beam. Lock and square the material together, then weld everything with 7018 welding rod.

The dimensions of the frame total 8 foot by 5 foot 1/2 inch.

Tongue

The tongue of the welding trailer is made from 20 wall 2 inch by 4 inch rectangular tubing. The tongue projects 4 foot out from the trailer and the hitch is a bull dog hitch. With the tongue and hitch attached to the mainframe, the trailer is 12 foot long.

Using 7018 welding rod, weld the two pieces of rectangular tubing that create the tongue at the tips, under the bull dog hitch. Also weld the hitch to the tongue.

Axle

Use a single, 6 foot long axle (the type with springs, not a rubber torso) that will hold 3,500 pounds. Center the axle under the trailer to balance the weight of the weld and other items to be carried on the trailer. Weld the axle to the outside beams of the mainframe.

To install the axle, turn the trailer upside down so that the axle can be set properly. Using 7018 rod, weld the spring hangers to the mainframe in four places. Note: be cautious in rotating the trailer.

Bed

Use one full sheet of 8 foot by 5 foot by 1/8th inch deck plate to make the trailer bed. Weld the deck plate completely around the trailer using 6011 1/8th rod. The deck plate is attached to the outside beams of the mainframe of the trailer frame.

Toolbox

There are two 1 foot by 3 foot toolboxes, one on each side of the trailer. Make these with 1/8th inch deck plate, bent at three separate 90-degree angles. To make the toolboxes removable, do not weld them directly to the trailer. Instead, weld a separate piece of 1/8th inch flat, 1 foot by 3 foot, to the bottom of each toolbox, using a Mig welder for this step. Drill holes in the bottom of each toolbox and in the bed of the trailer. Then attach each toolbox to the trailer bed with bolts.

Make the toolbox lids using 1/8th inch deck plate. Bend the plate to a 90-degree angle in four places, using a hand bender, to make a lip over all sides. Weld on the hasps using a Mig welder. Also weld on the hinges using a Mig welder.

Headache Rack

Make the headache rack using 2 x 2 inch square tubing. The rack will measure 3 foot 4 inches tall, and 3 foot inside to inside, with the top bar at 5 foot 3 inches.

On both sides of the top bars, make the side metal holders from 2-inch strap, but to 7 inches. Also make two welding lead hangers, consisting of two pieces of 2 inch pipe, set 5 inches apart. Weld a piece of strap 6 inches long by 3 inches wide to the outside of the pipe.

Weld the entire rack to the back of the tongue and the front of the bed using 6011 rod.

Cutting Torch Hanger

Use two pieces of 2-inch pipe and a piece of 3-inch flat to make the cutting torch hanger. Cut and clean the metal with a chop saw, plasma torch and grinder. Weld the cutting torch hanger to the front of the headache rack, on the center bar, using a Mig welder.

Oxygen/Acetylene Bottle Holders

To fabricate the bottle holders, cut the 10-inch pipe casing using an oxygen/acetylene cutting torch. The oxygen holder will be 14 inches tall and extend beneath the trailer bed. The acetylene holder will not extend beneath the trailer bed, since the cylinder is shorter and will set on top of the trailer bed. Both cylinders will be held upright against the headache rack with a bracket, which rotates, allowing the bottles to be released. Weld the pipe casing all around using a Mig welder.

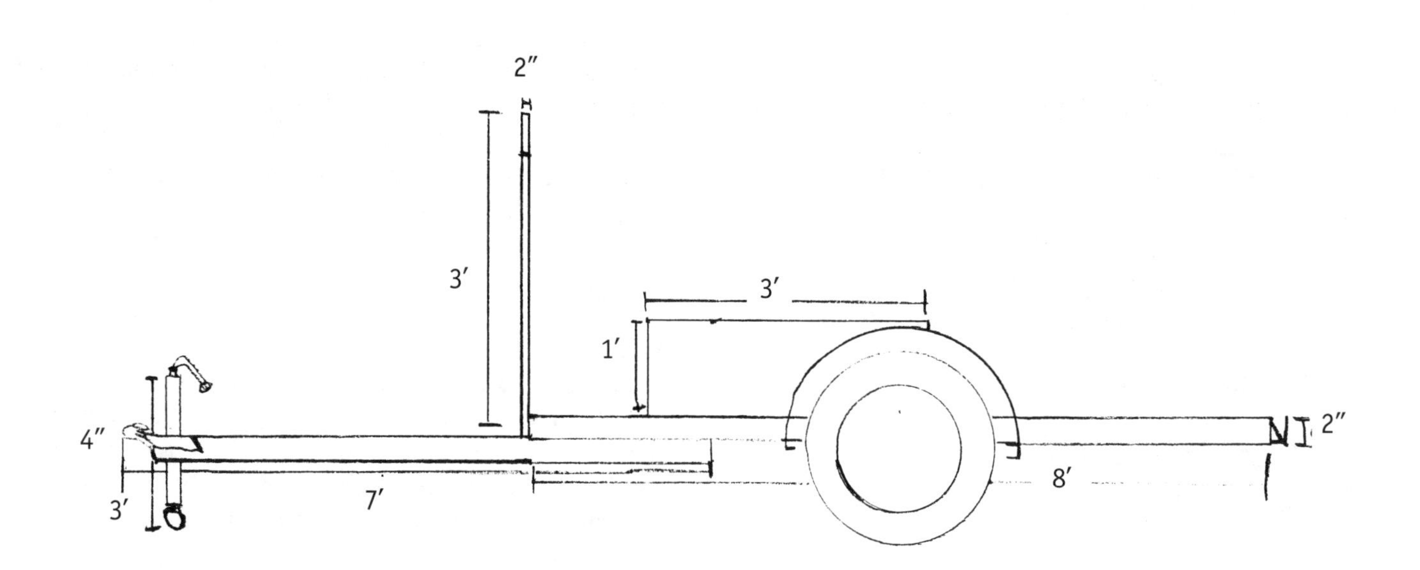

Heavy Duty Mower Trailer

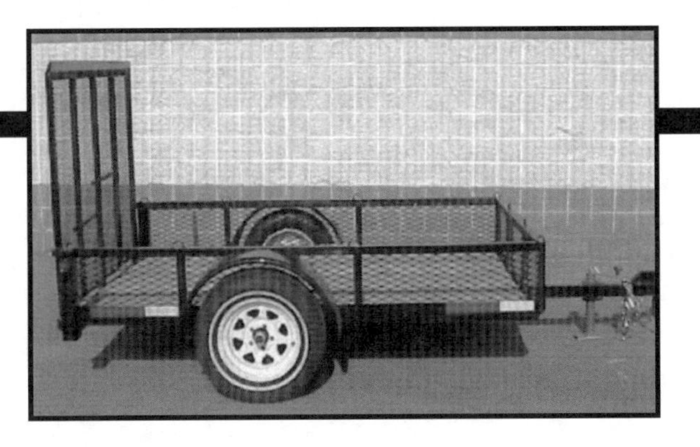

Author: Eric Hardel
Instructor: Dan Saunders
School: Monroe High School
City & State: Monroe, WI

Bill of Materials

Qty.	*Description*
2	1/2″ x 3″ x 4″ Angle – side frame members
1	1/4″ x 3″ x 4″ Angle – front frame member
1	3/16″ wall x 2″ Square tube – back frame
3	3/16″ x 2-1/2″ Angle – bottom frame members
1	3/16″ x 2″ x 3″ Tube – tongue
1	3/16″ x 2″ x 3″ Tube – tongue piece
1	1/4″ x 1-1/2″ x 7″ Flat bar – tongue cap
1	1/4″ x 1-1/2″ x 3-1/2″ Flat bar – tongue cap
1	14 gauge x 1″ x 5′6″ Square tube – wiring protection
2	1/8″ x 3″ x 4″ Light posts
2	3/16″ x 2″ x 8″ Angle – top side rails
1	3/16″ x 2″ x 6′ Angle – front rails
8	3/16″ x 1-1/2″ x 15-1/2″ Angle – side vertical supports
1	14 gauge x 1″ x 70-1/2″ Square tube – top tailgate
4	14 gauge x 1″ x 14″ Square tube – tailgate verticals
1	3/4″ schedule 40 Pipe – 70-1/2″ – bottom tailgate
1	10 gauge x 1-1/2″ x 67″ expanded metal – tailgate
1	10 gauge x 1-1/2″ x 14″ x 68″ ex. metal – front trailer
2	10 gauge x 1-1/2″ x 15″ x 96″ ex. metal – side trailer
5	14 gauge x 1-1/4″ Sq. tube x 48″ – ramp gate
1	14 gauge x 1-1/2″ Sq. tube x 68″ – ramp gate
1	3/4″ schedule 40 Pipe x 68″ – ramp gate
1	3/4″ x 10 gauge x 48″ x 68″ Raised ex. metal – ramp
2	11″ x 32″ fenders
1	2000# Torsion axle – 0 degree
2	Trailer tires and rims
1	2″ Coupler
1	2K Top wind jack
1	Package of torx deck screws
2	Red clearance lights
2	Oval tail/turn lights
1	Set of safety chains
1	Flat four trailer connector

Qty.	*Description*
1	6′ – Reflective tape
1	17′ 4-wire
1	6′ 3-wire
16	Heat shrink wire connectors
6	Red oxide primer
6	Satin black truck bed liner protector
6	Gloss black spray paint
6	2′ x 12″ Treated boards

A 5 x 8 utility trailer can haul about everything an ambitious homeowner needs. Whether it is lawnmowers with the attached ramp tailgate or drywall for a new addition, this trailer can handle it all. The trailer is equipped with the appropriate reflective stickers, tail-lights and marker lights and with legal safety chains.

First, cut the frame and tongue. Weld the frame together, checking constantly for square. After the frame is together, make the tongue and attach it, welding at all the seams.

Next, comes the axle, which in the case of this project was purchased from a local custom trailer shop. Weld the axle onto the frame slightly toward the rear to give the correct tongue weight. With the axle on the frame, flip the frame over to start work on the sides.

The sides are a project in themselves. Trying to keep them level and square is not easy, but it can be done. Start with the light posts and then put on the front corners. With these on, put the top angle iron on and put in the middle supports accordingly. Repeat this step on the other side and front.

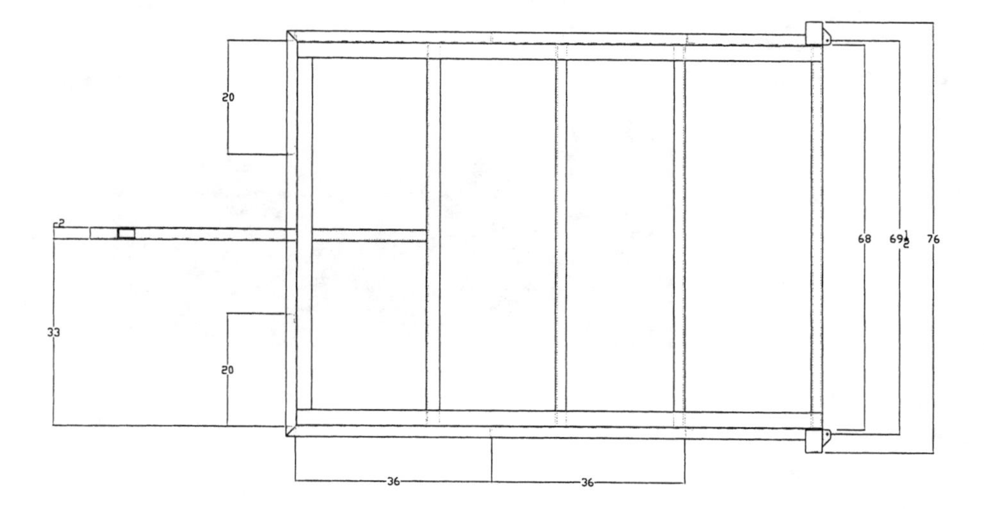

Notes:

All welds to be made using GMAW

Fenders have been omitted for clarity.

Expanded Metal has been omitted for clarity.

2000# Torsion axle.

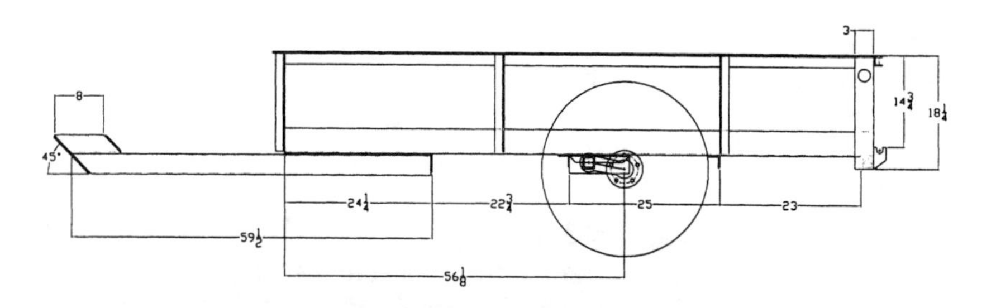

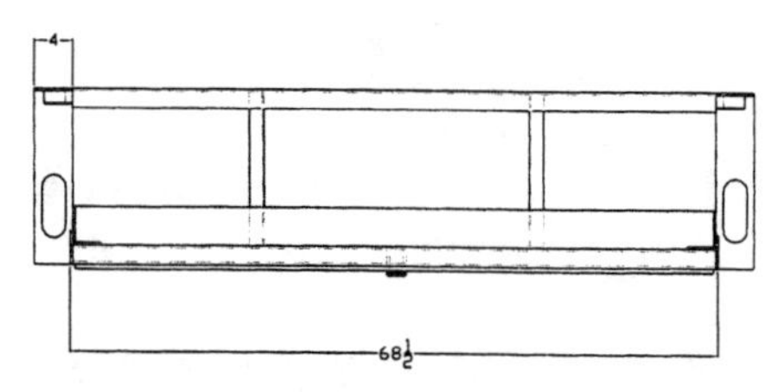

UTILITY TRAILER

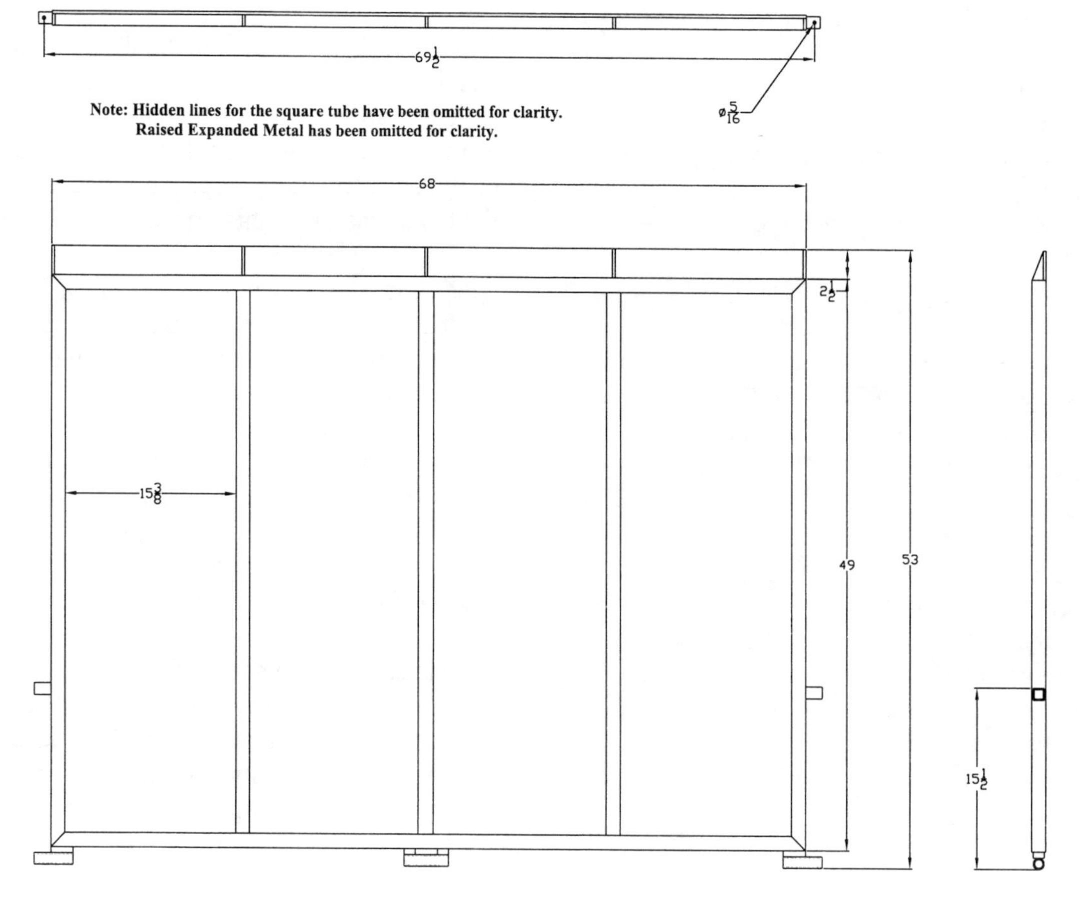

TAILGATE RAMP

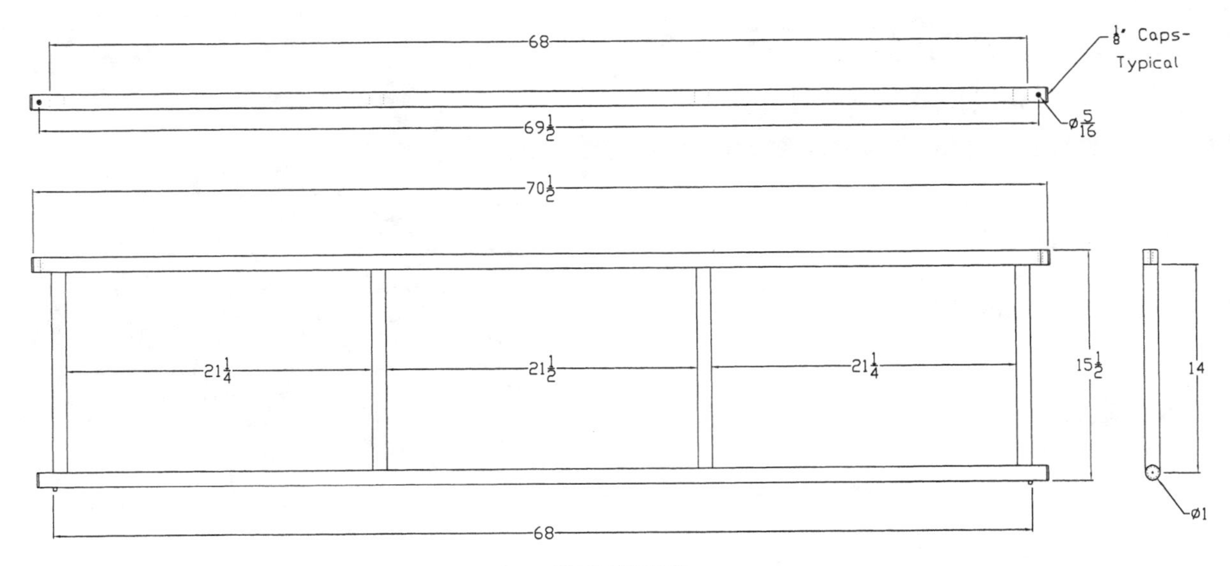

SIDE VIEW

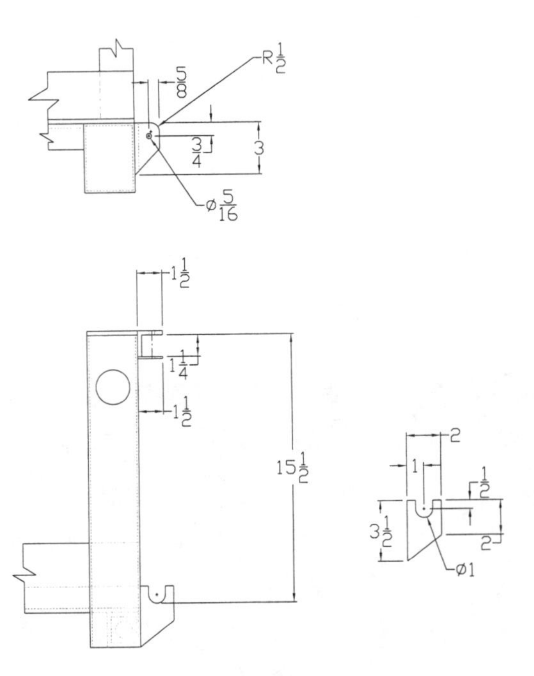

TAILGATE DETAIL

With the sides completed, attach the pre-fabricated fenders. (The fenders came from the same place as the axle.) For the fenders, get the correct spacing around the tire and weld them onto the frame.

With the fenders on, cover the side with expanded metal. The metal comes in 4 x 8 sheets, so use a plasma cutter to cut it to fit. Then spot weld the metal to the frame. One caution: this metal is the hardest thing to weld — since it is so thin, it is very easy to burn through.

With the side now covered, move on to build the tailgates. The ramp is a simple rectangle with some support in the middle, covered with raised expanded metal to provide grip. The smaller tailgate is a bit more difficult to fabricate since it has to line up with the top of the sides. It requires some patience and a good tape measure.

With the tailgates built, the trailer is ready to go to the paint shop. (Since my father works for an auto body shop, the trailer was sand blasted and painted in one day and done very well.) After the paint dries, take the trailer back to the shop to add the 2 x 12 treated wood floor and the wiring for the tail lights. (The taillights were purchased from the same shop as the axle and the fenders.) Run the wire from the tongue to the back, then split and run it to the tail lights and marker lights. With the lights finished, and reflective stickers attached, the trailer is road ready.

Off-Road Monster Grill-BBQ
(Not for Road or Highway Use)

Author: Bill Stevens
Instructor: David H. Murray
School: Ferris State University
City & State: Big Rapids, MI

Materials

A.	0.125″ plate
B.	0.25″ plate
C.	0.375″ plate
D.	0.5″ plate
E.	2″ x 3″ tubing
F.	2″ x 2″ tubing
G.	1″ x 1″ tubing
H.	6″ round stock
I.	2″ round stock
J.	0.75″ round stock
K.	1″ round stock
L.	0.25″ bar stock
M.	Purchased axle

Fabricating the Trailer Bed

A. From the 2″ x 3″ tubing, cut two pieces that are 10′ long, and two pieces that are 55.25″ long.
B. Cut a 45 degree angle on all of the corners to make a box with no open ends.
C. Again, using 2″ x 3″ tubing, cut the cross members for the trailer bed.
D. Begin assembling the trailer bed by first setting the two 10′ pieces parallel to each other, and then setting the 55.25″ pieces at the ends of the long pieces. They should fit up well and there should be no holes showing.
E. Using a framing square and CC welder with an E6010 electrode, make sure the frame is square, then tack the outside corner. Repeat with the other corners. Flip the frame over and tack all of the corners on the other side.
F. Put the cross members into the trailer frame, spacing them 24″ apart. After making sure they are square, tack them in.
G. Using GMAW, weld the trailer bed together. There should be no gaps or holes anywhere in the trailer bed.

Assembling the Axle

A. Take the axle out of the box and set it on a table.
B. Remove all of the hardware and read the instructions.
C. Take the springs out of the plastic and use the u-bolts to attach them to the axle.
D. Attach the trailer mounts onto the springs.
E. Take the bearings and pack them with a high temperature bearing grease.
F. Insert the bearings into the hubs, and slide the hubs onto the axle.
G. Thread the holding nut onto the end of the axle and put the cotter pin through it.
H. Put the dust caps onto the axle.

Attaching the Axle to the Trailer Bed

A. Set the axle on the trailer bed.
B. Measure the difference in width between the trailer bed and the axle to determine what size to make the axle clearance blocks. (They should be 4" x 6".)
C. Measure halfway down the trailer bed, and then go back from that half the width of the spring on the axle.
D. Tack the axle blocks on and then weld them up solid.
E. Set the axle on the axle clearance blocks and make sure that it is square.
F. Tack the axle to the clearance blocks.
G. Weld the axle on solid.
H. Bolt the fenders on a piece of 0.125" plate that has a 90 degree bend in it so that the legs are 2" x 6". Then weld the custom angled piece onto the trailer frame.

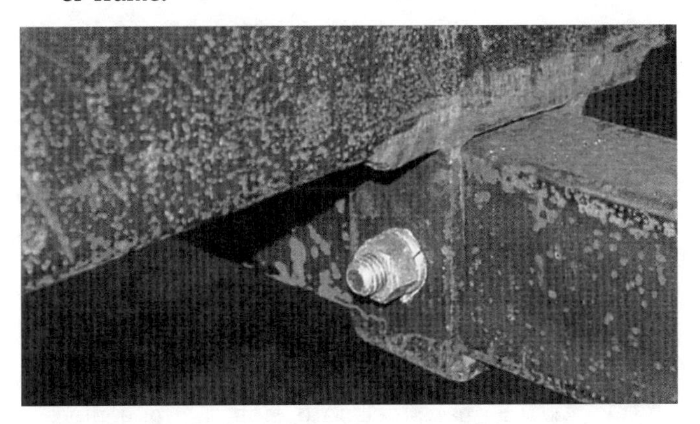

Fabricating the Tongue Brackets and the Tongue

A. Cut out four pieces of 0.25" mild steel plate that are 2" x 3.25", two pieces that are 2.25" and four that are 1" x 2".
B. Take the four longest pieces and the two second-longest, and tack them together so that they look like a (u). Now tack the two pieces that are 1" x 2" to the ends of the long pieces so that the tabs hand out from the inside of the hang out from the inside of the (u).
C. Once all of the tabs and pieces are tacked together as described, GTA weld them with ER60 electrode.
D. Place the tongue on the trailer frame with the brackets on the tongue. The tongue should be in the middle of the trailer with the brackets on the first two cross members.
E. To make sure the trailer rolls smoothly, square up the tongue to the axle.
F. Put two bolts through the tongue and tighten them.
G. Put the wheels on the hubs.

Installing the Ball Receiver and the Trailer Jack

A. Place the ball receiver over the 2" tubing and put a 3" clamp on it.
B. Take the tongue and the ball receiver over to the drill press and drill out the two holes that are needed for installation.
C. Decide on the most convenient spot, and install the trailer jack.

Fabricating the BBQ

A. Cut out two squares on the plasma burner. The top of the BBQ will be 55.25" x 56" and the bottom will be 55.25" x 60".
B. Bend the top square at a 45 degree angle at 6", 14", then 42", 50", 56". Make a 3" radius hole for an exhaust that will make the top shell of the BBQ.
C. Make the bottom of the BBQ the same way, but make the bends at 8", 18", 42", 52", and at 60". There are now two trough shaped pieces that will be the top and bottom of the BBQ.
D. To cap the end, cut out the parts to seal the trough. Put two baffles in the trough.
E. Weld the caps onto the troughs with GTAW and grind them flush. Use a very fine grind paper to finish the grind.
F. For each half of the BBQ, make a frame out of 1" tubing to fit inside the rim and make it rigid. Since the halves have different measurements, each frame will have to be custom fit.
G. To fabricate the hinge for the BBQ, cut out nine L shaped pieces with a 3/4" hole in each one. Space them evenly in groups of three across the back of the BBQ. Weld the hinges solid with a GMAW process. When welding them to the BBQ, put the rod that will be used inside the L-shaped parts so that they will be receptive to the rod after they are welded.
H. Weld a washer on one end of the hinge and put a cotter pin in the other end of the rod so that it will not slide around. The hinges are now complete.

Installing the Handle and the Brace for Counterweights

A. To install the handle, first cut two pieces of 3/8" bar stock, 3" wide and 48.5" long.
B. Weld them so that the bottom of the bar stock lines up with the bottom of the top of the BBQ.
C. Make the handle of 1.5" or 2" OD pipe. It will be 55.25" long.
D. The pipe will extend to 1/4" from the ends of the handle mounting bars, where it will be welded solid.

E. The counter weight support bar and the braces for it will be mounted on the 6″ of bar sticking out behind the BBQ.
F. Make the 55.25″ long counter weight mounting bar out of the same material as the handle mounting bars.
G. Make two 7.75″ long gussets out of the 3/8″ bar stock. Space them evenly behind the BBQ and weld them solid.
H. The amount of weight and location of the weight to be used for counter balance will be decided by the builder of the BBQ.

Making the BBQ Cooking Surfaces

A. Shear out three pieces of expanded steel, 33″x 17.25″ each, and attach them to a steel frame made of 1″ x 1″ tubing.
B. Fabricate the cooking surface adjusting tabs using 36 1″ x 2″ pieces of 0.25″ steel plate. Place them in layers 3″ apart inside the BBQ on the bottom half to make the BBQ surfaces adjustable for different cooking applications.
C. Create a chimney and adjustable vents for the BBQ to control air flow. The chimney can be made with 6″ ID pipe and a baffle can be made out of a disk of 6″ round stock.

Fabricating the Chili Cooking Arms and the Cooking Units

A. Cut two pieces of 2″ x 2″ tubing 70″ long, and one piece 53.25″ long. Nine pieces of 2″ OD pipe will also be needed.
B. Mark out one 2.5″ in from the end of all of the pieces of tubing, and mark the middle of all three, also. Weld a piece of the 2″ OD pipe on all of the marks so that there are three pieces of pipe on all of the pieces of tubing. Now set the short piece of tubing on the back of the trailer bed, tack it in place, and weld it.
C. Take the two longer pieces and put a 2″ notch in one end of each piece for the hinge.
D. Put a 3″ radius hole on the other end of the two pieces for the chili cooker support arm.
E. For the hinge, use three 2.25″ pieces and one 6″ piece of 0.75″ round stock.
F. To make the hinge, weld a 1″ thick disk to the end of the 0.75″ round, then weld that piece to the back corner of the trailer bed, one on each side. Take another section that is 2″ thick and weld it into the chili cooker arm with the 2″ notch in it. Then slide the section just welded onto the cooking arm over the 0.75″ round. Do this for both cooking arms. Now slide the next section of the 2.25″ round over the 0.75″ round.
G. The cooking arms are now assembled.
H. The next step is to make the burner supports.
I. Cut 2″ x 2″ tubing into eight 2″ long pieces. Cut 0.25″ x 0.5″ bar stock into eight 4″ long pieces. Cut eight pieces of 0.125″ thick plate that are 1″ x 1″ and drill a 0.25″ hole in each of these.
J. Weld the pieces of bar stock to the pieces of tubing at a 45 degree angle so that the end of the bar stock hangs out of the tubing by 2″. Weld the 1″ x 1″ tab onto the bar stock so that the plate is parallel with the top of the tubing.
K. Weld the burner supports to the longer chili cooker arms at 12″ from the end, and also at 36″ and at 60″.
L. On the short piece of tubing, weld them on at 12″ and 36″.
M. To make the supports for the cooking surface, cut eight 10″ lengths of 1″ x 1.5″ tubing, and eight 6″ lengths of 1″ x 1.5″ tubing. Weld them into a T-shape above the previously welded pieces of pipe. Do this for all of the supports on all of the pieces of tubing.
N. Put on the burners and the burner shields using a 0.25″ bolt and nut.
O. Weld the chili cooker support arm mounts onto the tubing.
P. Weld a piece of 3″ OD pipe into the 2″ notch that was made earlier.

Q. There will also be a 3/4″ set bolt in the pipe.
R. Make the support arm out of a 36″ length of 2.75″ OD pipe. Weld a 3″ x 3″ piece of 0.25″ pipe onto the bottom to slide up and down to adjust for different types of terrain.
S. Next, fabricate the cooking surface for the chili cookers.
T. Burn out three shapes out of 0.25″ plate; 2 rectangles measuring 72.25″ x 12″, and another rectangle measuring 53.25″ x 12″.
U. On the two longer pieces of plate, burn out three 8″ diameter circles; on the short plate, burn out two 8″ diameter circles.
V. The parts that fall from the plate also will have holes burned in them.
W. Use 0.125″ steel plate to make the chili cooking arm shield. Bend an 8″ piece of plate with a 90 degree angle at 2″ so that it forms an L-shaped piece. Set the guard on the T-shaped parts so that it is centered and then make and drill 0.5″ holes where the burners are.
X. Now, set the guard onto the T-shaped supports and then place the cooking surface onto the T-shaped supports. Mark and drill 5/32″ holes on all of the T-shaped supports and put 0.25″ self-tapping bolts into each of the holes. This completes the cooking arms and the monster grill.

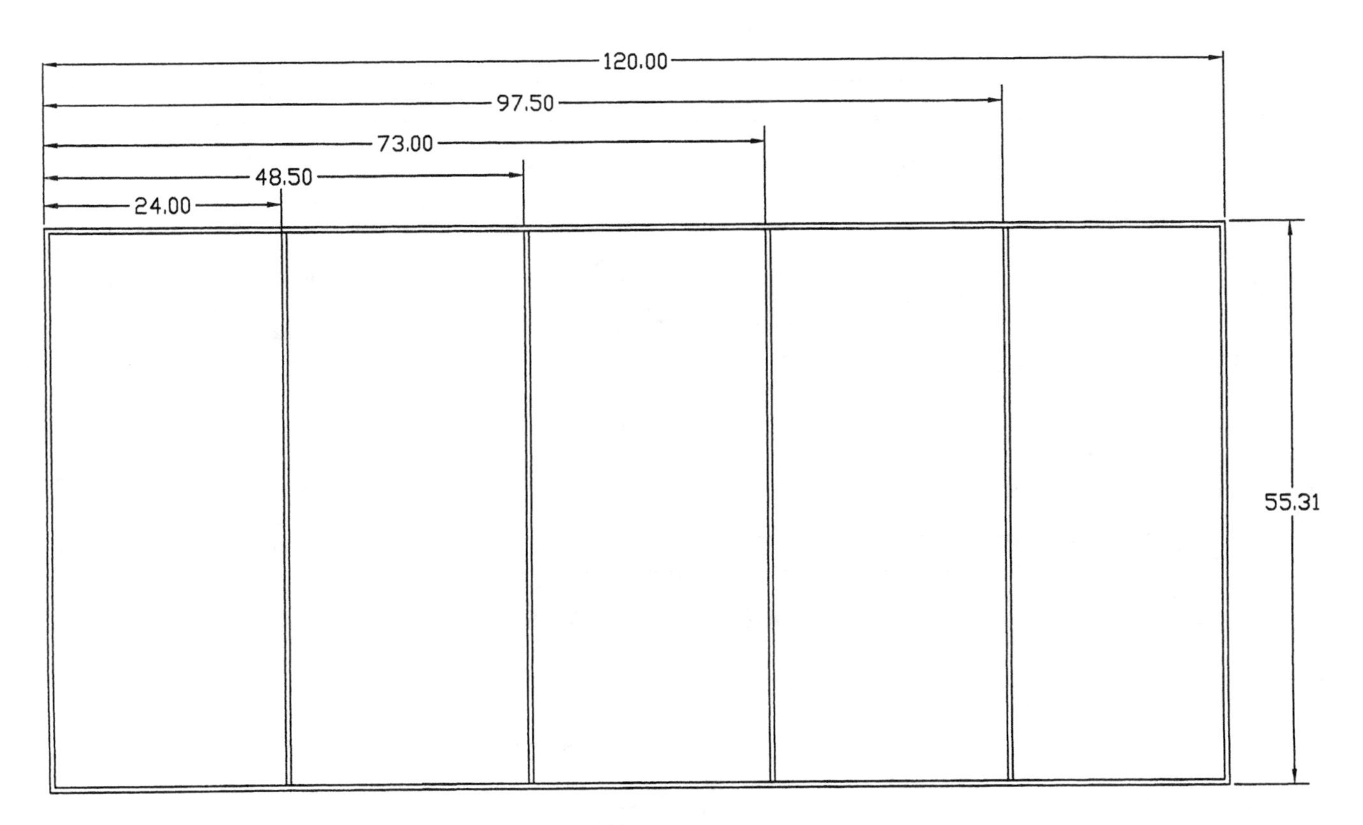

TRAILER FRAME

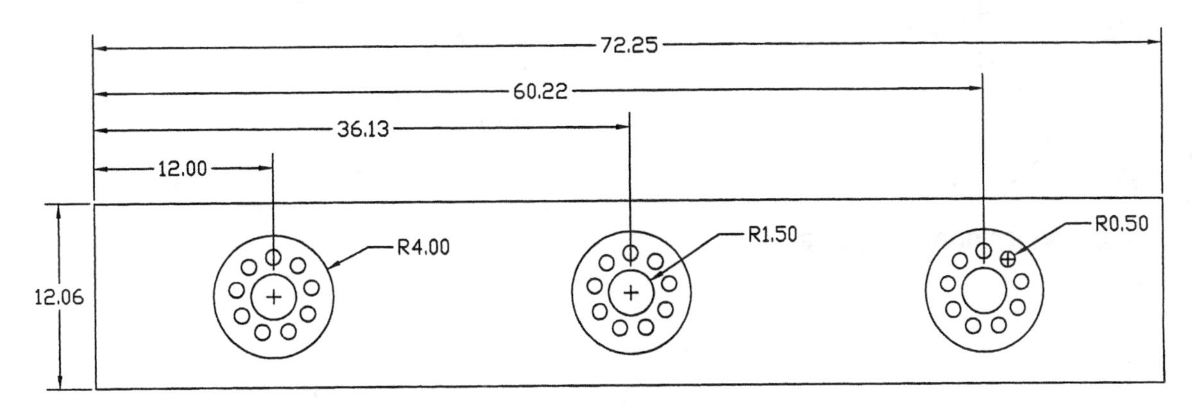

CHILI COOKER TOP SURFACE

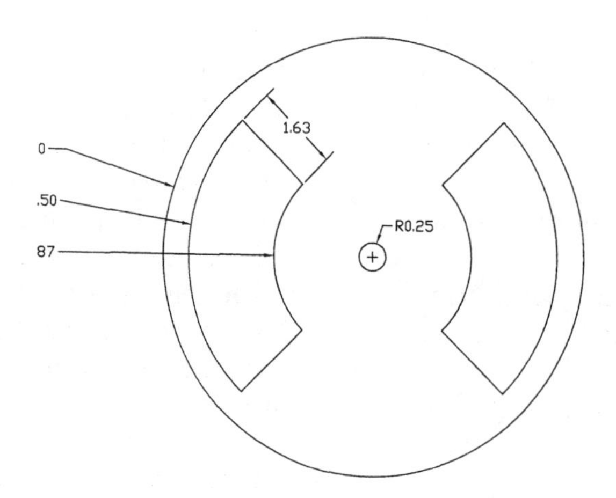

AIR VENT DOOR

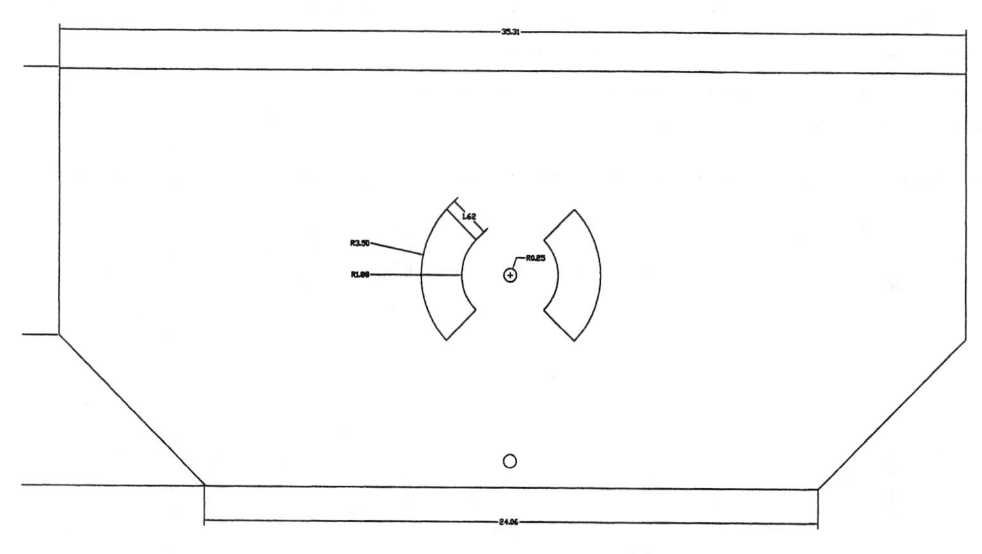

BBQ BOTTOM SIDE PLATES

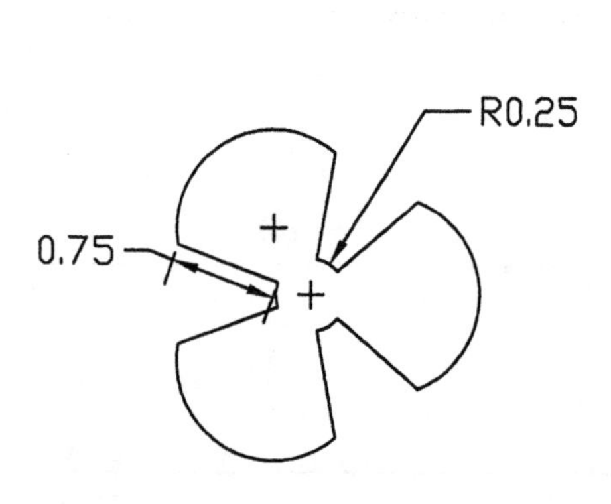

AIR VENT DOOR SPINNER

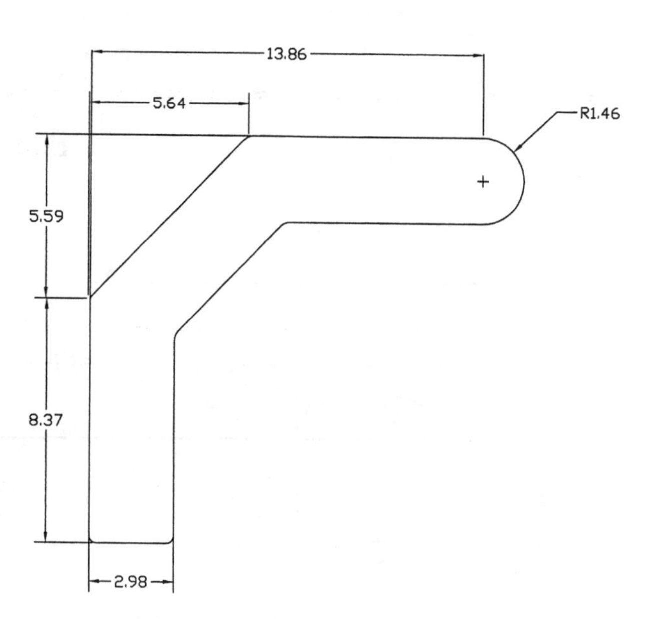

COUNTERWEIGHT BRACKETS

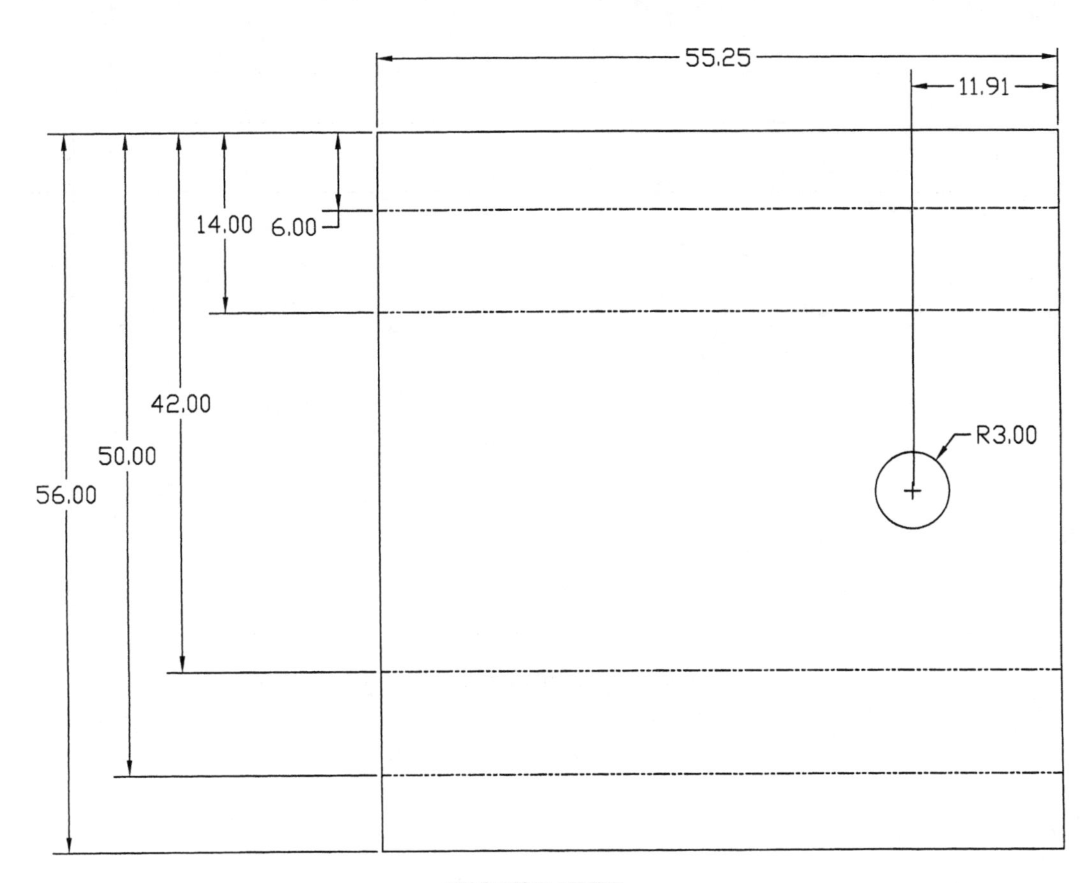

BBQ TOP UNIT

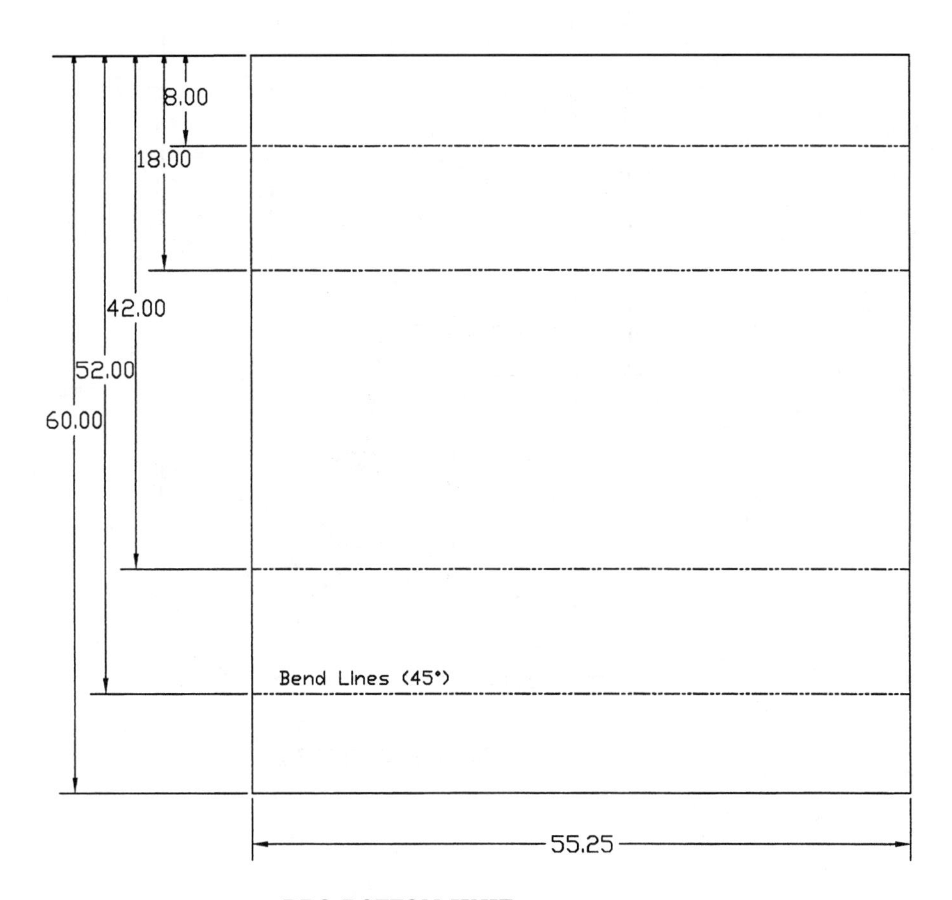

BBQ BOTTOM UNIT

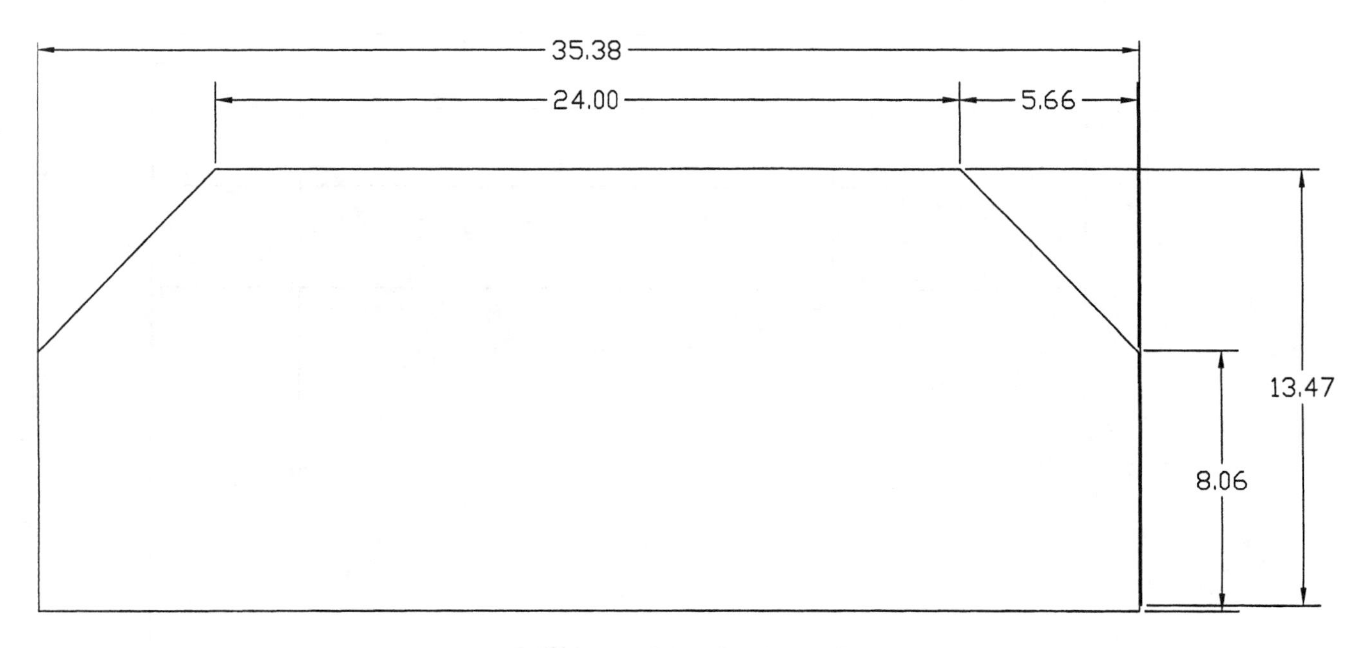

BBQ TOP UNIT SIDE PLATES

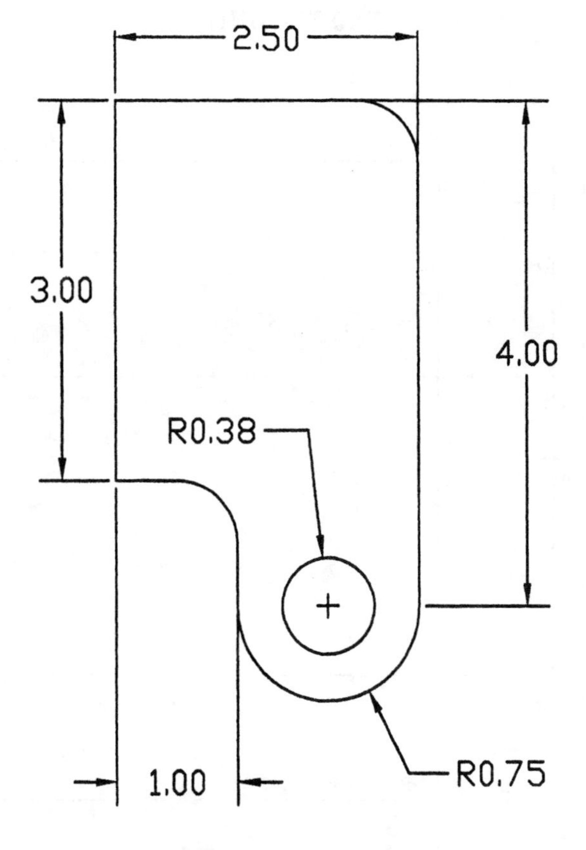

BBQ HINGE PIECES

20-Foot Gooseneck Cattle Trailer

Author: Daryn Hanley
Instructor: Dan Froneberger
School: Winona High School
City & State: Tyler, TX

Bill of Materials

Qty.	Description
11	1-1/2″ x 1-1/2″ x 20″ Tubing
5	1″ x 2″ x 20′ Tubing
7	1″ x 3″ x 20′ Tubing
1	2″ x 6″ x 20′ Tubing
5	Tire and wheel
1	Bulldog jack
1	Bulldog coupler
1	Set of fenders
700 sq. ft.	16 ga. Sheet metal
4	2″ x 2″ x 1/4″ x 20′ Angle iron
1	2″ x 3″ x 1/4″ x 20′ Angle iron
1	1-1/2″ x 1-1/2″ x 20′ Angle iron
1	1/2″ x 20′ Pipe
1	2″ x 20′ Pipe
5	1-1/2″ x 20′ Pipe
4	1″ x 20′ Pipe
2	Set of axles
10	Gussets
1	4″ x 20′ Channel iron
2	6″ x 20′ Channel iron
1	3″ x 20′ Channel iron
1	Brake away system
2	Rain deflectors
3	2″ x 1/8″ x 20′ Flat bar
1	3″ x 1/2″ x 20′ Flat bar
1	1-1/4″ x 1/8″ x 20′ Flat bar
1	1″ x 1/8″ x 20′ Flat bar
1	3/4″ x 20′ Cold roll
1	1″ x 20′ Cold roll
1	1/2″ x 20′ Cold roll
4	Oval brake lights
4	Armor plated lights amber
2	Armor plated lights red
1	Rear clearance light
60 ft.	Light wire
6 ft.	Whip cord
35 ft.	Brake wire
1	Carpet
4	Rubber floor mats
1	Floor
3	Gallons of paint
3	Gallons of primer
30 ft.	Reflector tape
10	Chrome strips
2	Light boxes
2	Safety chains
1	License plate light

The first step in building this cattle trailer is to connect two 20′ fenders together. To accomplish this, weld six pieces of channel iron inside the fenders for the floor bracing. For this job, use two pieces of 6″ x 6′5″ channel iron for the front and the back. Weld these two pieces flush with each end of fenders. Then weld three 4″ x 6′5″ pieces of channel evenly spaced between the two pieces of channel.

The second step is to place the six bows that you have created in a tubing bender. These bows are attached to the floor braces to stabilize the fenders and form the roof. They are made of 1-1/2″ x 1-1/2″ square tubing. Insert them into the sides of the fenders after carefully computing the proper placement and cutting of the insertion holes for them. Then, weld the bows in place and insert the proper bracing between the bows to complete the upper support for the trailer.

The third step is to put the sheet metal on the roof. Use 16-gauge sheet metal cut to fit. Bend the sheet metal around the bows with well-placed ratchet straps. These straps are also used to hold the metal in place while it is welded. Now weld all of the sheet metal on the roof to make a solid roof.

The fourth step is to stabilize the outside of the fenders to keep them from eventually warping. This can be prevented by welding four gussets under each of the fenders. Cut these gussets at a ninety-degree angle and in a triangular shape designed to add support.

The fifth step is to mount the axles under the trailer. Use two 8000 lb torsion flex axles mounted 2/3 from the front and 1/3 from the back. These require very careful welding, as they will support the entire weight of the trailer and its entire load capacity.

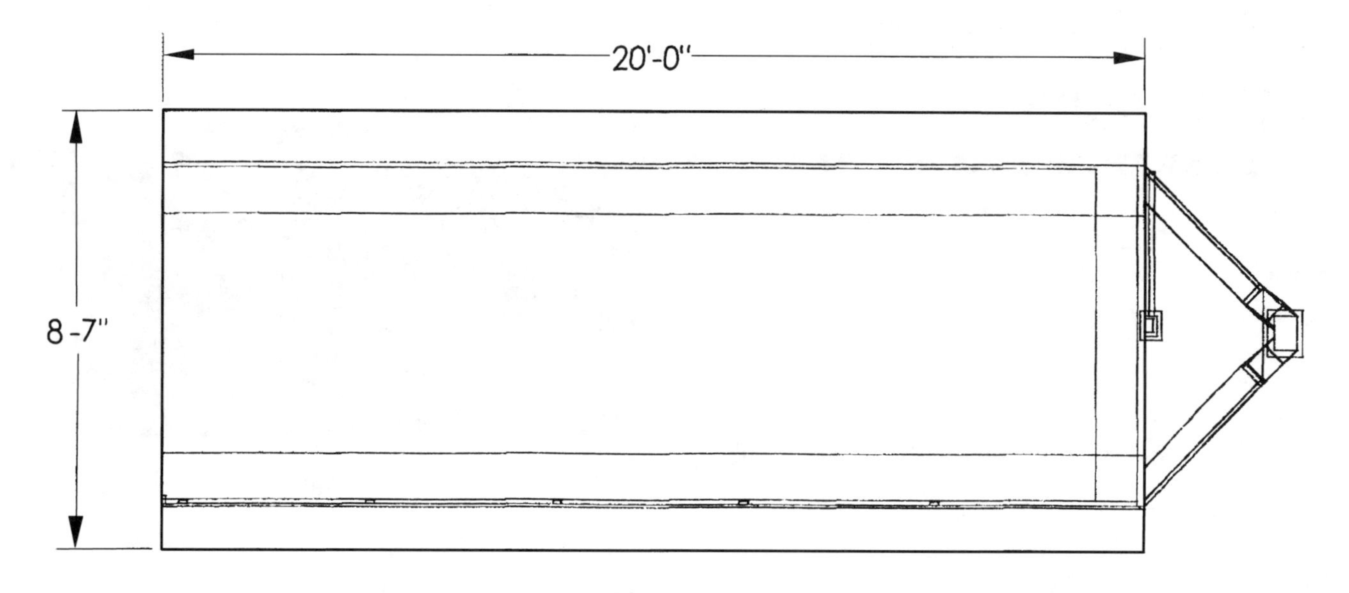

TOP VIEW

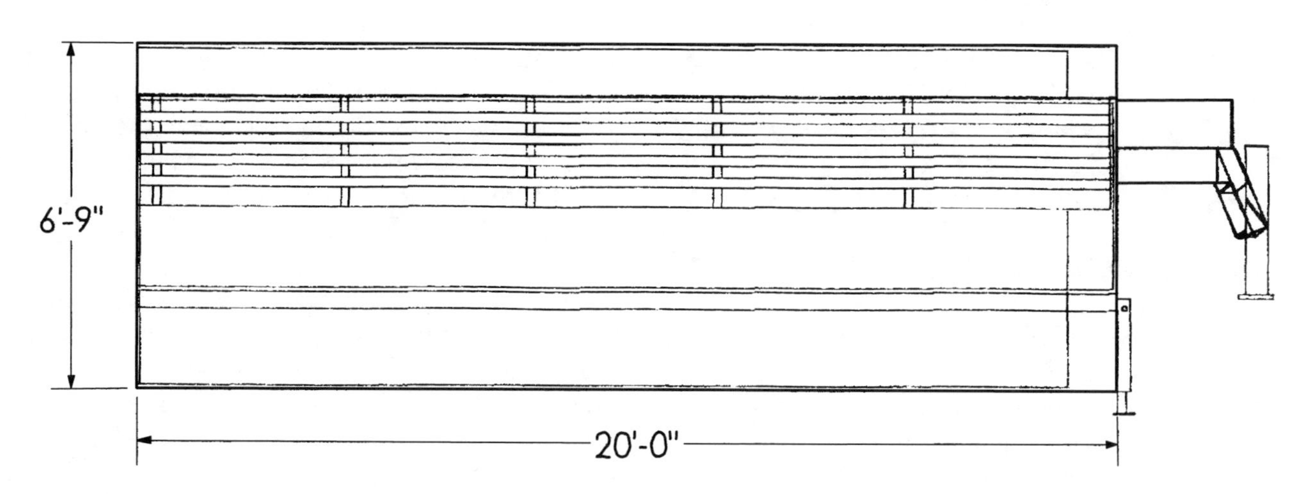

SIDE VIEW

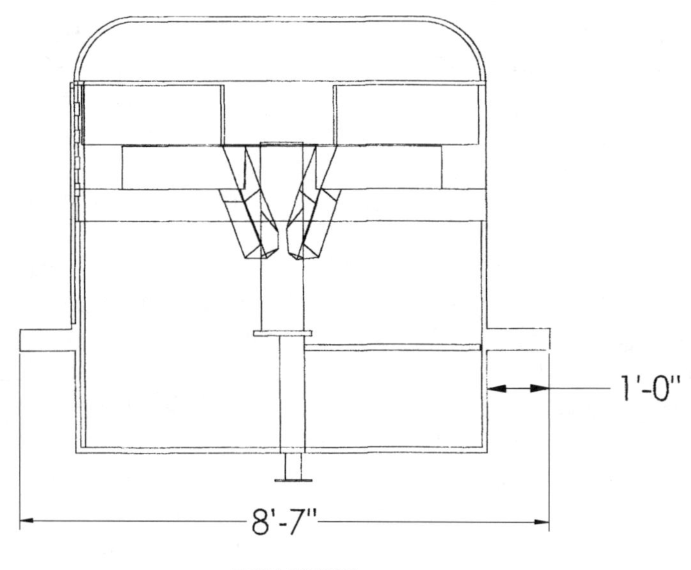

END VIEW

The sixth step is to create the front wall panel of the trailer. This wall panel is made of 12-gauge sheet metal and welded to the frame, fenders and the supporting bows.

Then the seventh step is to design the neck of the gooseneck trailer. It is made of 8″ tubing. Cut two 8′ pieces and then cut them down to where you can bend them by using a torch. Connect the two arms together by welding three pieces of 3″ channel iron across them to brace them and give support. Then hoist these arms up into the air with a cherry picker so that you can weld them in place to the front of the trailer.

The eighth step is to weld a bulldog coupler dead center of the arms of the trailer. This coupler is necessary to connect to the truck and must be welded strong as it is the life of the trailer.

Then the ninth step is to attach the bulldog jack dead center of the front wall of the trailer. Also mount an arm and a handle that will operate the jack that extends from the jack to just inside the fender. This is required to be extended to the ground when the trailer is at rest.

The tenth step is putting on the walls of the trailer. The walls of the trailer are made from 16-gauge sheet metal cut to fit. They are braced by pieces of 1-1/2″ x 1-1/2″ tubing that also are welded in between each of the bows. Then weld the pieces of sheet metal in place.

The eleventh step is to weld the four pieces of 3″ x 20′ rectangular tubing that are used for railings on each side of the trailer. Also, you can use pieces of 2″ x 20′ rectangular tubing welded in place to help stabilize the walls a bit.

The twelfth step is putting in the floor. Use #2 wolmanized pine for the flooring. Cut the boards with a skill saw and place them in the trailer. Then you can come right behind that and weld in place two 2″ x 2″ x 20′ angle iron in on top of these boards to hold the floor down.

The thirteenth step is to make the gates and latches and install them properly on the trailer. Start by making the back gates out of 2″ pipe. Bend the pipe with a pipe bender to make a door shape. Then make the cross braces of the gates with 1/2″ pipe to support and complete these gates. After welding them together, weld them into place on the trailer.

Next is the middle gate, which is constructed with the same type of material and created the same way. Weld it, then mounted it onto the trailer.

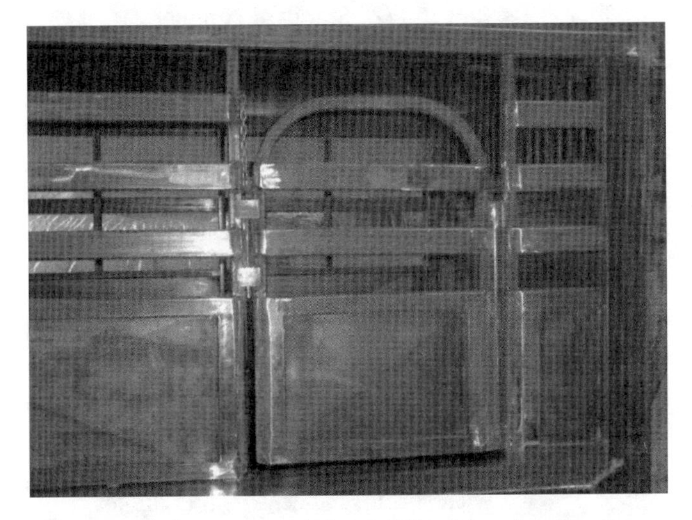

The escape door uses 1/2″ x 1/2″ tubing for the frame part of the door and cross bracing to match the same effect as the outside look of the trailer.

Finally, make a gate to fit up in the nose of the trailer to be able to open and close for a storage area. This gate/door is made of 2″ rectangular tubing and 1/2″ pipe for the cross bracing also.

Then, weld all of the latches into place so that all of the gates can be secured.

The fourteenth step is to wipe down the trailer with a degreaser to get all the dirt and oil off the metal so that it can be painted. This will take some time.

The fifteenth step is the priming and painting.

For the final step of the project, allow the paint to dry thoroughly before installing reflective tape and lighting. Wire all of the lights, brakes and brake-away system up at once, install the wheels, and take the trailer for a test drive.

16-Foot Lowboy Trailer

Author: Sarah Risse Mayfield
Instructor: Ty Anne Bixler
School: Winona High School
City & State: Winona, TX

Bill of Materials

Quantity	*Description*
44′	1/4″ x 3″ x 4″ Angle
128′	1/4″ x 2″ x 2″ Angle
18′	1/4″ x 2″ x 2″ Angle
6′	1/4″ x 1″ x 1″ Angle
10′	1/8″ x 3″ x 3″ Square tubing
82″	11 Gauge x 2″ x 2″ Square tubing
13′	3″ x 1-1/2″ Standard channel
22′	1/4″ x 2″ Flat bar
2 sq. ft.	1/4″ Flat plate
5 sq. ft.	1/4″ Diamond plate
1 pair	66″ x 9″ Tandem axle fenders with fender backs
1 ea.	7000 lb. Drop leg coupler
1 ea.	7000 lb. 2″ Ball coupler 3″ Square channel mount
1 ea.	50″ + 3/8″ Proof safety chain with hooks
2 ea.	1-1/2″ x 2″ U-bolts, Sae grade 8, with nuts and washers
1 ea.	7000 lb. Axle tandem axle assembly with spring, hanger kit, and brakes
107 sq. ft.	2 x 8 Treated lumber
4 ea.	16″ wheels and tires
1 ea.	D.O.T. approved light kit
1-1/2 gallon ea. – metal primer and paint	

Steps

First, cut two pieces of 3″ x 4″ x 1/4″ angle 192″ long with a 45 degree angle on both ends and two pieces of 3″ x 4″ x 1/4″ angle 78″ long with a 45 degree angle on both ends. Place the pieces on the ground in the shape of a rectangle with the 4″ flange facing up. Before tack welding together, measure across from opposite corners and move pieces until the same measurement is achieved. Then double-check these measurements after tack welding.

Cut nine pieces of 1/4″ x 2″ x 2″ angle 77-1/2″ cut square on both ends. These are the cross members for the trailer. Starting at the front of the trailer, the first piece is placed 3″ behind the 4″ flange. Number two is 21″ from the first, and pieces three through eight are 24′ apart. Piece nine is 19″ away from eight and 5″ in front of the 4″ flange at the back of the trailer. Square these cross members, remeasure, and tack weld. Double-check the measurements.

Next, cut sixteen pieces of 11 gauge x 2″ x 2″ square tubing 16″ long with a square cut on one end and a saddle cut on the other. Set these pieces vertically at a 90-degree angle on the outside of the frame. Starting from the back, the first piece is flush with the back of the trailer. The second piece is 24-1/2″ from the very back of the trailer – not the back of the tubing. The third piece is 22-1/2″ from the back of the second tubing. The fourth tubing is 66″ from the back of the preceding piece. Piece number five is 29″ from the front of the fourth piece and piece six 25″ from the back of five. The last piece is 25″ from piece six and flush with the front of the trailer. The opposite side of the trailer is done the exact same way. Place the two remaining pieces in the front of the trailer. They are both 25-1/2″ from the nearest side and 27″ apart from each other. Double-check the measurements and tack weld.

Cut two pieces of 11 gauge x 2″ x 2″ square tubing 194″ long with one end cut square and the other at a 45-degree angle. Also, cut a piece of 11 gauge x 2″ x 2″ angle 82″ long with a 45 degree cut on both ends. These are the rails that fit across the pieces in step 3. Tack the rail into place flush with the back, but overlapping the front by two inches. Then weld all the joints on the entire trailer completely, going in an alternate sequence of welds from top to bottom and side to side to minimize heat distortion. Using CNC, fabricate the gusset used in the front corners of the trailer. It is 15-1/2″ tall and 22″ wide.

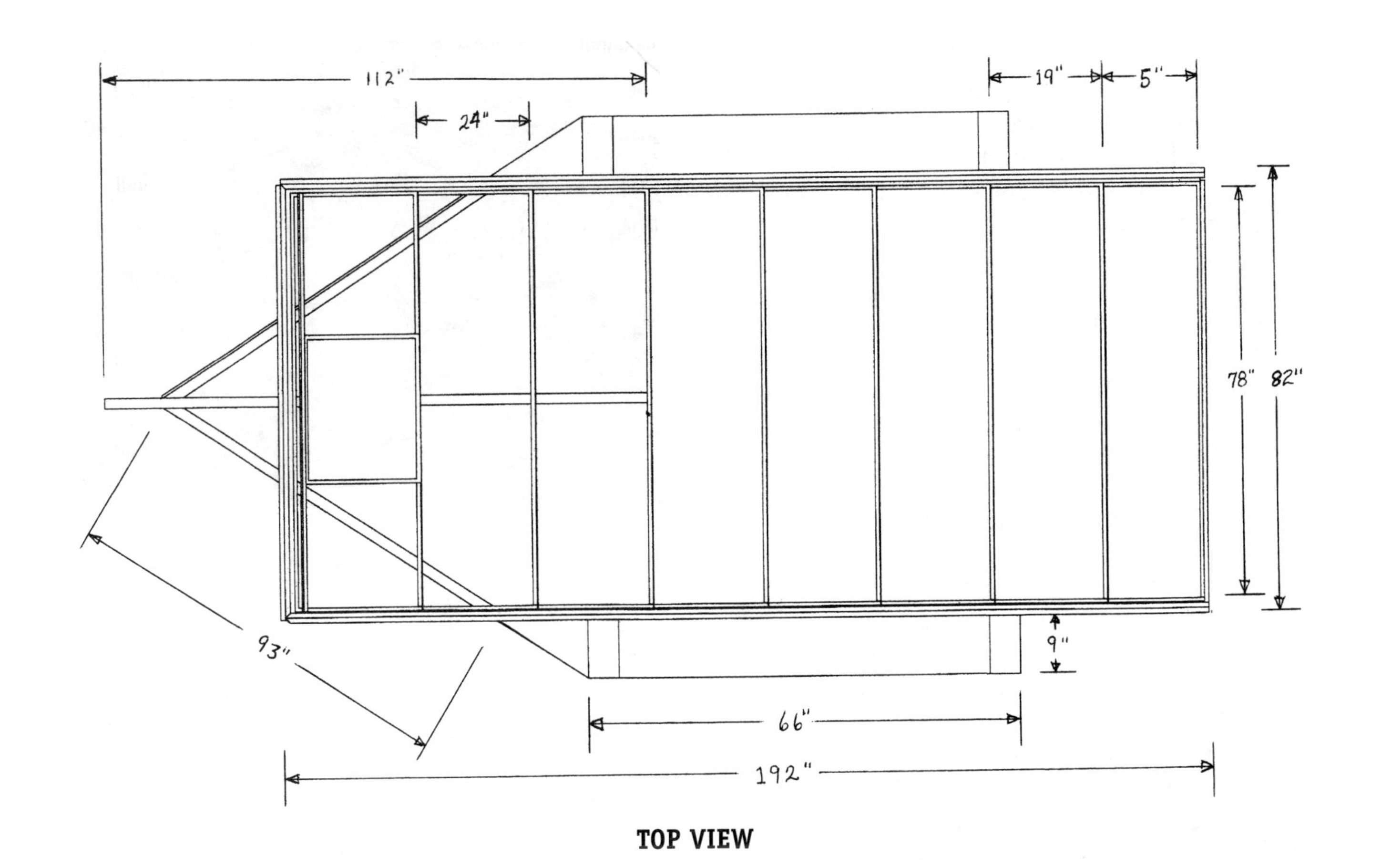

TOP VIEW

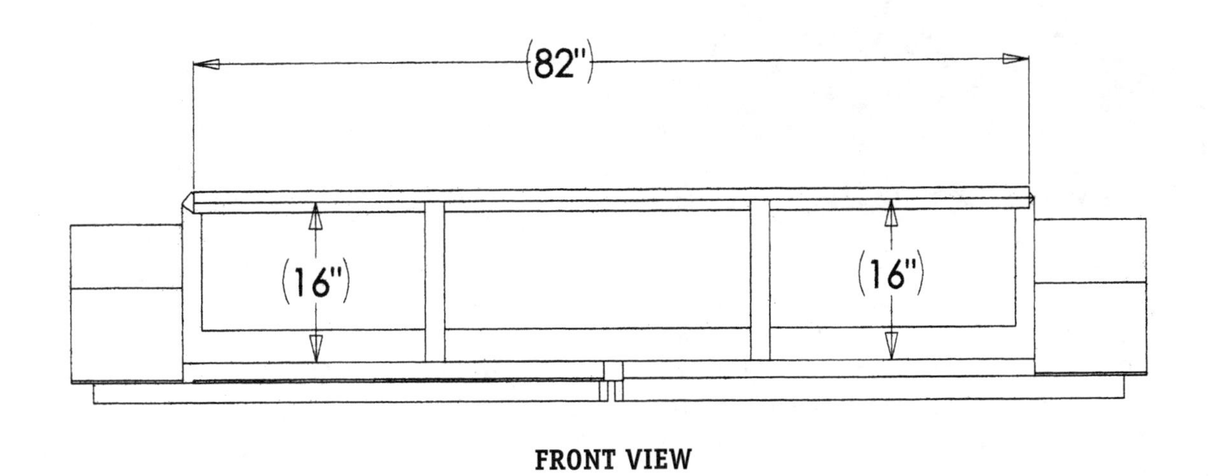

FRONT VIEW

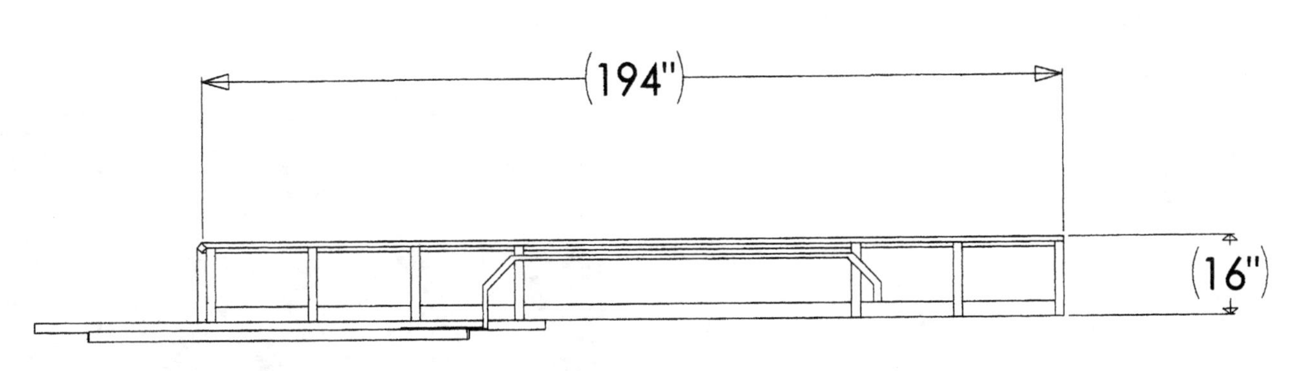

SIDE VIEW

Next, cut one piece of 1/8″ x 3″ x 3″ square tubing 112″ long square on both ends, two pieces of 1/4″ x 2″ x 3″ angle 93″ long with a square cut on one end and a cut to fit at the other, ten pieces of 1/2″ x 2″ x 2″ angle 5″ long with a square cut, four pieces of 3″ x 52″ standard channel with square cut on both ends, and six pieces of 1/4″ x 2″ flat bar 15″ long. Flip the trailer to make adding the ramp pockets and the tongue easier. To build the ramp pockets, place the channel 15″ apart, measuring from the outside to the inside. Lay the 15″ flat bar perpendicular to it on top. The flat bar should be flush with the outside of the channel. Two pieces at opposite ends are flush with the edges as well as the channel's sides and one directly in the middle. Tack weld these on and remeasure. Make two of the pockets. Measure 5″ from the outside of the frame to the outside of the channel, line the ramp pocket up flush with the back, and tack weld them on. Weld a 1/4″ x 1″ x 1″ locking tab with a 3/8″ hole to the back flat bar to provide a lock for the ramps. Tack weld four pieces of the 5″ long angle to the cross member the ramp pockets cross. Weld one side to the flange and the other to the ramp pocket.

Now make the tongue of the trailer. Lay the square tubing on the bottom of the trailer in the center. Tack weld the 5″ long angle to the second through fourth cross members to where one side is welded to the cross member and the other is lying next to the tubing. Then weld the end of the square tubing to the angle pieces where the tubing protrudes out the front 40″. Lay the braces (93″ angle) to where it sticks out the side of the trailer 9″ and intersects with the side of the trailer, approximately 41″ from the front towards the back. It attaches to the tongue 24″ from the front of the trailer to the outer edge of the braces. Tack weld them on.

Attach the suspension hangers. The centerline of the axle assembly is 112″ from the front of the trailer and 80″ from the back. Tack weld the center suspension hanger in place. Measure from the tip of the tongue at the center to the center point of the center suspension hanger. Repeat on the other side. When both meas-

urements are equal, tack weld the other hangers in place 29-1/2″ from the centerline of assembly. All the hangers are centered on the 3″ flange on each side of the main frame. After double-checking measurements, weld all the hangers, the tongue, and the braces completely, alternating the sequence from top to bottom and from side to side to minimize heat distortion.

Grind the 3″ square tubing at the tip of the tongue to provide space for the coupler to fit. Tack weld the coupler onto the tongue. Then measure from the coupler's tip to the centerline of axle assembly on both sides to make sure they are even. The distance between the front of the trailer and the tip of the coupler should be no more than 40″ maximum. Completely weld the coupler into place.

Next, flip the trailer and put the axle assemblies on. Install brakes on both axles using self-locking nuts. The centerline of axle assembly is where the axles meet.

Cut four pieces of 1/4″ x 2″ x 2″ angle 6-3/4″ long, two pieces of 1/4″ x 3″ flat plate five inches long, and two 3/16″ x 38″ x 9″ diamond plates trimmed to fit. CNC-fabricate the fender. Tack weld the fender on, check for fit, and weld with intermittent welds. The top of the fender will be 2″ under the top rail. Weld a 3/4″ x 20″ round bar from frame to inside corner of

fender for tire clearance. Place the 6-3/4″ angle at the outside and the bottom of the fenders for more support. On top of that angle and the tongue brace, weld the diamond plate on.

Fabricate the A-frame using CNC. The wide base is 29″ wide and the narrow base is 3-1/2″ wide. Cut notches for the uprights.

Install the tires and level the trailer. Then install the jack on the right side of the trailer at the back edge of the coupler, perpendicular to the ground to optimize road clearance and give adequate jack performance.

For the light and license plate brackets, cut one piece of 3/16″ x 14″ x 12′ plate and one piece of 3/16″ x 8″ x 6″ plate. Weld the bracket on the inside of the vertical uprights. Place the bigger piece on the left side of the trailer and the other on the right. Drill holes as needed for light mount and wire clearance and for the license plate. Grind a 1″ radius on all exposed corners.

Drill tow holes in the braces near where they intersect the tongue. Bolt the U-bolts and attach the safety chain.

To build the ramps, cut four pieces of 1/4″ x 2″ x 2′ angle 52″ long with a square cut on one end and cut to fit on the other, eighteen pieces of 1/4″ x 2′ x 2″ angle 13-1/2″ long cut square on both ends, and two pieces of 1/4″ x 2″ x 3″ angle 14″ long and cut square on both ends. Lay the 52″ angles on the ground 14″ apart and parallel to each other. The first 14″ piece is flange down with the outer edge flush with the end of the 52″ angle. The next eight pieces are flange up in a "V" look and 2-3/4″ apart from edge to edge. Tack weld all these pieces. Cut a slot in the last piece for the locking tab to fit through. Then put it on the ramp at the end with one side perpendicular to the ground and the other parallel to it. It also has an inch that hangs below the ramp. After tack welding and remeasuring, weld all joints completely.

Next, cut one piece of 1/4″ x 1″ x 1″ mangle 78″ long. Weld it on the back of the trailer 1″ above the bottom of the mainframe.

Fabricate a box out of 1/4″ sheet metal and insert it into the frame of the trailer. It is 31″ x 16-3/8″ and 6-1/2″ in depth. There is a 1-1/2″ rise in the middle for where the tongue is.

Before installing the floor and wiring, paint the trailer. Wire brush the trailer completely over every reachable inch and wipe all surfaces down with degreaser. Apply one coat of primer and two coats of top grade enamel for the final application.

Wire the trailer according to the instructions included in the lighting kit.

Cut eight pieces of 2″ x 6″ treated lumber 191″ long, five pieces of 2″ x 6′ treated lumber 4-3/4″ long, five 169-5/8″ long, and one piece of 2″ x 4″ cut 4-3/4″ long and one piece of 2″ x 4″ 169-5/8″ long. Also cut 1/4″ x 2″ x 77-1/2″ metal strap. After placing lumber on the cross members, weld the metal strap on each end of the trailer.

BBQ Trailer

Author: Benjamin Roustio
Instructor: Robert Daiber
School: Triad High School
City & State: Troy, IL

Materials List

Qty	*Description*
75′	1/8″ x 1″ Angle iron
27′	1-1/2″ scheduled 40 Carbon steel pipe
30 sq. ft.	Expanded metal grating
10′	2 x 10 #2 construction grade
8	1/4″ 20 x 2 Round head screws, nuts and washers
4	1-1/2″ U bolts
10′	1/2″ schedule 40 Carbon steel pipe
10′	1/2″ Carbon steel rod
30′	1/8″ x 1/2″ Flat iron
8′	1/8″ x 1″ Flat iron
3′	3″ schedule 40 Carbon steel pipe
3 sq. ft.	1/4″ Plate
4′	2″ schedule 40 Carbon steel pipe
1 qt.	Black paint
1 qt.	Metallic aluminum paint

Parts List
250 gallon oil drum
2″ Ball hitch
Rear axle from an S-10 pickup
2 - 205-70-r15 Tires

Steps to Complete Project
Make drawings and materials list for project

A. Build the trailer for the grill.
Get 1/2″ scheduled 40 pipe.

Cut pipe to length.

Lay out frame and weld it together.

Grind down weld on trailer frame.

Purchase a rear axle of an S-10 pickup from a junkyard.

Take brake assemblies off rear axle.

Remove leaf springs from rear axle and move them in to fit the trailer frame.

Weld 1/2″ pipe on ends of leaf springs for mounting.

Drill holes in the frame to mount frame.

Make shackles to mount axle.

Mount axle onto frame using 1/2″ x 2-1/2″ bolts and lock nuts.

Cut off excess pipe on front of trailer frame to mount the hitch.

Cut a piece of 2″ pipe to weld the hitch to.

Weld a round plate to the 2″ pipe to close it up.

Buy a hitch.

Weld the hitch to the 2″ pipe.

Slide the hitch assembly onto the trailer frame and drill a hole to attach the hitch.

Clean the frame for painting.

Paint the frame.

Remove cover from the rear axle to drain the gear oil.

Install the rear cover with new gasket.

Put gear oil in the rear end.

Put a new vent tube on the rear axle.

Make a bracket for the vent tube.

Install new wheels and tires on trailer.

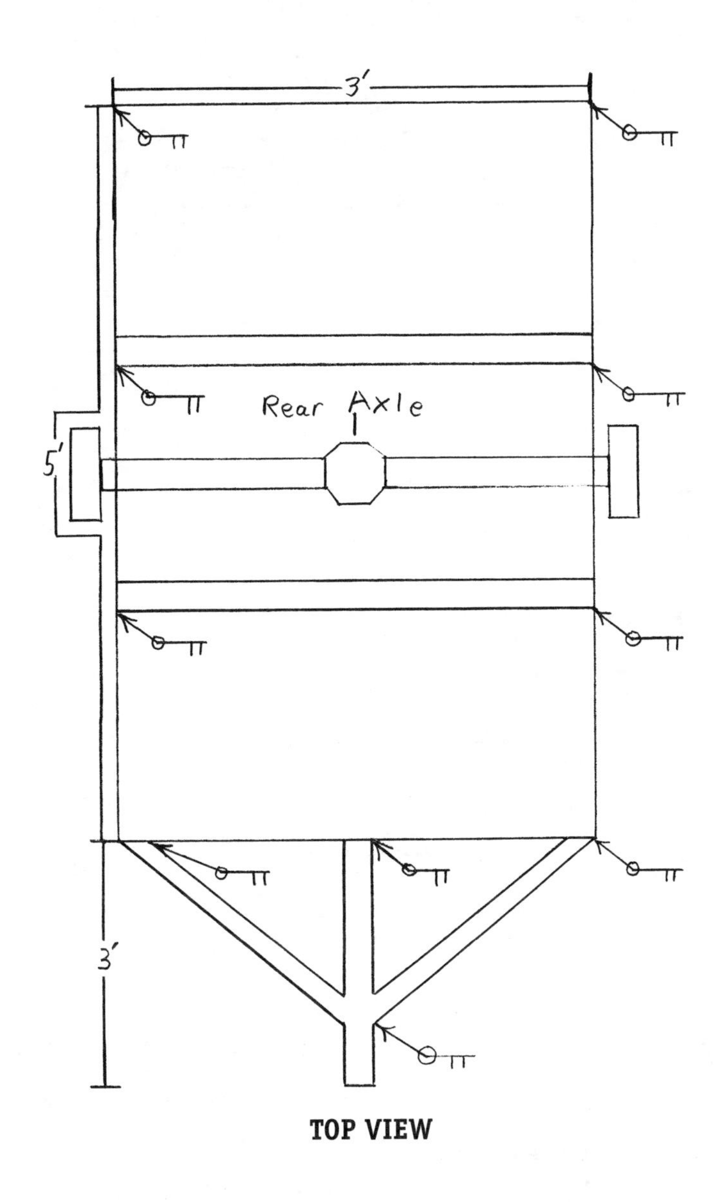

TOP VIEW

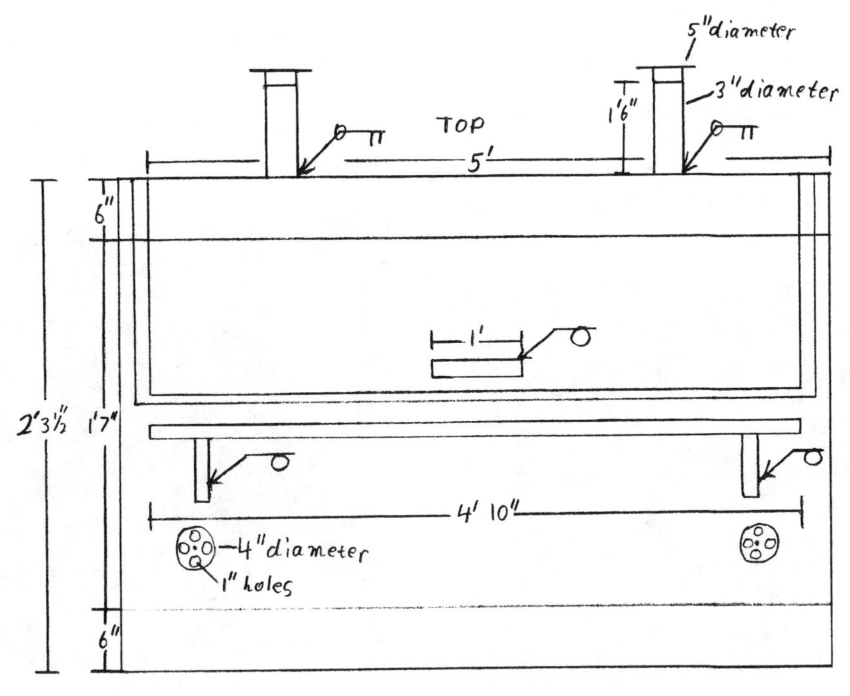

FRONT VIEW

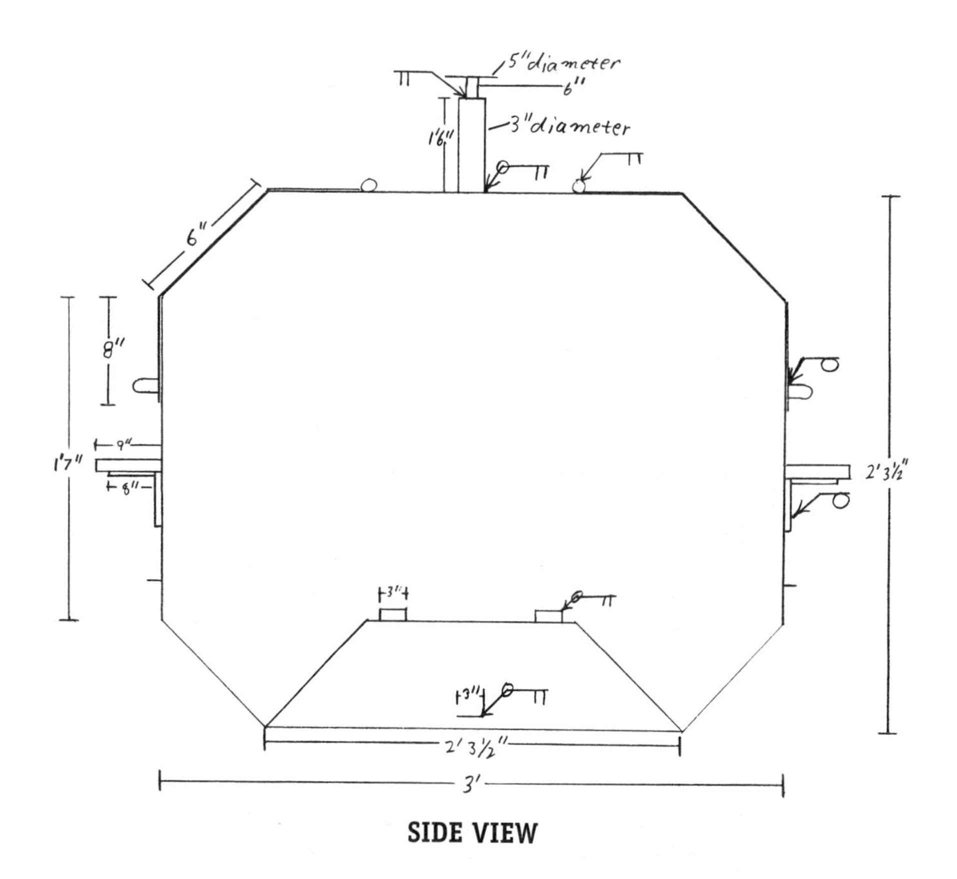

SIDE VIEW

B. Build the Grill

Locate and pick up a 250 gallon oil drum.

Drain the oil from the drum.

Cut doors out of the drum using a reciprocating saw.

Put a fire in the drum to burn off excess oil.

Wash drum and doors with a power washer.

Cut 1/8″ x 1″ angle iron for the grating supports and weld them into place.

Cut out ash door on the back of the drum using the reciprocating saw.

Make the ash door out of a piece of 3/16″ plate steel.

Make the hinges for the ash door out of 1/2″ rod, 1/2″ pipe and 1/8″ x 1″ angle iron.

Weld hinges to the drum and attached door.

Make a handle for the ash door out of 1/2″ rod and weld it to the door.

Frame the door openings with 1/8″ x 1″ angle iron and stitch weld it to the drum.

Trim the doors to fit into the framed openings using a cut-off wheel.

Pack the doors into the openings.

Frame the doors with 1/8″ x 1″ and 1/8″ x 1-1/2″ flat iron so the doors close against the drum.

Make hinges for the doors with 3/16′ x 2″ flat iron, 1/2″ rod, 1/2″ pipe, and 1/8″ x 1″ angle iron.

Weld hinges onto drum and doors.

Cut tacks on the doors and open them to check for alignment.

Make brackets for door handles out of 1/8″ x 1-1/2″ flat iron and 1-1/2″ dowel rods.

Drill sixteen 1″ holes in the side of the drum for air vents to the fire box.

Make air vent louvers out of a 5″ round piece of 20 gage stainless steel plate and drill four 1″ holes.

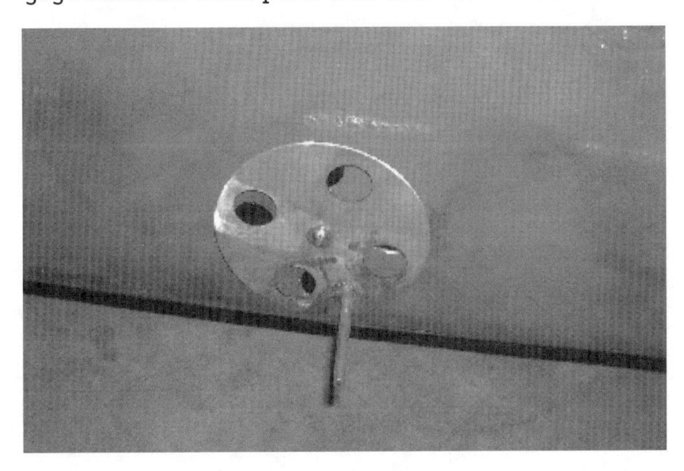

Attach louvers to the drum with 1/4″-20 x 1″ round head screws, nuts and washers.

Cut two 3-1/2″ holes in the top of the drum using a hole saw.

Insert a piece of 3″ scheduled 40 pipe into each hole and seal weld it.

Weld a 5″ round piece of 1/4″ plate on top of the 3″ pipe, leaving a 1 1/2″ gap between the plate and the pipe for the exhaust to exit.

Cut 1/8″ x 1″ angle iron to length to make L brackets for the shelves.

Weld brackets together.

Weld L brackets to the drum.

Prepare drum for paint using a wire wheel.

Wipe drum off with paint thinner.

Paint drum with metallic aluminum paint.

Purchase a 2 x 10 x 10 #2 construction grade piece of lumber for the shelves.

Cut the 2 x 10 to size for the shelves.

Attach the 2 x 10 to the L brackets by drilling holes and using 1/4″-20 x 2″ long round head screws, nuts and washers.

Paint the shelves black.

C. Assemble Grill to Trailer

Lift grill onto the trailer.

Center grill onto trailer.

Strap grill to trailer using tied down clamps.

Locate and drill holes in the bottom of the drum for 1-1/2″ U bolts.

Install U bolts with a backer plate made out of a piece of 1/4″ x 2″ flat iron and tighten them down.

Weld a center support for the bottom grating.

Install bottom grating and tack it down.

Weld a center support for the top grating.

Install top grating.

Make a support to hold the doors open out of 1/2″ rod and 1/2″ pipe.

Weld the door supports onto the grill.

Shop Projects

Table of Contents

Hydraulic Work Station

Author: Bret Parker Kennemer
Instructor: Dan Froneberger
School: Winona High School
City & State: Winona, TX

The Hydraulic Work Station project has five distinct components: the base and roller; the brake; the hydraulic table; the bender; and the convenience table and chair. The bill of materials for the entire project follows.

Base and Roller

1	49″ x 96.5″ Diamond plate
33.5″	1.5″ Square tubing
2	4′ x 8″ Channel
1	2′ x 8″ Channel
2	1′ x 4″ x 0.5″ Flat bar
2	18″ x 4″ x 0.5″ Flat bar
6	Bearings
1	Enerpac hydraulic cylinder
3	Rollers
2	24″x1″ Cold roll
1	12.5″ x 1″ Cold roll
1	8″ x 50.75″ Channel
1	8″ x 4″ Flat bar
4	7″ x 4″ x 0.5″ Flat bar
4	6″ of 0.5″ x 0.5″ Angle iron
2	8″ x 4″ x 0.5″ Flat bar
3	Sprockets
1	3′ Chain
1	Gear motor
4	0.5″ x 3″ Cold roll
4	1.75″ x 14″ x 0.177″ Springs

Brake

30″	6″ Channel
74″	8″ Channel
4″	8″ x 2″ x 0.25″ Tubing
18″	3″ x 3″ x 0.5″ Sheet metal
8″	8″ x 11″ x 0.25″ Sheet metal
12.5″	8″ x 5.5″ x 0.5″ Sheet metal
2	6″ x 4″ x 0.5″ Sheet metal
4	Bolts
1	Enerpac hydraulic cylinder
26″	26″ x 2″ Pipe
25″	4″ x 4″ Sq. Tubing
34″	4″ x 4″ Sq. Tubing
1	14″ x 33″ x 0.5″ Sheet metal
1	Brake

Hydraulic Table

4	21″ of 1.5″ Sq. Tubing
1	17″ x 18″ x 0.25″ Sheet metal
1	Enerpac Hush pump (hydraulic pump)
4	Nipples
4	Valves
4	Hydraulic hoses
4	Gauges
4	Gauge receptacles
1	Manifold

Bender

1	Hydraulic cylinder
2	52″ x 8″ Channel
1	8″ x 4″ x 0.5″ Flat bar
2	12″ x 4″ x 0.5″ Flat bar
2	24″ x 4″ x 0.5″ Flat bar
1	3.5″ x 4.5″ Pipe
4	8.25″ x 1″ Cold Roll

Table & Chair

4′9″	2″x 2″ x 25″ Tubing
1	13″ x 16″ x 0.25″ Plate
1	34″ x 16″ x 0.25 Plate
1	Piece of wood, 18″ x 11″ x 1″
1	20″ x 3″ Pipe
4	Bolts

Finishing Touches

2	25′ Electrical cords
10	Hose clamps
3	Electrical boxes
2 qts	Omni epoxy primer
2 qts	Omni black paint
1 pint	Omni yellow paint

Base and Roller
First, weld the square tubing to the bottom of the diamond plate to create the base of the Work Station. Next, weld pieces of diamond plate to the sides of the base like the ramp, to prevent tripping. To make the roller, weld a 1′ x 8″ piece of channel to the top of two pieces of 4′ x 8″ channel, leaving an 8″ gap from center to center. Then weld two pieces of 1′ x 4″ x 0.5″ flat bar to the bottom for support and to keep the channel at the proper width. Further support this with the addition of two pieces of 18″ x 4″ x 0.5″ flat bar. Weld a piece of 8″ x 4″ flat bar in between the channel and the base to form the base for the hydraulic cylinder. Cut a piece of 8″ channel. Drill four plates to match the bearings, then weld them 2.25″ from the ends of the channel, one on each of the four sides. Weld a link to the four 6″ pieces of angle iron to connect to the springs 17″ from the base. Weld cold roll 3″ long on all four sides to connect the springs to. Add four springs on each side, to help pull the roller base back to its normal position after bending the metal.

Brake
To build the brake, start by creating the base, using two pieces of 15.25″ long channel, and two pieces of 24″ long 2″ x 8″ tubing. Weld up the tubing inside the channel, then weld a piece of channel on top of the channel on the base. Weld on one 0.5″ plate, 18″ x 8″, to make the surface for the brake. Next, weld the "v" of the brake to the surface. To make the brake function properly, weld two 6″x 4″ x 0.5″ plates vertically, 3.25″ apart, on the flap, to hold the hydraulic cylinder. Then brace the plate with 0.5″ plate cut in a triangle. Cut four holes in the 6″ x 4″ plate, two on each side, to support and hold the hydraulic cylinder.

Hydraulic Table
Cut the 4 pieces of 1.5″ square tubing to make the legs of table, and one 17″ x 18″ piece of sheet metal to serve as the tabletop. Weld the table to the base, 2′ from the corner so it can be accessed easily from all the machines on the project. Center and place the hydraulic pump under the table for protection. On top of the table, weld a piece of channel 13″ long, 5″ from the side, onto the top to place the manifold on. Position the manifold and connect all of the gauges, receptacles, valves, and hoses to complete the hydraulic table.

Bender
To add a bender to the Work Station, start by cutting two pieces of 8″ channel, 52″ long, and welding them to the Work Station floor, 8.25″ apart. Brace the channel with the 12″ x 4″ flat bar, positioning it 16.25″ from the outside of the channel on both sides. Cut holes (four on each side) through the flat bar to hold the cold roll. The four pieces of cold roll are used to roll the metal through. Brace the channel again at the hydraulic cylinder, capping the channel at the top of the brace with the 8″ long piece of flat bar. Bolt the hydraulic cylinder to the brace going downward. Using the pipe, make an end for the hydraulic cylinder, welding it to a bolt that goes into the cylinder and can be taken out easily.

Table & Chair
For convenience and to have a place to put things, create an L-shaped table using 0.25″ plate for the surface and 2″ x 2″ square tubing for the legs. Weld it to the channel on the side of the bender and the channel on the side of the brake.

To fabricate the chair, weld a piece of 3″ pipe to the base and support it with two pieces of 0.5″ flatbar. Then weld a swivel to the top of the pipe, and bolt on the piece of wood to serve as the seat. Sand the corners smooth.

Finishing Touches
Prime and paint the Work Station. Next, do all of the wiring. Mount the four electrical boxes, all of the hydraulic hoses, gauges, and the electric hydraulic pump. Finally, screw the hose clamps on, and install the light.

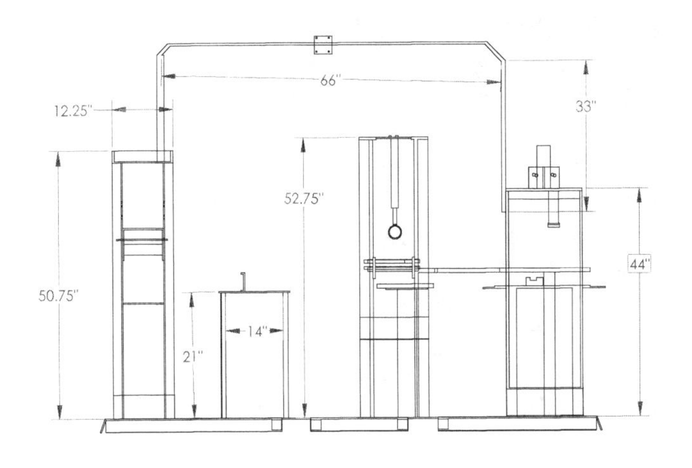

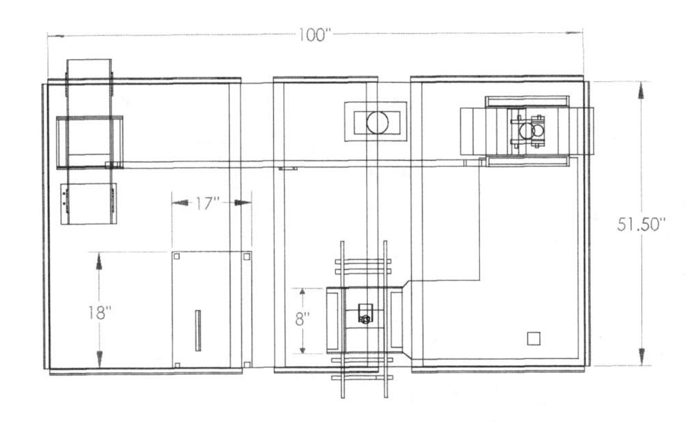

Engine Hoist

Author: Jason Scott Sanders
Instructor: Bill Gleason
School: University of Montana, College of Technology
City & State: Missoula, MT

Bill of Materials

Qty.	*Description*	*Length*
1	3″ x 3″ x 0.25″ sq. tubing	58″
4	0.25″ x 1″	2.5″
1	2.5″ x 2.5″ x 0.25″ sq. tubing	41″
2	0.25″ x 3″	8″
1	3″ x 3″ x 0.203 sq. tubing	41″
2	0.25″ x 3″	12″
1	8-ton hydraulic jack	
2	0.25″ x 6″	6″
4	caster/swivel 3″ steel wheels	
1	3″x 3″x 0.25″ sq. tubing	29″
2	C4″ x 7.25	16″
1	3″ x 3″ x 0.25 sq. tubing	37″
2	2.5″ x 2.5″ x 0.25″ sq. tubing	65″
8	3/16 x 1	1″
2	1″ x 3″ x 0.25 rect. tubing	16″
1	0.25″ x 12″	14″

Welding Procedure Specification

Job Title: Engine Hoist-1
Welding Process: FCAW, Semiautomatic
Material: A-36
Joint design: Tee
Position: 2F
Root opening: None
Land width (W): NA

Joint Design

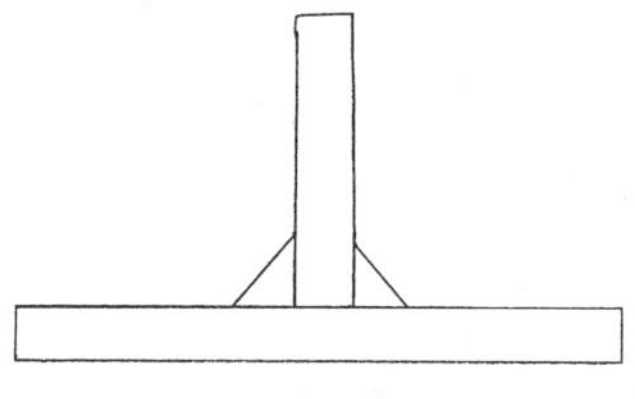

Backing Type: None
Electrode #1: E-71T-1, Size: 0.45″ Amperage: 170
Filler metal: Same
Wire feed speed: 260 ipm
Voltage range: 23-25
Polarity: DCEP
Electrode stickout: 1/2″
Shielding gas (type): CO2 30 p.s.i

Technique

Stringer
Travel angle: Push 80° Work angle 45°
Single Pass
Initial cleaning: Chip & Wirebrush

Welding Procedure Specification

Job Title: Engine Hoist-2
Welding Process: FCAW, Semiautomatic
Material: A-36
Joint design: Flare-Bevel
Position: 1G
Root opening: None
Land width (W): NA

Joint Design

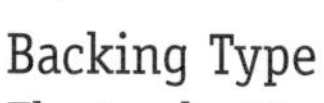

Backing Type: None
Electrode #1: E-71T-1, Size: 0.45″, Amperage: 170
Filler metal: Same
Wire feed speed: 260 ipm
Voltage range: 23
Polarity: DCEP
Electrode stickout: 1/2″
Shielding gas (type): CO_2 30 p.s.i

Technique

Stringer
Travel angle: Push 75° Work angle 90°
Single Pass
Initial cleaning: Chip & Wirebrush

Introduction
Making an engine provides good practice in cutting with different machines, cutting certain angles, drilling holes in the right place, fitup and assembly. The engine hoist described here is an original design by the author

Fabrication
Order the steel according to the dimensions shown on the blueprint drawing. When it arrives, measure all the pieces to make sure they conform to the order.

First, cut the flame cut edges off the edges of the channel. Measure and mark out one 16″ piece of C4″ x 7.25″ and cut it on the ban saw. Then measure, mark and cut the other 16″ piece. Make both cuts at a 90° angle.

Take the 174″ section of 2.5″ x 2.5″ x 0.25″ square tubing to the ban saw and cut the flame cut edges off. Measure, mark and cut one 65″ length at 90° and do the same for the other 65″ piece.

Again using the ban saw, measure, mark and cut both pieces of 3″ x 1″ rectangular tubing to 16″ lengths. Also cut the 58″, 41″, 29″ and 37″ pieces of 3″ x 3″ square tubing, as well as the 41″ piece of the 2.5″ x 2.5″ square tubing. Mark out the appropriate angle for the 58″ piece, then set the angle at 80° for the first cut, then 60° for the second.

Using the shear, cut the 0.25″ x 12″ x 14″ plate for the bottom assembly. Then cut two 0.25″ x 1″ x 2.5″ pieces, and two more 0.25″ x 1″ x 3″ pieces. These are for the stopper for the boom extension.

Mark out both the 29″ and the 37″ lengths of 3″ x 3″ square tubing for fitup of the bottom assembly. Use two bar clamps to hold pieces B7 and B3 in between B4 and B2, while centering the pieces. Tighten both clamps a bit to hold all the pieces together. Check to

be sure everything is square by measuring from one corner of each square made by the pieces to the opposite corner. If needed, make slight adjustments with the hammer and then re-measure. If everything is square and level, these pieces of the bottom assembly are ready to weld.

Begin by tack welding the outside, bottom, right corner, and then the inside, top, right corner. Continue in this fashion until all the corners are tacked on the topside of the bottom assembly. Turn the assembly over to the right, and tack weld the bottom side in the same manner as the top.

According to W.P.S. Engine Hoist-1, weld the 4″ channel down the inside of the flange and a little bit around the corner. This is where the channel fits up with the 3″ x 3″ tubing on both sides. These welds can be done in the same manner as the tack welds (criss-cross) and are done in the flat position. After every two welds, stop and let the whole piece cool.

After welding the channels on, use W.P.S Engine Hoist-1 on the 1″ x 3″ rectangular tubing with a one inch weld on the 1″ side. Weld the two opposite corners on each side, and then the other two sets of corners. Allow time for the metal to cool after each pass. Weld the three inch sides of the 1 x 3″ tubing according to the same W.P.S. Use the same criss cross and cooling pattern that was used when tacking.

Next, following W.P.S. Engine Hoist-2, weld the bottom ends of the 4″ channel to the 3″ x 3″ square tubing. Then, weld down the outside flange of the channel where it connects at the 3″ x 3″ tubing.

Pre-fitup the 65″ piece of 2.5″ x 2.5″ square tubing to the 37″ piece of 3″ x 3″ tubing; use an angle grinder to grin out the outside corner welds to make a better fitup.

Use two bar clamps to fitup the two 65″ pieces. When they are level and square, tack them together. Stagger the welds (W.P.S. Engine Hoist-2) connecting channel with the square tubing and weld all three sides that connect the 2.5″ x 2.5″ tubing to the 3″ x 3″ tubing. After each pass, let the whole piece cool.

Following W.P.S. Engine Hoist-1, weld the 0.25″ x 12″ x 14″ plate onto the top of the bottom assembly. Next, cut two 0.25″ x 3″ x 8″ plates, and mark out the 10° angles. Make the cuts on the shear. Mark the 3″ x 3″ tubing, clamp both pieces on with clamps, and tack weld them. After removing the clamps, use W.P.S. Engine Hoist-2 to weld all outside edges.

Using W.P.S. Engine Hoist-1, weld the stopper on the extension. Tack weld the extension to the main boom to maintain alignment when drilling all holes. Mark and center punch the holes that need to be drilled. Using the drill press with a 41/64″ drill bit, drill all the holes for the boom and the boom extension. C-clamps can be used to hold the pieces in place. Grind off the tack welds holding the extension to the boom, and test to make sure the holes align properly all the way down. Next, using the same drill bit, drill the holes in the 0.25″ x 3″ x 8″ plate where the boom connects to the main support piece.

Using the shear, cut both 3/8″ x 3″ x 12″ pieces that will be welded to the boom. After welding the pieces onto the boom using W.P.S. Engine Hoist-2, mark out the holes and drill them on the drill press using an

11/64″ bit. Check the fitup and make any necessary adjustments to ensure the piece will rotate properly when the hoist is jacked up.

Measure the 0.25″ x 6″ plate 6″ down one side, then 5″ down the parallel side. Connect the two points to make the angle necessary for both pieces with one cut. Cut out both pieces on the shear using vise grip c-clamps. Use a compass to make the arc for the pieces, use the shear to create the basic shape of the arc, then clamp the two pieces and grind down the arc the rest of the way. After welding the pieces onto the 3″ x 3″ tubing for the lower hydraulic brace, measure, mark and drill the two 41/64″ holes needed.

If needed, get help to center the piece before welding it on, using W.P.S. Engine Hoist-2. Finally, weld the wheels, using W.P.S. Engine Hoist-1. Take care to quench each weld so as not to distort the wheel bracket.

Finishing
First, remove all spatter with a chisel and hammer. Sandblast with 70 grit sand followed by 30 grit sand. Spray the hoist off with compressed air to remove most of the sand, then wash off the remainder and wipe down all the pieces to be painted. Use two coats of primer and three coats of paint. Reassemble the pieces.

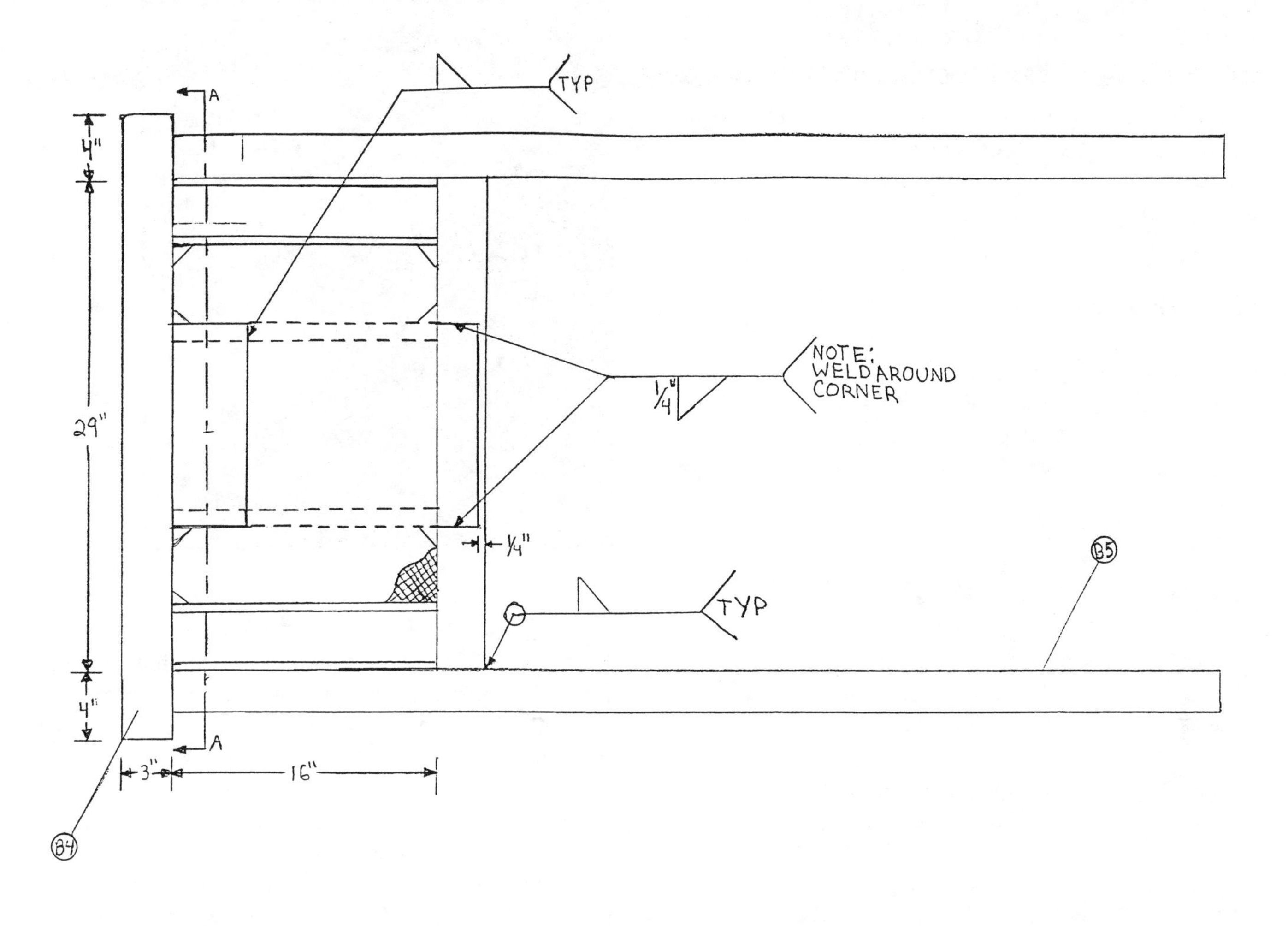
TYP
A
4"
29"
4"
NOTE:
WELD AROUND
CORNER
1/4"
1/4"
B5
TYP
A
3"
16"
B4

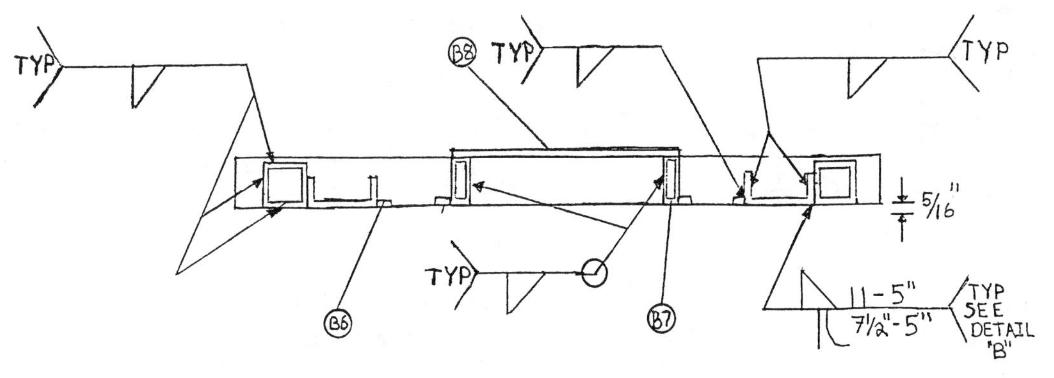
TYP
B8
TYP
TYP
5/16"
TYP
B6
B7
11-5"
7 1/2"-5"
TYP
SEE
DETAIL
"B"

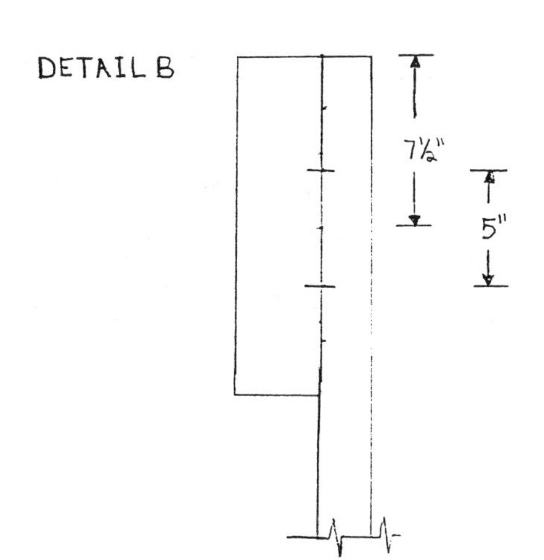
DETAIL B
7 1/2"
5"

Flat Bar Steel Roller

Authors: Geoffrey H. Akin & Justin Eckeberger
Instructor: Dan Froneberger
School: Winona High School
City & State: Winona, TX

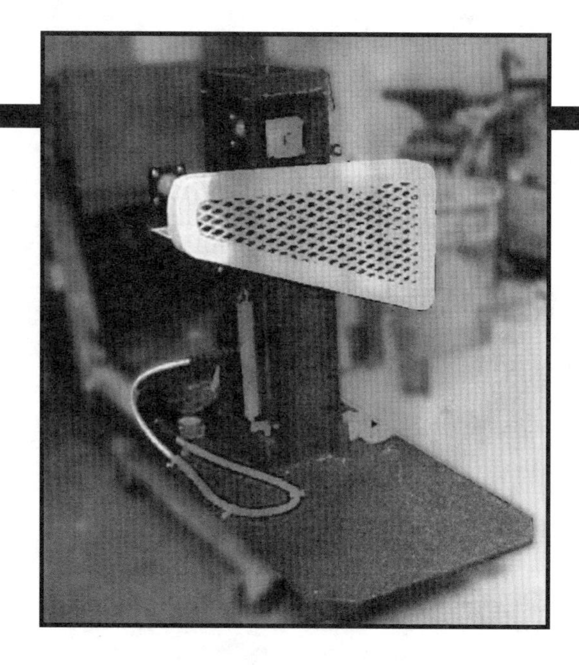

Bill of Materials

Quantity	*Description*
3	Rollers
1	Foot pedal
1	Motor
1	Pump
1	40″ chain
2	Sprockets
1	Motor sprocket
6	Flange bearings
24	HYK bolts
2	37.5″ x 8″ channel
1	2′ x 6″ channel
5	7′ x 4″ channel
1	61″ x 24″ diamond plate
4	24″ x 3″ square tubing
4	1/2″ x 1/2″ x 6″ angle
1	1″ x 1″ x 6″ angle
1	8″ x 4″ flat bar
2	12″ x 4″ flat bar
1	12″ x 8″ channel
1	25′ electrical cord
3	1″ shafts
1qt.	Primer with hardener
1qt.	Blue paint
1qt.	Clear gloss

Project Description
Begin building the flat bar roller by cutting two pieces of channel iron 37.5″ long. Drill four holes 1.5″ apart from each other and a center hole 1.5″ in diameter with a plasma cutter.

Cut another piece of channel 2′ x 8″, grind off the sharp edges, and space the two existing pieces the width of the 2′ x 8″ channel.

Cut four pieces of flat bar in 7″ x 4″ sections, drilling four 7/16″ diameter holes in one.

Attach the channel to the 1/4″ diamond plate that has been cut into a 61″ x 24.5″ piece, and notch the edges.

Weld a 3″ x 3″ x 3/16″ tubing on to the bottom, leaving a 3″ overhang on two sides, and two of the 3″ x 3″ x 3/ tubing in the center, 24.25″ long, parallel to each other.

Attach the 37.5″ channel onto the diamond, welding the four pieces of angle iron to the 2″ x 8″ channel, allowing it to slide up and down on the four pieces of flat bar. Then, weld them to the movable channel and attach the flange bearings to the four holes in the flat bar.

Weld the braces onto the center of the channel and attach the sprockets. After running the 1′ shafts through the flange bearings, cover each shaft with a nylon shaft cover.

Weld a piece of flat bar between the channel irons to attach the hydraulic cylinder. Drill four holes in the diamond plate to attach the hydraulic hand pump.

Weld a 12″ piece of channel to the top of the two existing channels and cap off the two open ends with flat bar.

Attach the motor and cover it to protect it while painting the unit. To complete the project, prime and paint the unit, topping it with a clear gloss coat for protection. After allowing ample drying time, remove the plastic and test the project.

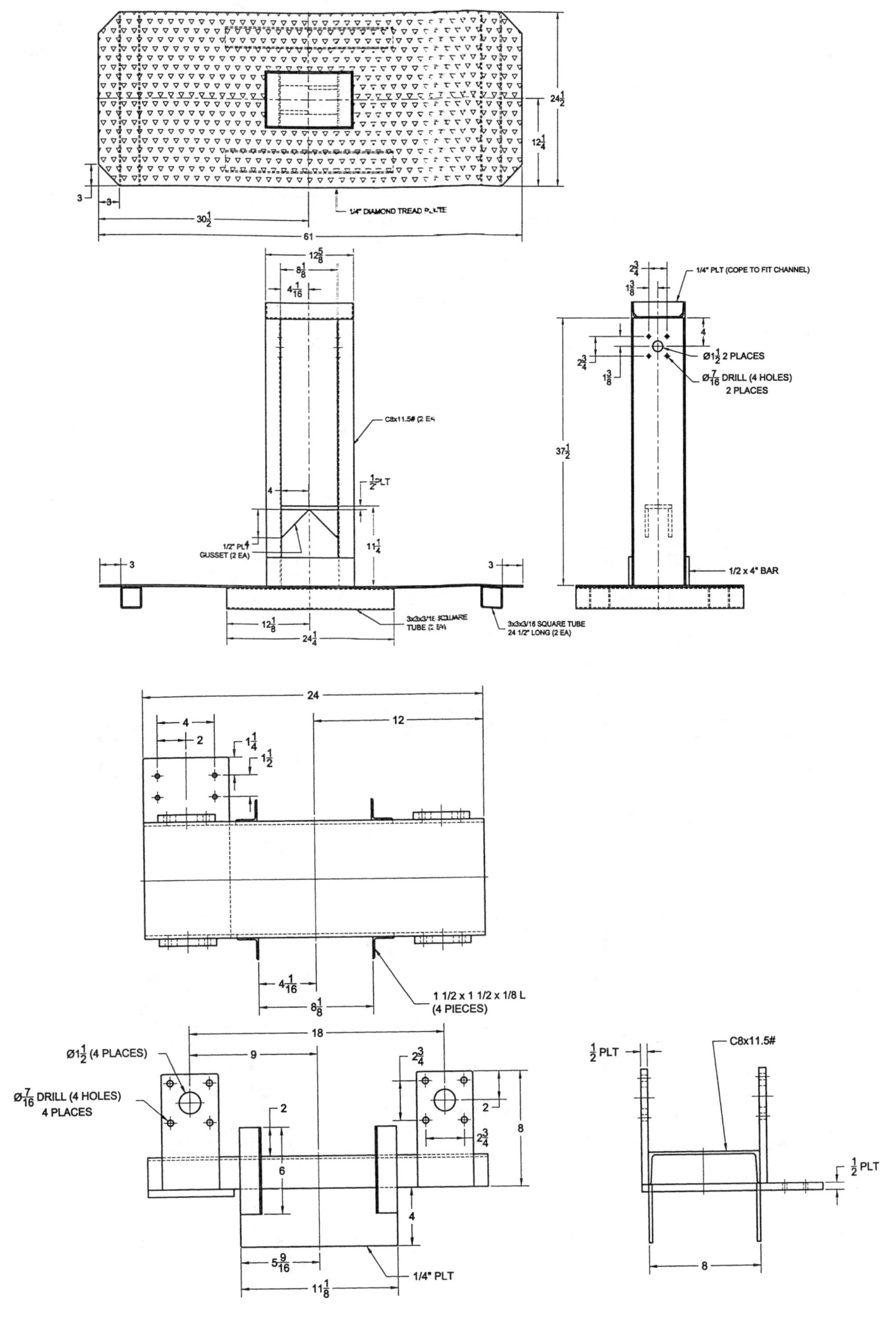
24 1/2
12 1/4
3
3
30 1/2
61
1/4" DIAMOND TREAD PLATE
12 5/8
8 1/8
4 1/16
C8x11.5# (2 EA)
1/2 PLT
4
4
1/2" PLT
GUSSET (2 EA)
11 1/4
3
3
12 1/8
24 1/4
3x3x3/16 SQUARE TUBE (2 EA)
3x3x3/16 SQUARE TUBE
24 1/2" LONG (2 EA)
2 3/4
1 3/8
1/4" PLT (COPE TO FIT CHANNEL)
4
2 3/4
1 3/8
Ø1 1/2 2 PLACES
Ø 7/16 DRILL (4 HOLES)
2 PLACES
37 1/2
1/2 x 4" BAR
24
12
4
2
1 1/4
1 1/2
4 1/16
8 1/8
1 1/2 x 1 1/2 x 1/8 L
(4 PIECES)
18
9
Ø1 1/2 (4 PLACES)
Ø 7/16 DRILL (4 HOLES)
4 PLACES
2 3/4
2
2
2 3/4
8
6
4
5 9/16
11 1/8
1/4" PLT
1/2 PLT
C8x11.5#
1/2 PLT
8

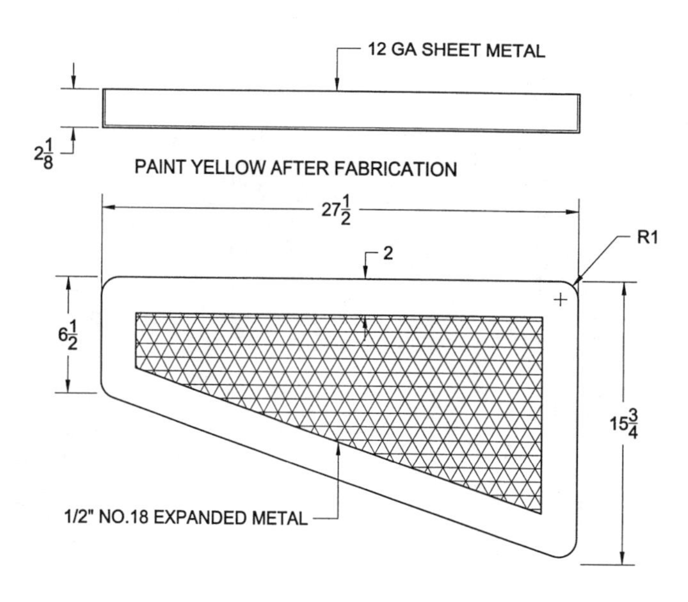
12 GA SHEET METAL
2 1/8
PAINT YELLOW AFTER FABRICATION
27 1/2
2
R1
6 1/2
15 3/4
1/2" NO.18 EXPANDED METAL

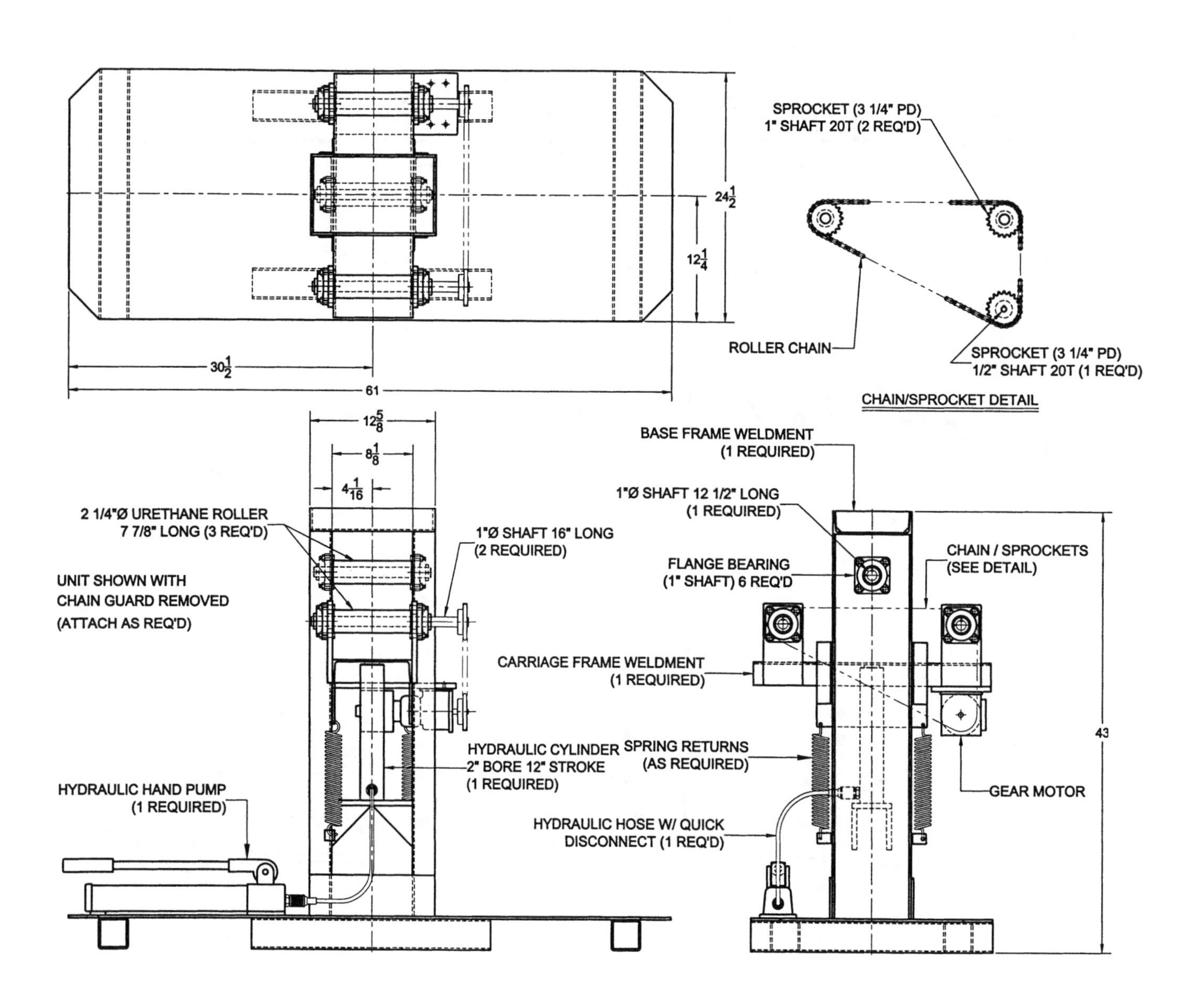
SPROCKET (3 1/4" PD)
1" SHAFT 20T (2 REQ'D)
24 1/2
12 1/4
ROLLER CHAIN
SPROCKET (3 1/4" PD)
1/2" SHAFT 20T (1 REQ'D)
30 1/2
61
CHAIN/SPROCKET DETAIL
12 5/8
8 1/8
4 1/16
2 1/4"Ø URETHANE ROLLER
7 7/8" LONG (3 REQ'D)
1"Ø SHAFT 16" LONG
(2 REQUIRED)
UNIT SHOWN WITH
CHAIN GUARD REMOVED
(ATTACH AS REQ'D)
BASE FRAME WELDMENT
(1 REQUIRED)
1"Ø SHAFT 12 1/2" LONG
(1 REQUIRED)
FLANGE BEARING
(1" SHAFT) 6 REQ'D
CHAIN / SPROCKETS
(SEE DETAIL)
CARRIAGE FRAME WELDMENT
(1 REQUIRED)
HYDRAULIC CYLINDER
2" BORE 12" STROKE
(1 REQUIRED)
SPRING RETURNS
(AS REQUIRED)
43
HYDRAULIC HAND PUMP
(1 REQUIRED)
GEAR MOTOR
HYDRAULIC HOSE W/ QUICK
DISCONNECT (1 REQ'D)

Exhaust Fan Housing

Author: Taylor Laskoski
Instructors: David Morgan and Anthony McIntosh
School: Franklin County Technical School
City & State: Turners Falls, MA

Material List
16 gauge coil stock
1/8″ x 1-1/4″ flat stock
1/8″ x 1″ flat stock

Procedure

1. Cut four pieces of 16 gauge steel to 6-7/16″ x 16-3/8″ using a hydraulic shear.
2. Cut two of the four pieces into strips. Make the first two strips 5″, the second two strips 5-1/4″, and the final two 6-1/8″.
3. Using a scribe, a ruler and a protractor, lay out the angles, 22.5 degrees, for the miter cut. (This is determined by using the formula of 90 degrees divided by the number of angles, in this case, four.)
4. With one of the remaining two pieces, lay out the bend lines. The first line is 5″. From that, mark the next line at 6-1/8″ and from that line, mark the next at 5-1/4″. For the last piece, mark out 2-1/4″, from there 1″, and finally 2-1/4″. Using a center punch, make marks at the ends of each line. These are reference points for bending.
5. Using the Finger brake, proceed to make the 22.5 degree bends for the top and bottom of the elbow. Use a protractor to check the angles.
6. Using clamps and a layout table, tack the side pieces together using a Tig welder.
7. Tack the top and bottom, stopping periodically to check for square and correct alignment.
8. Using 1/8″ x 1-1/4″ flat stock, cut four pieces using a hydraulic ironworker. Cut two pieces to 9-1/8″ long and two pieces to 6-5/8″. Use these to construct the flange, which will allow the fan to be mounted. Tack on the flange using the Tig welder.
9. Using 1/8″, 1″ flat stock H.R. steel, fabricate an internal mounting frame for a 120 volt fan. Cut two pieces 6-3/8″ and the other two 4.375″. Again, use the Tig welder to tack it together. Tack the frame inside the elbow using a Mig welder (set the fan frame back from the end 2-1/4″ inside before tacking it).

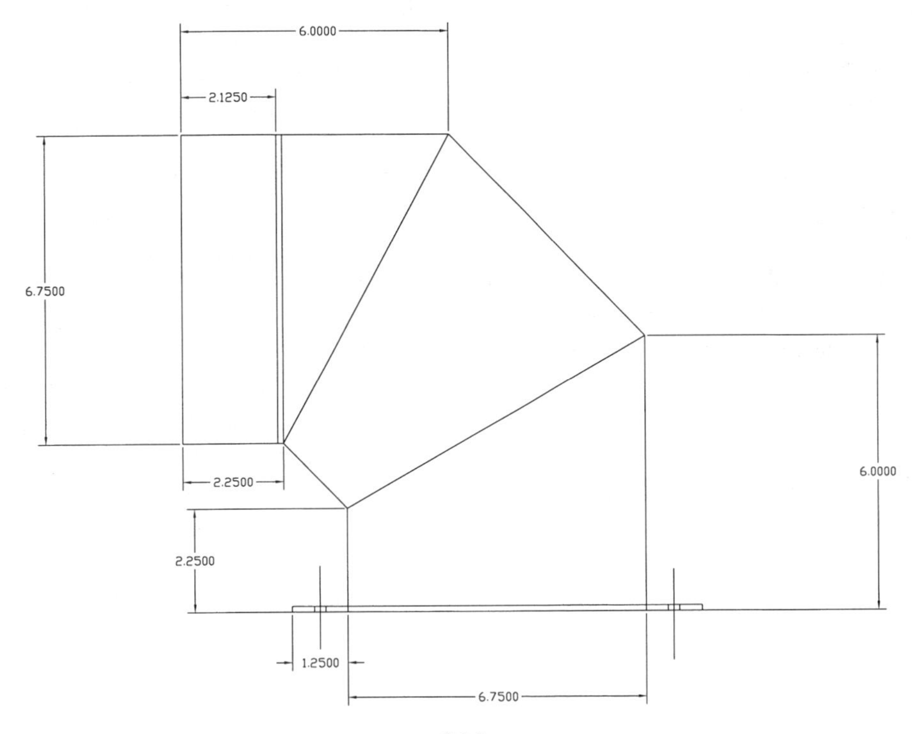

14-Inch Chop Saw

Author: Sarah Mayfield
Instructor: Dan Froneberger
School: Winona High School
City & State: Winona, TX

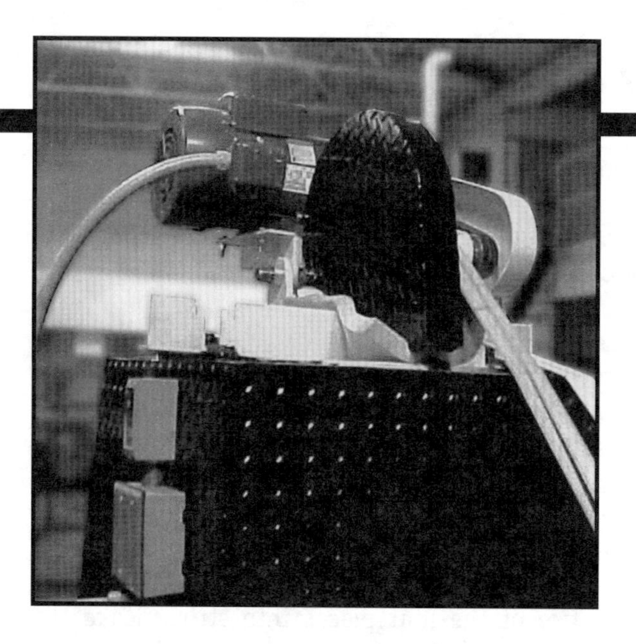

Bill of Materials

Quantity	*Description*
1	32-7/8″ x 27-1/2″ diamond plate
1	16″ disc
1	9-1/2″ x 8″ channel
1	7-1/2″ x 6″ channel
1	17″ handle, 1″ cold roll
1	5″ x 3″ oilfield tubing
1	28″ x 21-1/2″ steel plate
2	5-3/4″ x 1/2″ x 4″ flat bar
2	12-3/4″ x 2″ x 3″ angle iron
1	76″ x 90″ x 32″ diamond plate
2	31-1/2″ x 1-1/2″ x 1-1/2″ angle iron
4	1″ x 22″ x 1-1/2″ round tubing
2	2″ x 22″ x 1-1/2″ round tubing
2	2″ x 1-1/2″ x 1-1/2″ angle iron
2	2″ x 4″ x 1/2″ flat bar
2	Weld-on hinges
1	5-1/2″ x 8-1/2″ flat bar
1	1/2″ x 5-1/2″ x 8-1/2″ flat bar
2	Hi-Power II A40 belts
2	2-3/25″ x 2-3/25″ pulleys
1	NOM-117 switch box
2	17″ long chain
8	1/8″ bolts
2	1″ long Quicklinks
2	1641 DCTN bearings
1	Dayton 6 K 146 industrial motor
2	6-1/2″ x 1″ springs
1	1″ arbor 14″ x 3/32″ chop saw blade
2	11″ long springs
4	1/2″ x 3″ cold roll
2	Safety stickers
1	2′33″ x 2′27-1/2″ sheet diamond plate
1	3/4″ diameter x 6′ (L) wiring kit

Editor's safety note: *Be sure proper safety guards are installed along with proper warning signs.*

Base

Make the front and side panels of the chop saw base unit from a single piece of diamond plate steel bent at each corner with a 1-1/2″ lip on the back for attaching the back panel. Make the top of smooth surface sheet steel with a 15″ hole in the center. Weld the top plate to the base unit. Using a pad grinder, grind the welds to give a smooth, flat surface.

Turn the base onto its front side and mark the locations of the holes for the foot pedals. Cut the holes using a plasma cutter. Use the pad grinder to smooth the edges of the cutouts. A hammer and chisel can be used to remove the beads of excess metal produced by the welding and cutting procedures.

Drill four holes in the lip of the base unit, one near each corner. Place bolts in the holes from the inside, and weld. Mark places on the back cover to drill holes through, and drill them.

Place the back cover on the bolts, fit the nuts, and use a hacksaw to trim the bolts to the proper length—just outside the nuts. Then remove the nuts to rethread the bolts.

Pedals

Around the cutouts for the pedals, cut and weld pieces of 1/2″ cold roll to cover the edges. Weld a bolt to the inside of the base front to attach a return spring for the pedals. To attach the pedals, weld a piece of round tubing to a weld-on hinge, then weld the tubing to the lip on the back of the base. Repeat the process for the other pedal.

Cut two pieces of angle iron to extend through the foot pedal cutouts, and weld each one to each weld-on hinge. Fit each piece of angle iron with a 1″ quick link for attaching a spring to a hold-down chain. Also, attach a spring between the angle iron and the bolt on the inside of the front cover to support the pedals.

Form the foot pedals by cutting two pieces of 4″ x 6″ diamond plate. One at a time, put each piece into the hydraulic brake and bend 1″ of the end to a downward angle. Center the metal and weld one to each piece of angle iron.

Cutting Surface
To enable it to make cuts at various angles, the chop saw is designed with a circular pivoting base plate. The base plate is placed into a hole on the top of the base. Cut two pieces of flat bar and trim one end of each to fit the pivoting base. Weld the flat bar into place on each side of the base, forming a cutting surface across the chop saw, level with the pivoting base.

Drill holes in the flat bar to hold down a piece of angle iron. The angle iron will form a rear cutting fence and will extend over the pivoting base to hold it in place. To keep the angle iron aligned, bend a piece of 1″ cold roll and weld it in place. Weld another piece of angle iron to the chop saw base over the front edge of the pivoting base.

Drill a hole through the angle iron on the left side all the way through to the inside to make a place for the drop pin to keep the disc from sliding.

Chains
Using a plasma cutter, cut two 2″ x 3″ holes on top of the base for the chains to go through. Weld pieces of flat bar around three sides of the holes to give better support. On the fourth side, nearest the angle iron, weld a piece of 1/2″ cold roll to make the chains slide more smoothly. Grind the welds smooth.

On the angle iron that serves as a cutting fence, cut a 1″ space in the upright section on each side. Cut two more slits on each side of the space and bend the edges downward. Inside the base unit, attach the chains to the springs. Weld two pieces of 1/2″ x 3″ cold roll to the chains at different lengths. These pieces hook through the spaces in the angle iron and hold the metal in place.

Motor
Weld a piece of metal to the spinning disc. To this, weld a piece of metal on both sides with a 1-1/2″ hole for the arbor. This arbor will hold the motor mount.

To make the motor mount, weld a 9-1/2″ x 8″ channel to a 7-1/2″ x 6″ channel and pad grind the weld smooth. Bend a 17″ piece of cold roll to a 90 degree angle and weld it on for a handle near the front of the channel pieces. Weld a 5″ x 3″ piece of oilfield tubing to the front of the 6″ channel to hold the arbor for the chop saw blade.

Drill holes for where the motor will be bolted on. Also drill a hole slightly to the left and in front of the motor position, and weld a nut flush with the hole. A piece of all-thread welded to a 1/2″ x 3″ piece of cold roll can be used to screw into the hole to adjust how high the motor sits. A spring can be attached under the motor mount to help support the heavy motor.

Finishing Steps
Put the motor, the blade and the blade cover on. Attach the pulleys to the motor and the blade arbor. Put the belts on. Take the belt guard, and measure places for the holes to hold it up, and drill them. An extra hole can be drilled to add a 3″ piece of 2″ x 3″ piece of angle iron to attach a post-it note pad for recording measurements, etc. Put the belt guard on. Decide where to put the switch box and drill the necessary holes for it, and attach it. Wire the switch to the motor and add a 220 power cord.

Parts Cleaner Bin

Bill of Materials

Quantity	Name	Description
4	Legs	1x1x11ga. sq. tube, 42" long
4	Side braces	1x1x11ga. sq. tube, 28" long
6	End braces	1x1x11 ga. sq. tube, 16" long
2	Tank braces	1x1x11 ga. sq. tube, 28-1/4" long
1	Tank bottom	28-1/2 x 16-1/2 x 16ga. cold rolled
2	Tank sides	27-1/2 x 14 x 16ga. cold rolled
1	Tank end	16-1/2 x 14-1/2 x 16ga. cold rolled
2	Handles	4-1/2 x 5 x 16ga. cold rolled
1	Lid	28-1/2 x 16 x 16ga. cold rolled
1	Tank end	16-1/2 x 12-1/2 x 16ga. cold rolled

Project Description
This parts cleaning bin is designed to allow recycling of the cleaning solvent using a pump with a filter. The choice of the Mig welding process for the original project posed some problems with distortion from overheating. Tack welding a thick piece of angle iron to the side of the bin to serve as a chill bar cured some of the warping. This project provides good experience in welding on thin gauge material.

Procedure
First, make the frame. Weld the legs to the side cross members. Weld the tank braces in at an angle to compensate for the angle on the bottom of the tank. Now, weld all of the joints together.

Cut the material for the bottom of the tank. Notch each corner with a notching shear. Bend two of the edges on the bottom at a 90 degree angle. Bend one side at a 90 degree angle and one side at an 85 degree angle. This is so when it is welded together, the sides will be square and the bottom will be at an angle to make a sump in the tank.

Now, make the ends of the tank. Notch the corners and bend the two sides at a 90 degree angle. Put a hem on the top to get rid of any sharp edges.

Make the sides of the tank. One of the sides will be two inches longer than the other one to make the top sit level and still have the sump. Put hems on top of the sides to eliminate sharp edges.

Now, weld up the tank so it will not leak. Make the handles for the tank so it will be easier to lift out to drain. Finally, make the lid. Notch the corners and bend them at a 90 degree angle. Attach the lid with a small hinge. Paint the project.

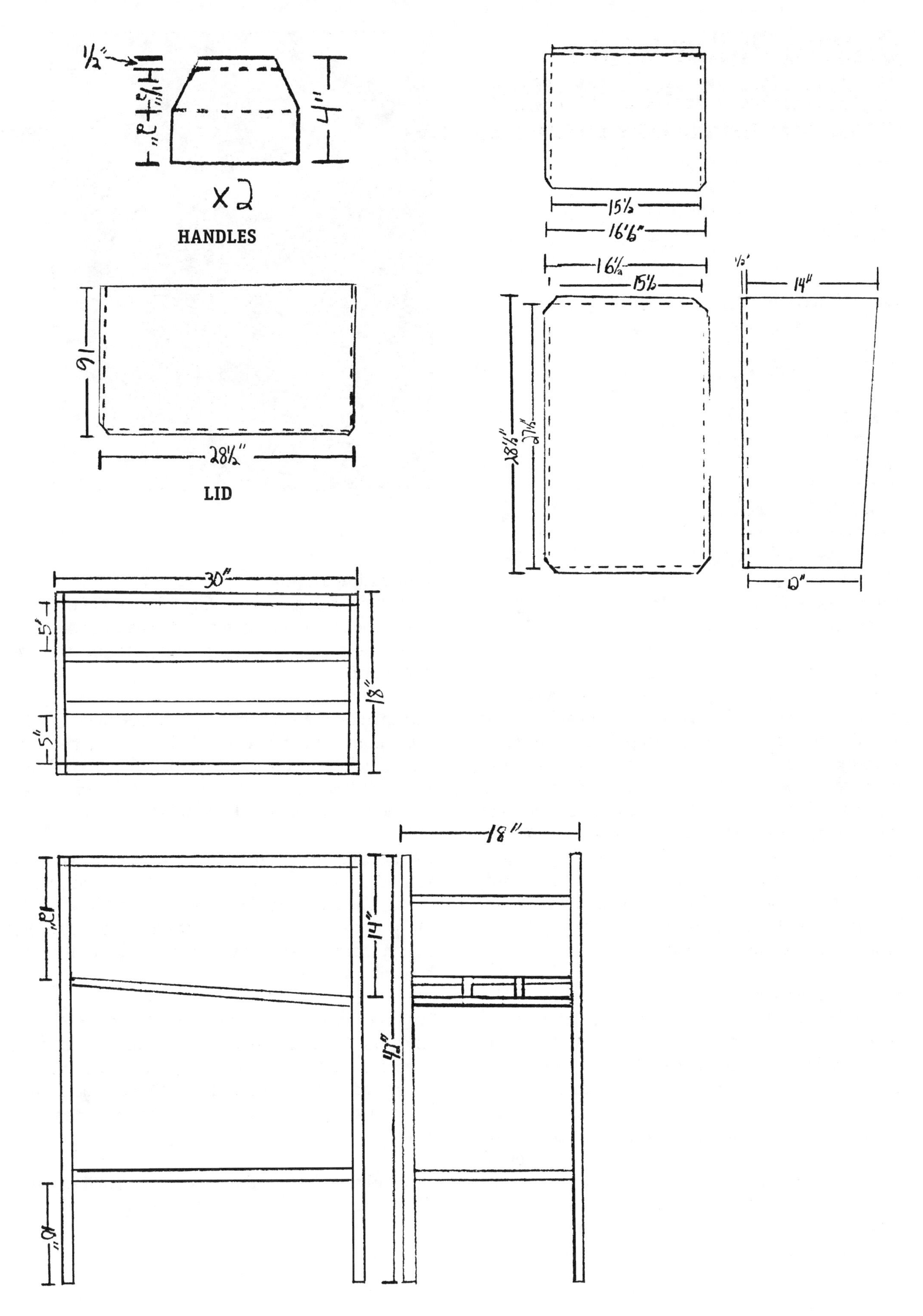

HANDLES
LID

Sleeve Puller with Hydraulic Power Station

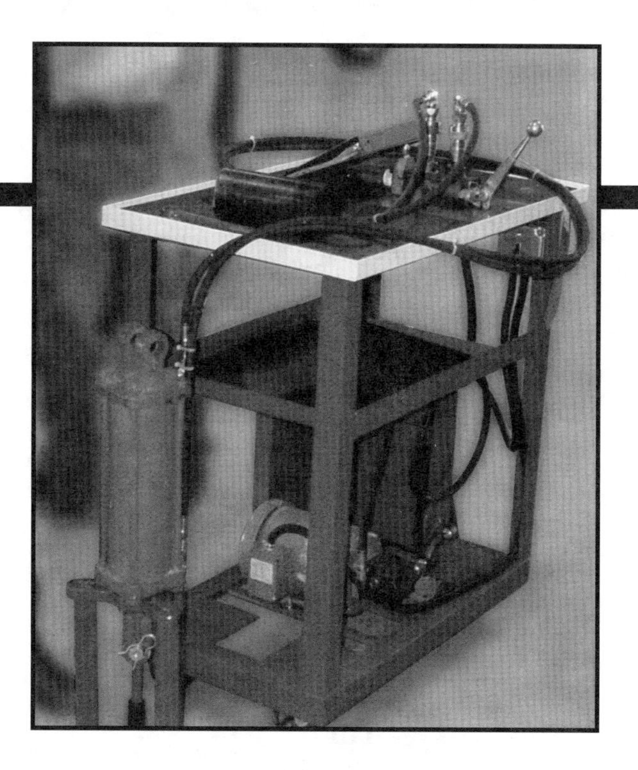

Author: Jordan Killingsworth
Instructor: Bob Killingsworth
School: Rains High School
City & State: Emory, TX

Bill of Materials

Used motor with pump
Used 4 wheel dolly
Used 4″ cylinder
Control valve
1/8″ plate
1/8 x 2 x 2″ angle iron
3/4″ plate
Hoses and fittings
Strainer/filler assembly
Quick couplers
3-phase switch and box
Plug and wire
Bolts
Primer, paint and thinner

Background

This project illustrates how an adverse event can inspire the creation of a useful piece of equipment at very little expense. A set of sleeves in a tractor block needed replacement, and the cylinder went bad. Presented with this challenge, the student and his instructor decided to create a hydraulic station that would work with the sleeve puller, as well as the press, hose crimping tool, track-pin remover, and portapower. The workstation was designed to incorporate storage for the sleeve puller, as well as a work table. The unit was also designed to be portable so that it could be moved around the shop.

Keeping Costs Down

The project relied on employing used components and materials wherever possible and practical. A four-wheel dolly, a 1 horsepower, 3-phase motor and pump, and a 4″ cylinder, were all purchased used?

A 3.5 gallon reservoir (dimensions: 5″ x 10″ x 16″) was fabricated from 16-gauge steel. Other sheet metal for the work surface was salvaged from an office structure. The control valve was purchased new and set to "relieve" at 2500 psi. The combination is calculated to develop just over 15 tons of pressure.

Fabrication Steps

First, build the reservoir, using a Lincoln PRO-CUT 55 to cut the metal, and then a Scotchman ironworker to break the sides. Cut holes for the filler neck, outlet and return.

Incorporate strainer assemblies at the filler and the suction. Drill and tap holes. Weld sides together using a Lincoln SP135 Plus with 0.023 ER70S6 electrode and Argon/CO2 shielding gas. (Settings used were H and 8 for the welder, and 24 CFH flowmeter.)

Arrange the cart to house the motor/pump and reservoir, provide for storage of the puller and attachments, and a work surface. The motor will be wired for low voltage 3-phase service, with a 3-phase electrical switch connecting power. Make all controls easily accessible for safety.

Cut attachments for the sleeve puller from 3/4″ steel using an old puller as a pattern. Build eyebolts to allow for variations of sleeve length; they will pin to the cylinder. Add quick couplers to the control valve to allow the power station to be used with the other tools.

Prep, prime and paint the power station and assemble the parts. Now it is ready for plumbing and wiring.

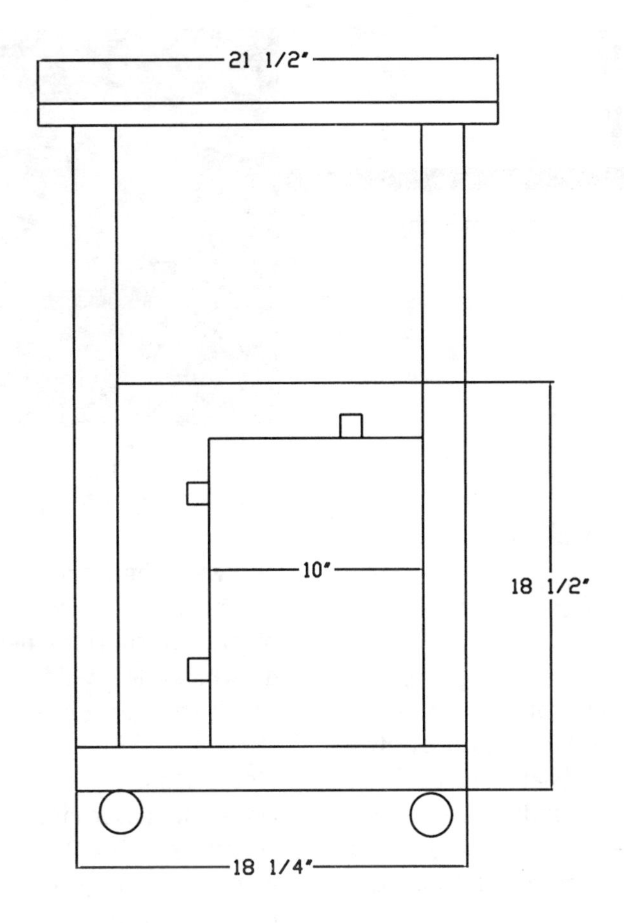
21 1/2"
10"
18 1/2"
18 1/4"

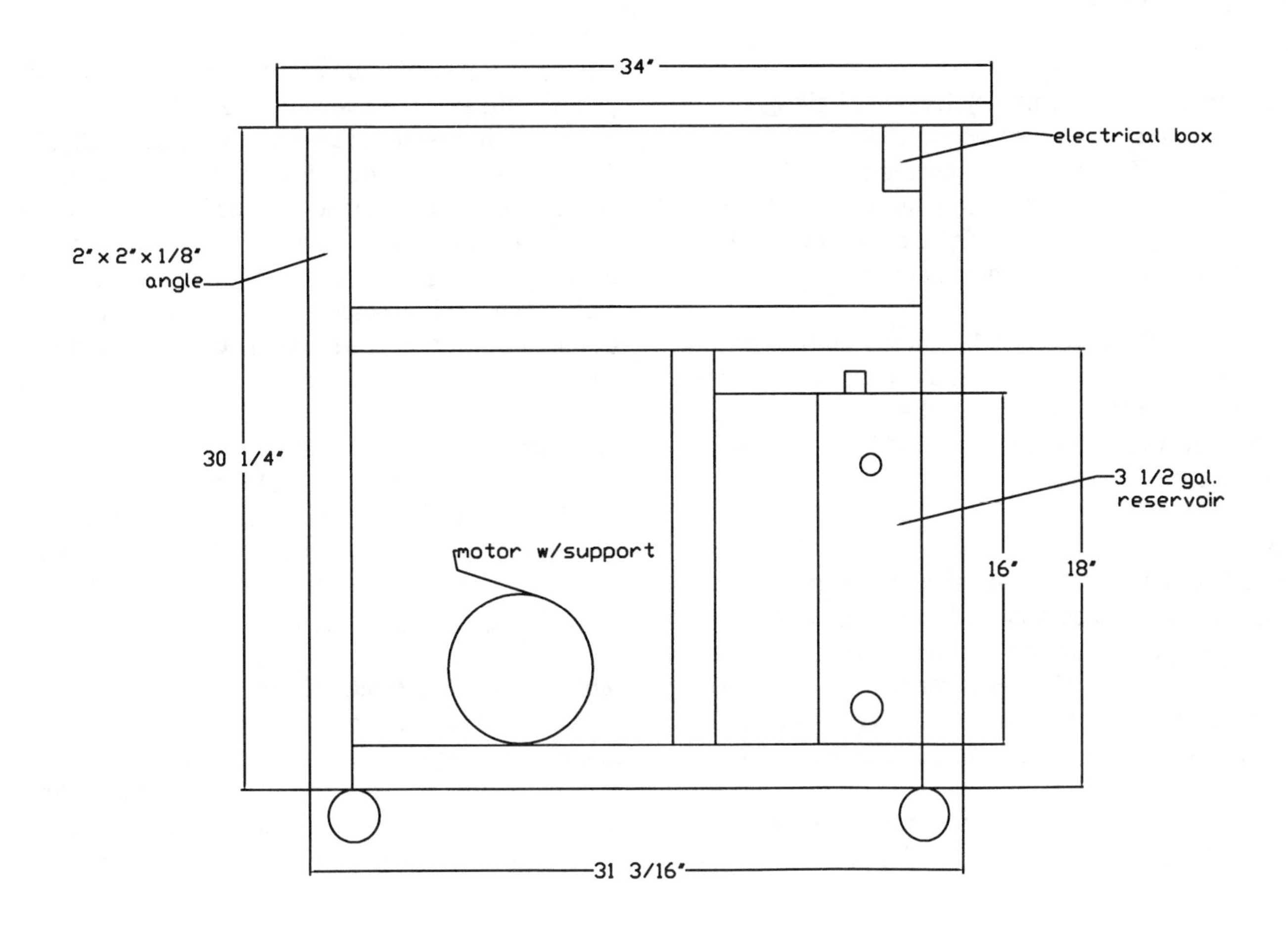
34"
electrical box
2" x 2" x 1/8"
angle
30 1/4"
3 1/2 gal.
reservoir
motor w/support
16"
18"
31 3/16"

Stainless Steel Pickup Sander

Authors:	Noah Mehaffy and Ron Brandenburg
Instructor:	James L. King
School:	Burlington High
City & State:	Burlington, IA

Bill of Materials

Quantity	*Description*
1	4′ x 10′ sheet of 11 gauge
1	4′ x 10′ sheet of 16 gauge
2	4′ x 8′ sheet of 16 gauge

Introduction

The sander a school district used to spread sand on icy roads became too rusty to be functional. School officials proposed that welding students at the local high school take on the project of building a new sander. The instructor and students decided to make the new sander out of stainless steel to protect it from rust and corrosion.

Step by Step Process

Base

Build the base by cutting (2) 12″ x 8′9″ pieces of 11 gauge stainless for the sides. Bend a 1″ tab on the bottom of both sides at a 90 degree angle. Cut slots 1.25″ x 7″ at both ends of both sides. Cut (2) 3″ x 8′9″ pieces of 11 gauge stainless and arc weld them tot he tops of each side at a 45 degree angle.

Next, cut a 13″ x 8′ piece of 11 gauge stainless for the drag plate. Cut 2″ x 3″ slots in all four corners. Bend a 1″ tab on both sides of the drag plate for easier welding. To finish the base, arc weld the drag plate between the two sides.

Frame

The entire frame of the sander is made of 16 gauge stainless. Cut (2) pieces 24″ tall, 94″ long at the top, and 88″ long at the bottom for the bottom side pieces. Next, cut (2) 14″ x 94″ pieces for the top side pieces and bend them to the correct angles (see drawings). Mig weld the top side pieces to the bottom side pieces.

Cut (2) triangular pieces for the ends, and bend the front end piece to the correct angles (see drawings). Bend a 1/2″ tab on all sides of end pieces for easier welding, and Mig weld both sides to both ends.

Mig weld the frame on top of the base. Cut (4) triangular pieces for the bracings and bend a 2″ tab on all of the bracings to help sturdy up the sander (see drawings). Mig weld the two bracings on both sides to finish the frame of the sander.

Motor Mount

Make the motor mount out of 1/4″ stainless. Start by cutting a 17″ x 27″ piece. Next, cut (5) odd shaped pieces and cut slots and drill holes where they are needed (see drawings). Arc weld those pieces to the 17″ x 27″ piece to finish the motor mount.

Motor Cover

The motor cover is constructed of (3) pieces of 16 gauge stainless. The two sides are 17″ x 22″ but slant downward in the back (see drawings). Bend a 1/2″ tab on all sides for easier welding. Make the top and back of the motor cover using one 23″ x 32″ piece, and bend it to the correct angle. Cut (2) 8″ holes in the back and an 8″ hole in the side for ventilation for the motor. Mig weld expanded steel in those holes. Mount the motor cover on the motor mount using a stainless steel hinge.

Sand Chute

Make the sand chute out of (2) triangular pieces of 16 gauge stainless for the sides (see drawings). Cut a 12″ x 10″ piece for the back and a 12″ x 8″ piece for the front. Mig weld it together to complete the sand chute.

Motor and Drive Mechanisms

First, mount the motor on the motor mount. Next, mount the gear reduction drive and connect it to the back conveyer sprockets. Install the clutch and connect it to the motor. Finally, install the sand thrower by using a chain to connect it to the gear reduction drive.

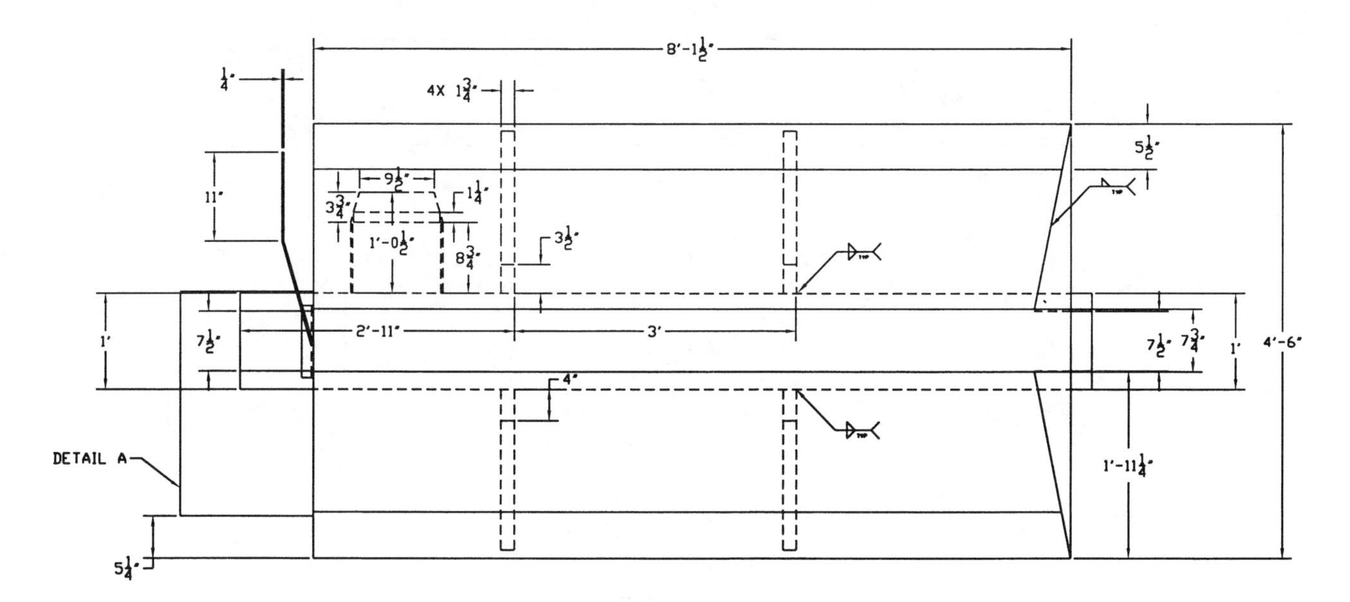

TOP VIEW

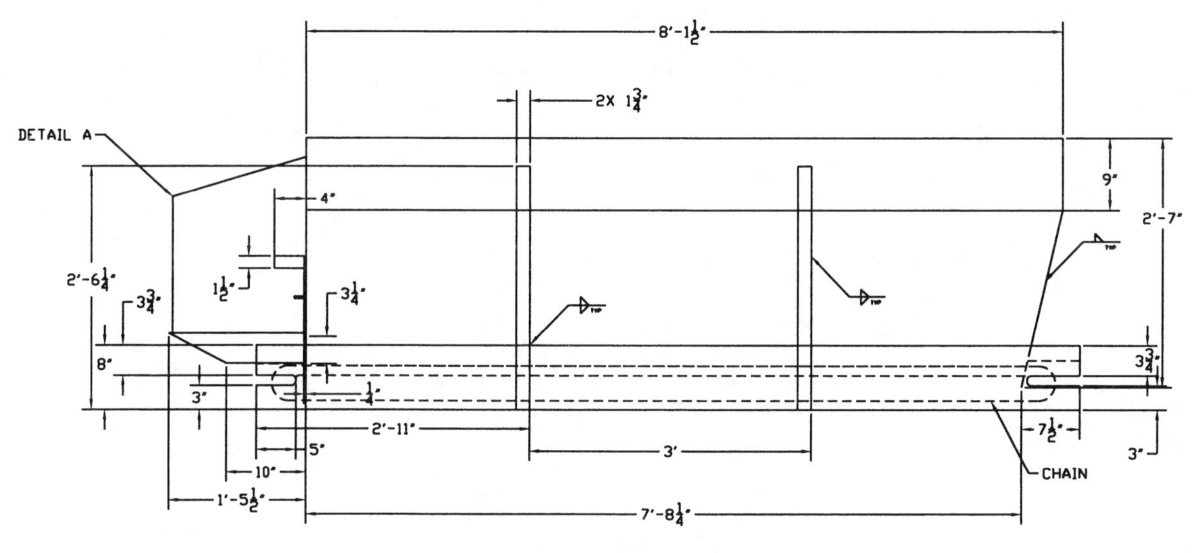

SIDE VIEW

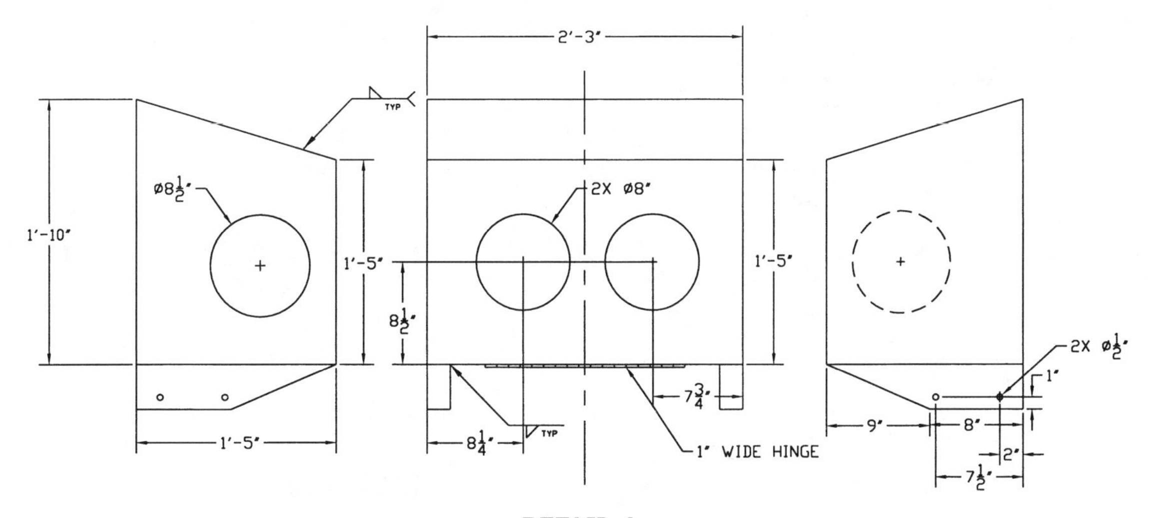

DETAIL A

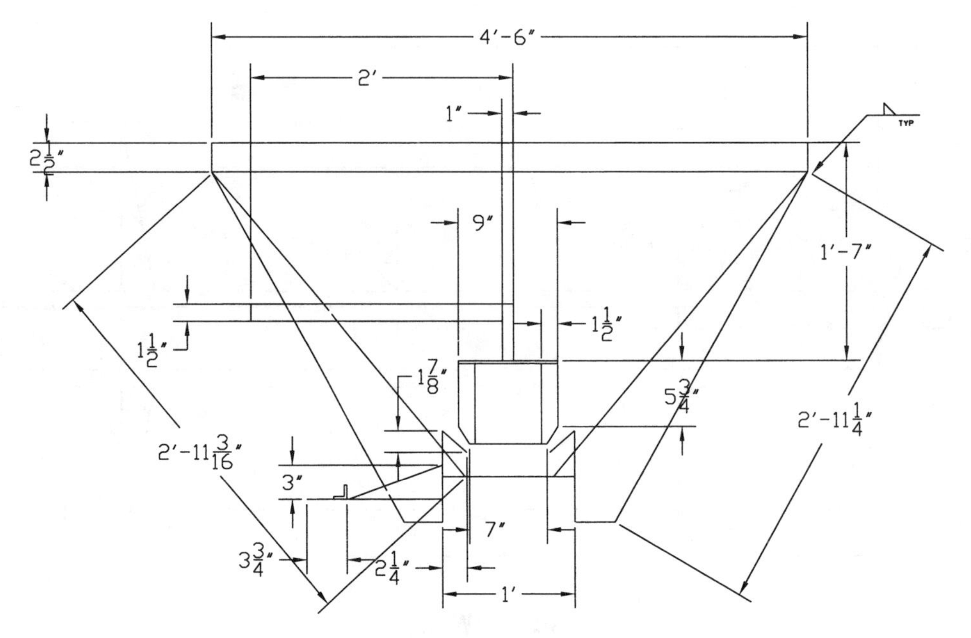

LEFT SIDE VIEW

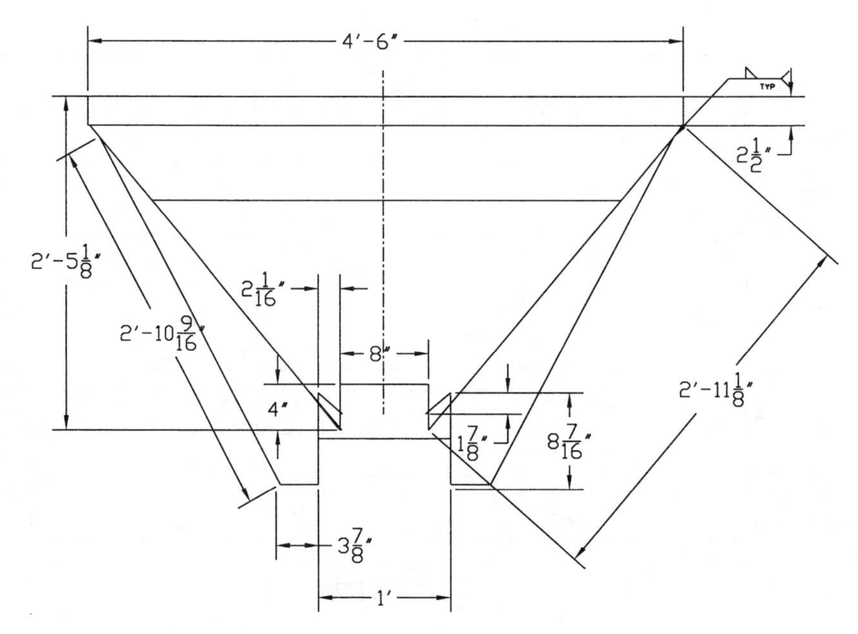

RIGHT SIDE VIEW

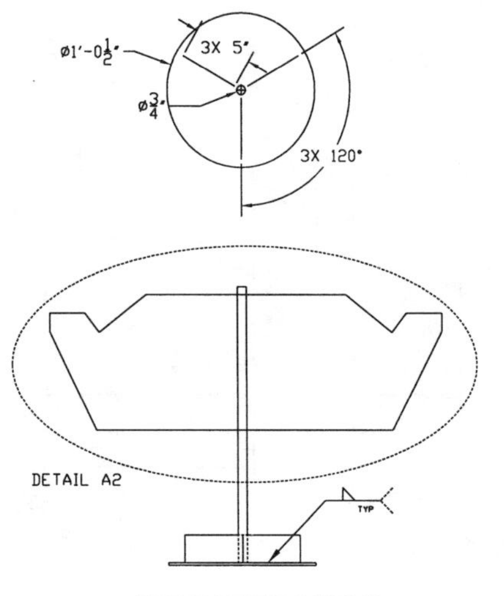

SPREADER VIEW

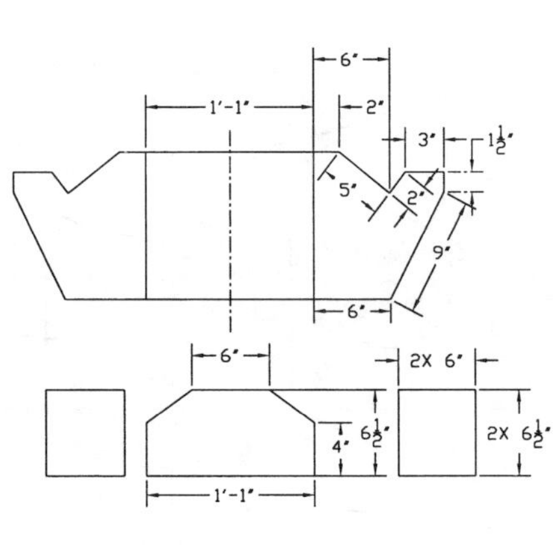

DETAIL A2

English Wheel

Bill of Materials

Quantity	*Description*
1	3" x 20′ – 3/16" wall, square tubing
1	2" x 20′ – 3/16" wall, square tubing
1	3" x 20′ – 3/16" wall, pipe
1	2" x 16" tube
1	3/8" x 6" x 5′ plate
1	machined wheel, 9" dia. x 3" width
1	machined wheel, 3" dia. x 3" width
1	machined handle
2	bearings, 3/4" ID x 1-5/8" OD x 5/16" width
2	bearings, 1/2" ID x 1" OD x 5/16" width
1	bushing, 2" ID x 3" OD x 3" length
1	2" rotating wheel
2	2" wheels
8	3/8" bolt & nut
4	3/4" ID washers
6	1/2" ID washers
3	10-32 bolts

Project Description

An English Wheel is a machine that makes compound curves in sheet metal by bending the metal when it goes in between the two wheels. An English Wheel can be used to make gas tanks, body panels, and bike fenders, among many other items.

Start construction by cutting a 3" x 3" square tube to 6′ long, to make the back piece. Then cut the top and bottom pieces out of 3" x 3" tube to 3′9". Next, tack the top piece down from the top of the back piece one foot, and tack the bottom piece two feet from the bottom of the back piece, as shown in Figure 1.

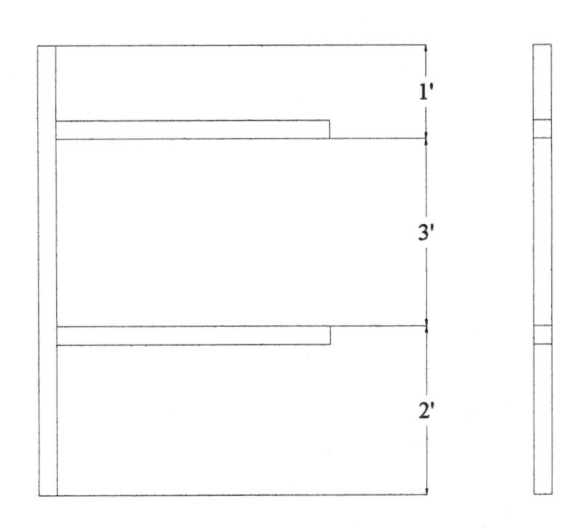

FIGURE 1

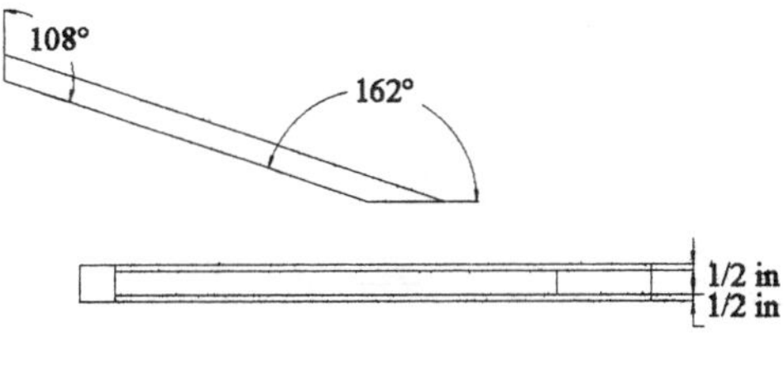

FIGURE 2

Next, cut the two support pieces of 2" x 2" tube to 3′ 3" with the angles as shown in Figure 2. Then tack them to the English Wheel as shown in Figure 3.

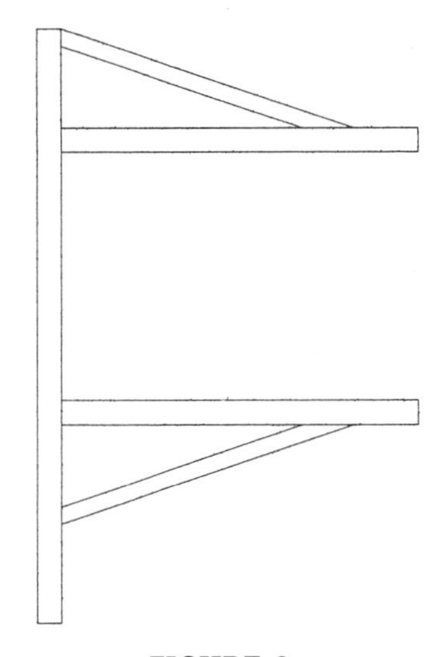

FIGURE 3

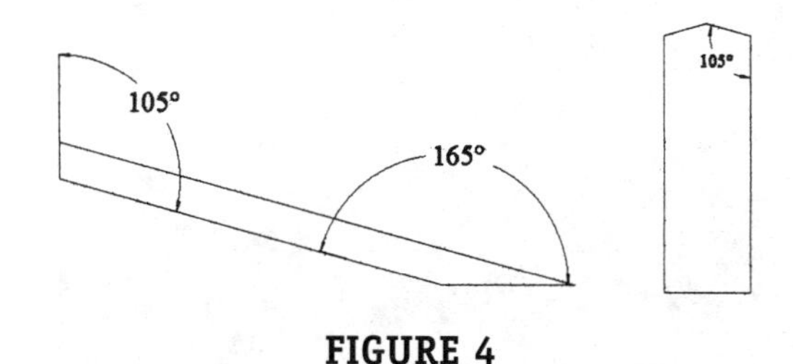

FIGURE 4

Now, make the back support by cutting two pieces of 2" x 2" tube to 6" with the angles as shown in Figure 4. Tack and weld them together as shown in Figure 5.

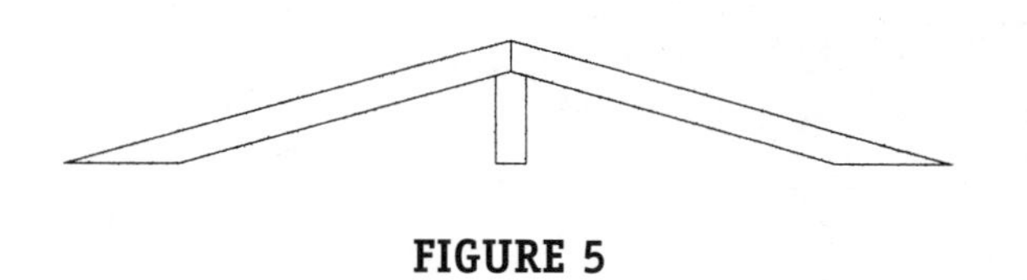

FIGURE 5

Tack and weld the back support to the English Wheel as shown in Figure 6. Now, make the front legs by cutting two pieces of 2" x 2" tube to 2'8", with the angles as shown in Figure 7. Tack and weld the piece as shown in Figure 7. Then, put the legs on the English Wheel as shown in Figure 8. Now, cap all the holes, and attach the wheels to the legs.

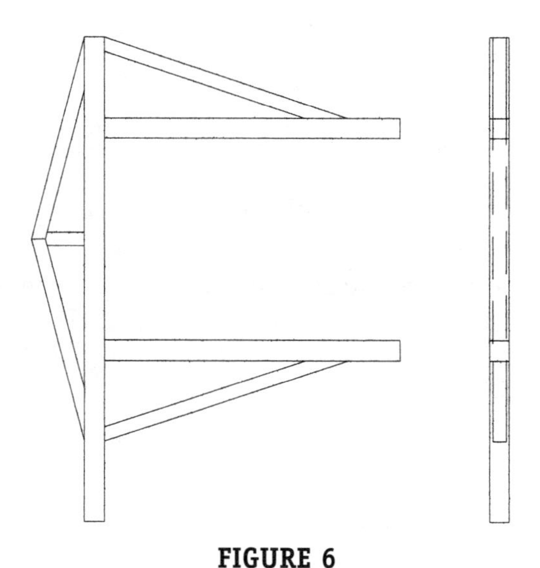

FIGURE 6

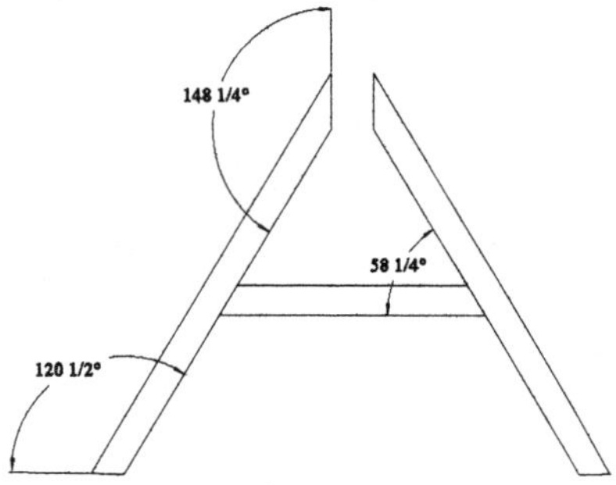

148 1/4°
58 1/4°
120 1/2°

FIGURE 7

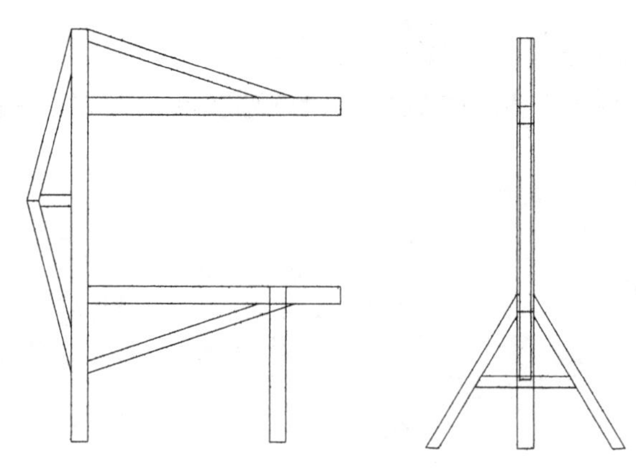

FIGURE 8

Make the lifting part by cutting a 3" pipe to 20" long, cutting a 2" tube to 16" long, and a 1" threaded rod to 16" long, with 3" of the bottom machined down to a 3/4" diameter. Now, center up and weld the threaded rod's nut to the 2" tube bottom. Cap the other end of the 2" tube, with a 3/8" plate cut to fit, and a hole drilled and tapped in the center of the circle to fit a 3/8" bolt.

Cap the 3" pipe with a 3/8" plate cut to fit. In addition, cut a 1" hole in the cap, drill three holes and tap them to fit a 10-32 bolt equally spaced on a 1-1/4" radius circle, from the center hole. Cut a 3/8" plate with a 1" radius, drill a 3/4" hole in the center, and drill three holes which fit a 10-32 bolt, and match the capped end of the 3" pipe.

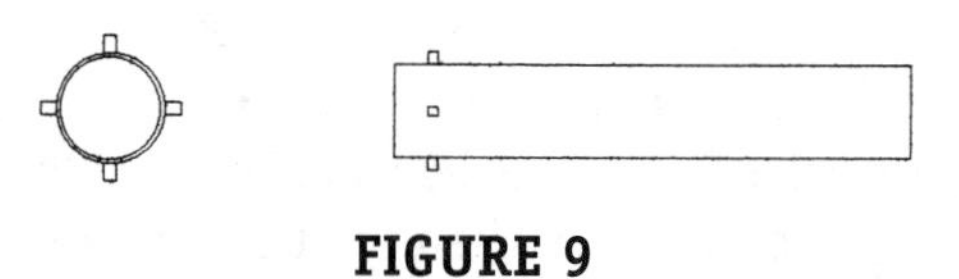

FIGURE 9

Drill the four holes, to fit a 3/8" bolt, equally spaced in the 3" pipe, 1-1/2" from the top, as shown in Figure 9. Now weld four 3/8" nuts on the 3" pipe in each hole that was drilled. Then assemble the lifting piece as shown in Figure 10.

Next, cut a 3" hole in the bottom support, 2-1/2" from the outside edge. Tack the lift piece in the hole and put the end piece back on.

Now make the brackets for the two wheels. Make the bracket for the top wheel by cutting one 4" x 3" piece of 3/8" plate, and two 4" x 4-3/4" pieces of 3/8" plate, as shown in Figure 11. Make the bottom bracket by cutting a 3" x 3" piece of 3/8" plate, and two 3" x 3" pieces of 3/8" plate, also shown in Figure 11.

Weld the brackets, and attach them to the English Wheel. Finally, paint the assembly and put on all the bolts and wheels.

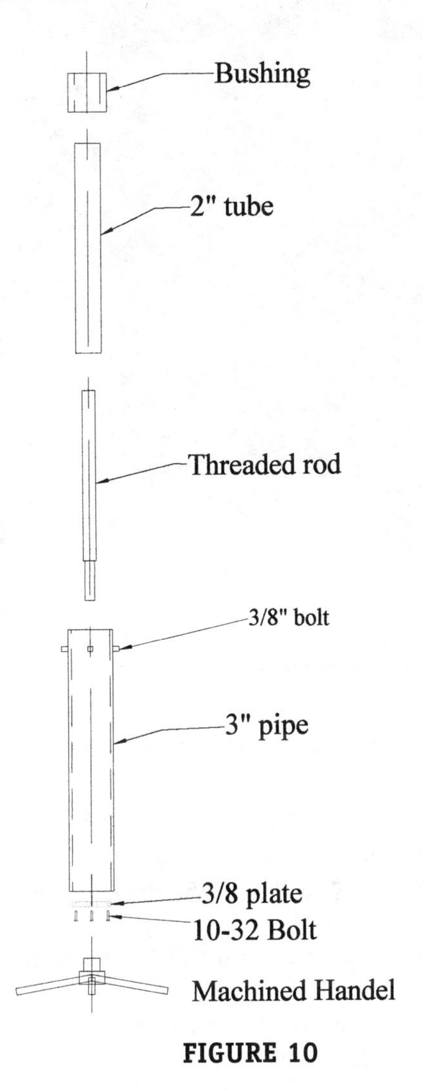

FIGURE 10

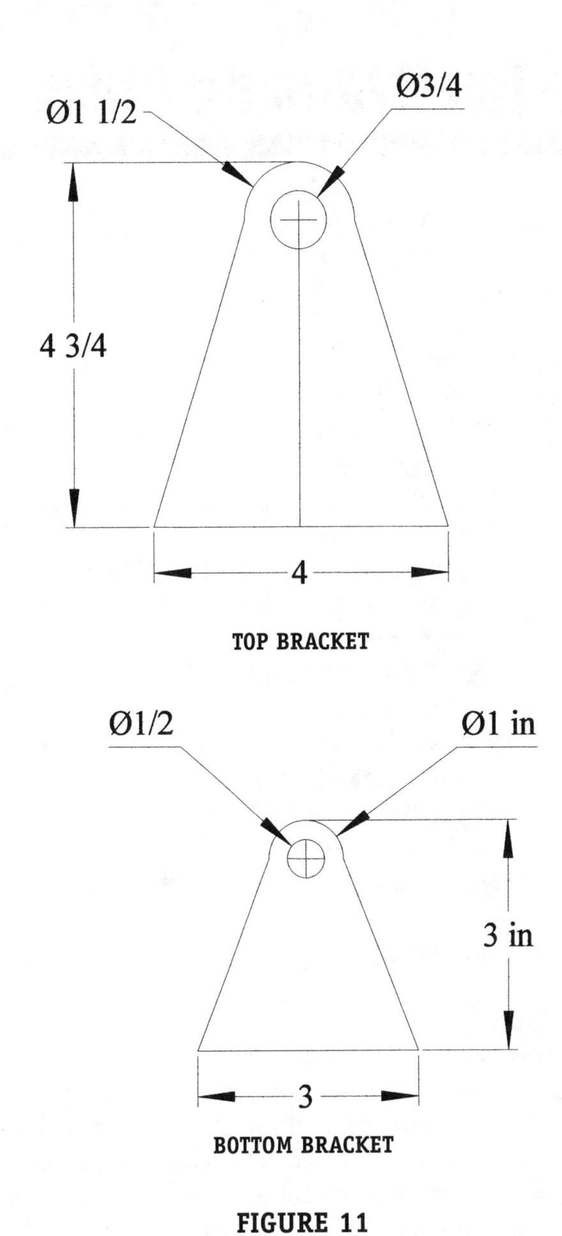

FIGURE 11

20-Ton Hydraulic Press

Author: Cody Madlung
Instructor: Thayer Davis
School: Tomahawk High
City & State: Tomahawk, WI

Bill of Materials

Quantity	*Description*
180"	8" channel x 1/4"
200"	10" channel x 1/4"
40"	1" steel round
110"	3" angle x 3/16"
1	jack holder and ram head, 1/2" plate (20 ton spec's)
6	1" pins, cold rolled, round
2	7/8" bolts, 12" long
2	7/8" nuts
4	7/8" washers
1	20-ton hydraulic jack
2	6" extension springs
36"	1" threaded rod
2	1" couplers
1 qt.	primer
1 qt.	flat black paint

Introduction

This press was designed and built for use at a construction shop. Its main purpose is to push pins in and out of bulldozer tracks, and bushings in and out of backhoe buckets. Its large size allows big items to fit, and the jack pushes these items out with ease. The springs are encased for safety in case they break; also, the press looks a little nicer with the springs out of sight.

Construction of the Framework

1. Cut all steel to required lengths.
2. Drill 1" holes every 10" on the 8" channel through both sides.
3. Make sure the holes are lined up straight so the table will sit square when it is resting in the hole.
4. Drill a 1" hole through two pieces of 50" channel.
5. Holes should be placed 1-1/2" from the outside of the channel.
6. These two pieces are the table.
7. Center the 8" channel on separate pieces of 30" channel.
8. These two pieces will be the posts.
9. Pin the table in place on the posts.
10. Bring another piece of 10" channel up to the top.
11. Make the top piece square and level on the inside. Tack weld in place.
12. Bring the other piece of 50" channel up to the top.
13. Make sure everything is still square and tack weld this piece opposite the other piece.
14. With everything still square, weld in the bottom cross member piece.
15. Center the 50" channel on the posts and tack weld in place.
16. Remove the table from the assembly.
17. Mark a hole 25" from each end.
18. Clamp the two pieces of the table together.
19. Drill 1" holes through both pieces.
20. Pin the table in place with 1" pins and place the 1" bolts through the holes just drilled.

Assembling the Ram

The assembly described here was made to fit with a combination of 8" and 10" channel. Each unit would have to be custom made if the dimensions are different. For this project, make the spring guards with 2" tube steel moving inside of two pieces of 2-1/2" channel welded together to form a piece of tube steel. Weld everything on the unit in place with 7024E electrode. The jack is held in place with 1/4" flat bar welded in place to conform to the shape of the jack. This keeps it from moving around.

This view shows the assembly without the piece of 2" round stock welded on the bottom to act as the push bar.

This view shows how the assembly will move side to side when it is installed on the framework. The couplings thread into the rod to hold the two pieces together.

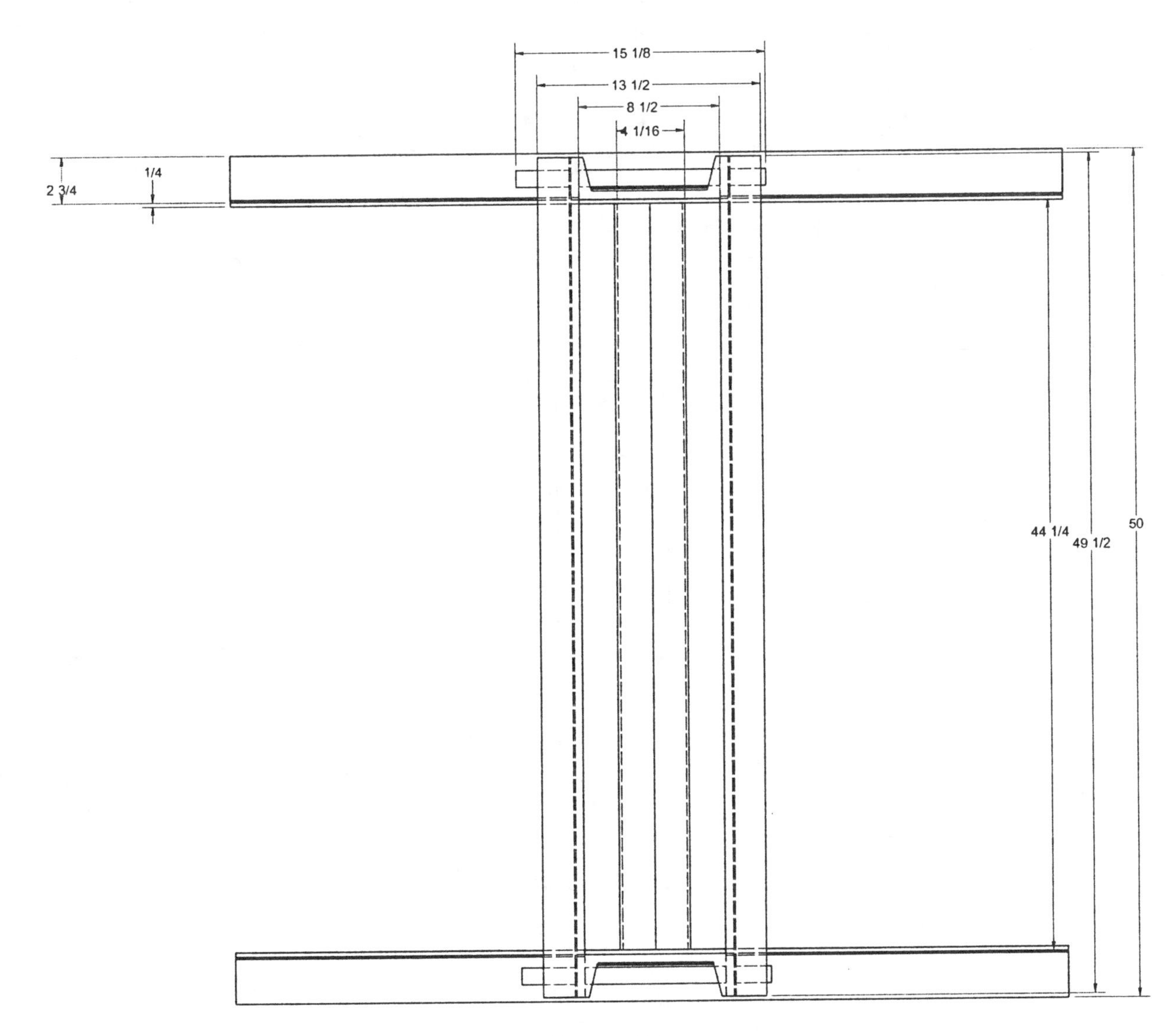

TOP VIEW

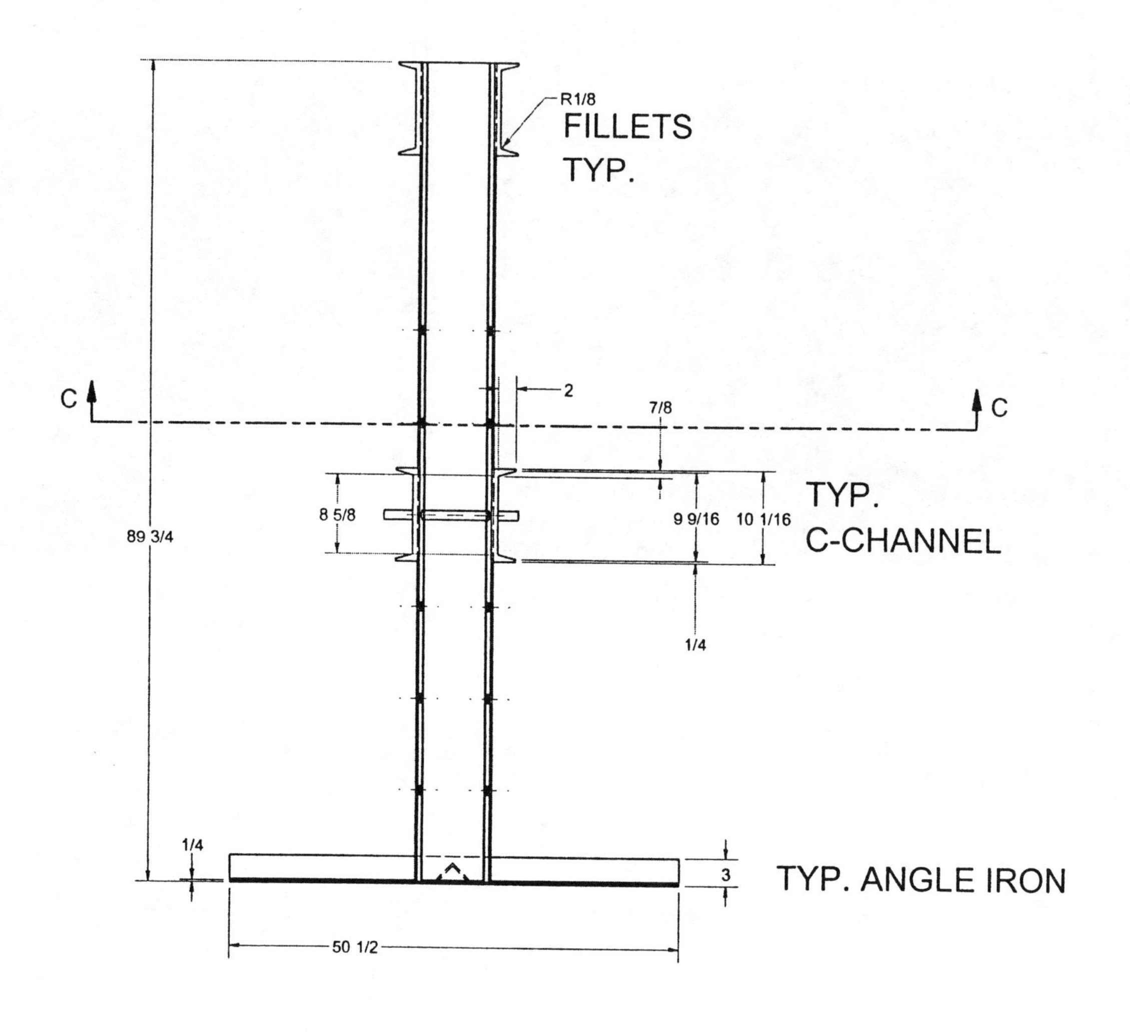
R1/8
FILLETS
TYP.
2
C
C
7/8
8 5/8
9 9/16
10 1/16
TYP.
C-CHANNEL
89 3/4
1/4
1/4
3
TYP. ANGLE IRON
50 1/2

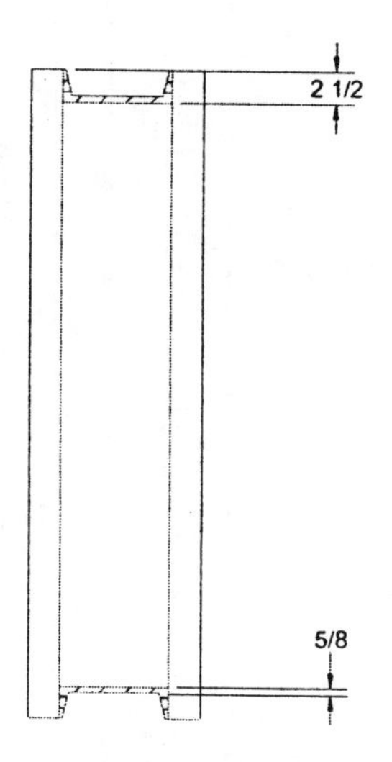
2 1/2
5/8

FRONT VIEW

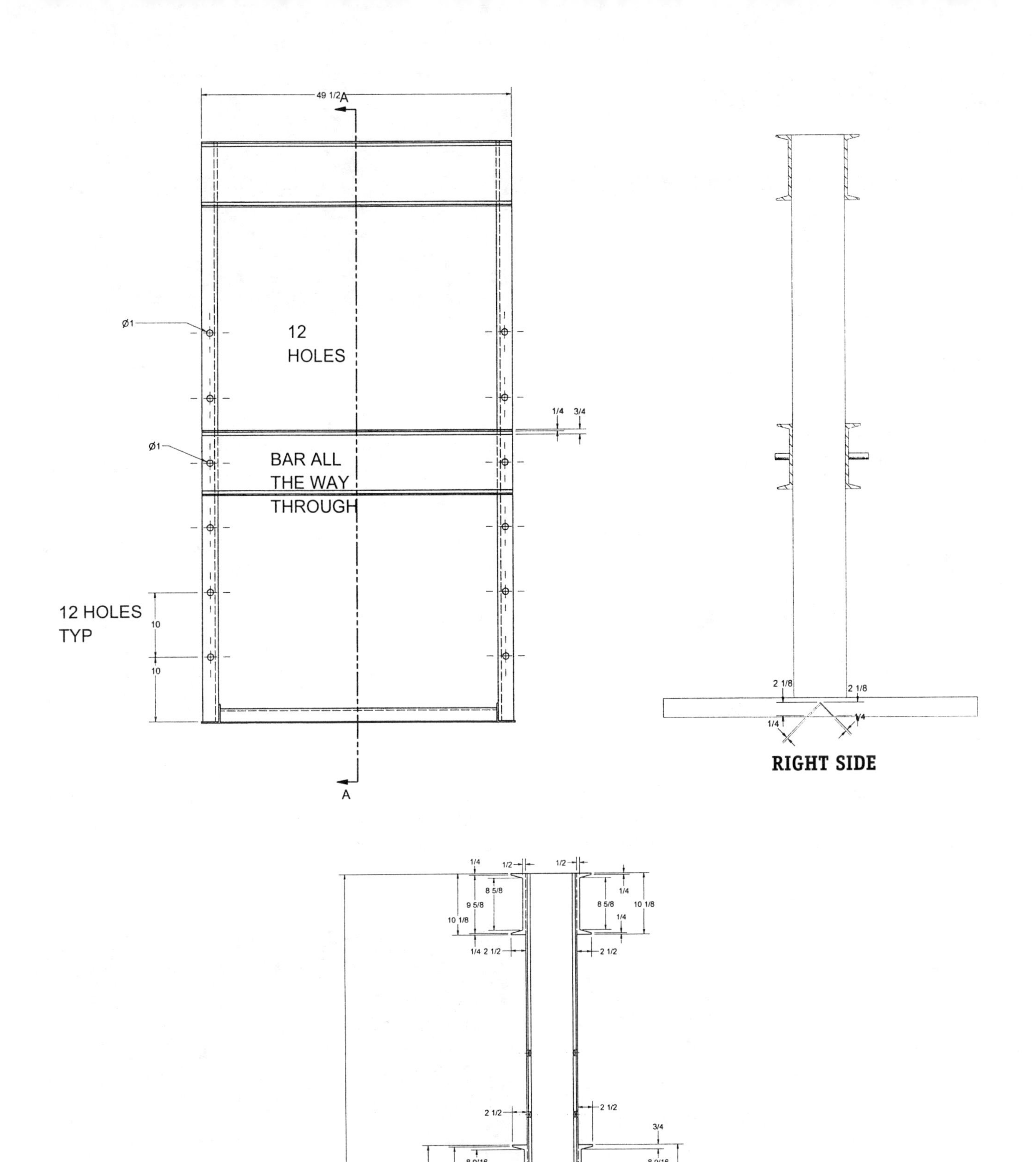

RIGHT SIDE

1/4 1/2 1/2
8 5/8
9 5/8
10 1/8
1/4
8 5/8
10 1/8
1/4
1/4 2 1/2
2 1/2
2 1/2
2 1/2
3/4
8 9/16
10 1/16 9 9/16
8 9/16
10 1/16
89 3/4
3/4 1/2
1/2
1/4
1/4
3
50 1/2

Adjustable Surface Cutting Table

Authors:	Sandy Spalding and Sheena Ann Schneider
Instructor:	Clifford Guyette
School:	Blackstone Valley Tech
City & State:	Upton, MA

Material List

16 feet 1-1/2" x 1-1/2" angle iron
21-1/4" x 1-1/2" diameter round stock
36-1/4' bars 1/2" x 3/4"
1/2- 13 x 1" bolts, 120 required

Item	*Qty*	*Description*
Fabricate frame of angle iron	1	46-1/2" x 46-3/16"
Machine pegs	120	from 1-1/2" round stock
Bolts	120	3-7/16" spaced on center
Bars	10	43-1/2" @ 4 inches on center

Introduction

The design and fabrication of an oxy-acetylene cutting surface involves making a frame with rows built into it. On the rows will be bolts that will be welded on. On the bolts will be pegs that will be screwed on. A CNC lathe, CNC milling machine, MIG welder and a grinder are needed. The hardest part of the project is welding on all the bolts so that they will be properly aligned. Practice pays off!

Procedure

First, cut pieces of A-36 stock. Write a program in G-code for the CNC lathe to machine the parts. After the parts are machined, build a frame out of angle iron and bar to hold the pegs and bolts.

There are ten rows of bars in which bolts will be welded using GMAW. Drill and tap the machined parts. Screw the parts onto the bolts. Align each bolt on the frame bars at 4 inches apart.

Put notches in the angle iron using a notch machine. Next, fabricate the 1/2" x 1-1/2" angle iron and cut it to size. Use the framing square to make sure that all the angles are accurate. Make sure it is square diagonally by measuring corner to corner. Tack it together using GMAW.

Lay out the 1/2" x 3/4" bar 4 inches from center. When tacking them, use the framing square to make sure that they are straight. Tack the bars on location using GMAW. Preheat the parts to be welded, then weld them solid using 1/4" welds. (We preheated because the steel we used was high carbon. This material was free, readily available, and suitable for the application.)

Lay out 1/2 x 13 bolt locations 3-7/16" on center using the framing square and the straight edge of the outside frame. Set the bolts, tack them on location, preheat, and weld.

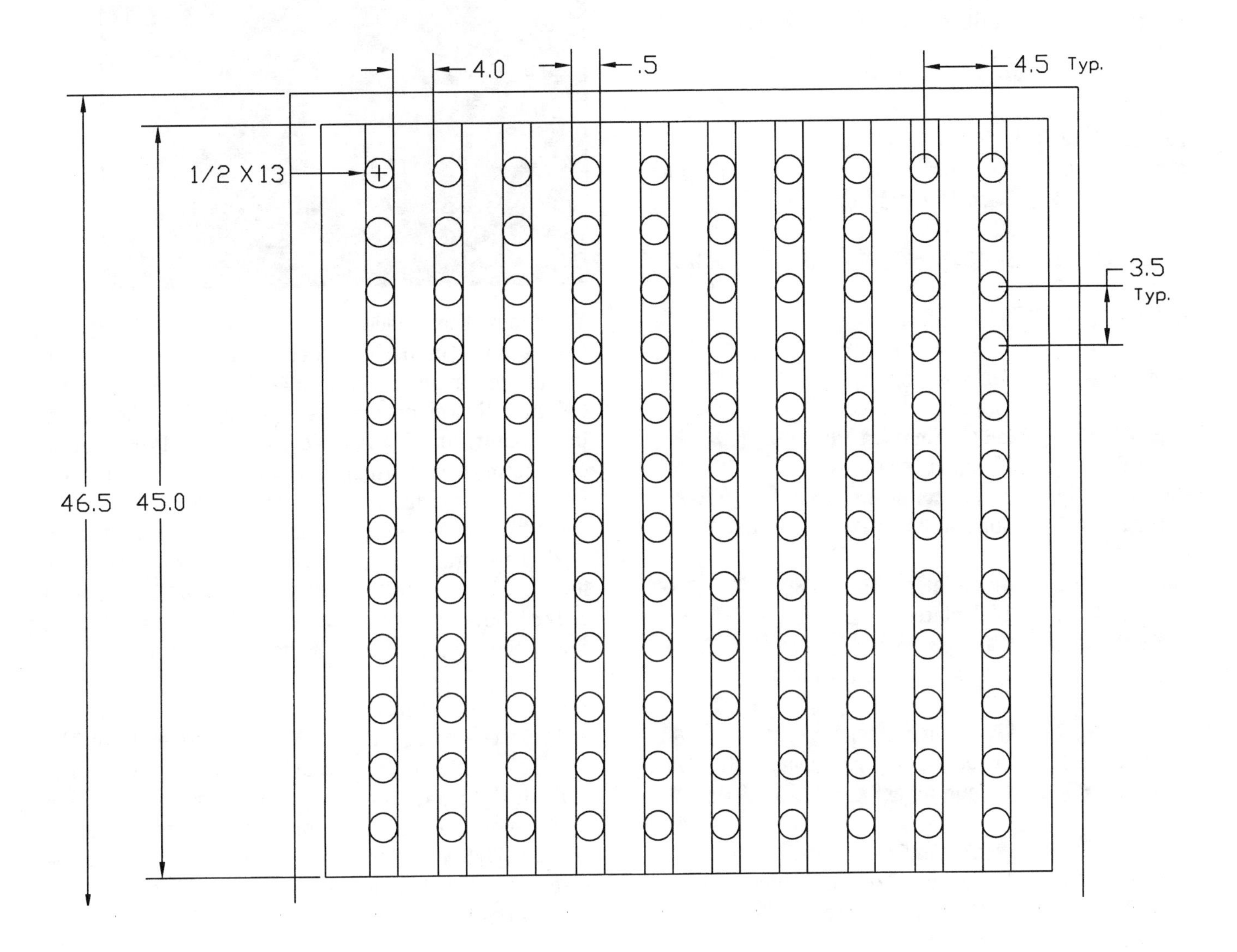
4.0
.5
4.5 Typ.
1/2 X13
3.5
Typ.
46.5
45.0

Guided Bend Tester

Author: Justin Cummins
Instructor: Bill Gleason
School: University of Montana C.O.T.
City & State: Missoula, MT

Bill of Materials

Mark/Pc	*Qty*	*Description*	*Length*
M1	1	Base Assembly	
M1-A	1	C6 x 10.5	20.5"
M1-B	1	5/16" plate	19.875"
M1-C	2	5/16" plate	5.25"
M1-D	2	5/16" plate	5.25"
M2	1	Jack Assembly	
M2-A	1	12-ton bottle jack	
M2-B	1	2-1/2" round stock	2"
M2-C	1	2" round stock	2"
M2-D	1	1-1/2" round stock	2"
M2-E	3	Std 1-1/2" pipe	1.25"
M3	2	Guided Bend Bracket	
M3-A	4	1/2" plate	6.5"
M3-B	2	1-1/2" wrist pin	2"
M3-C	2	1/2" x 1-3/4" sq. tubing	2"
M4	2	Pivot Point Assembly	
M4-A	2	11 gauge 1-3/4" sq. tubing	10.375"
M4-B	2	1" round stock	
M5	2	Height Adjustment Assembly	
M5-A	2	11 Gauge 1-3/4" sq. tubing	6.5"
M5-B	2	1/2" x 1-1/2" plate	1.5"
M5-C	2	5/8" bolt	4"
M5-D	2	5/8" nut	

Introduction

This bottle jack driven guided bend tester, designed to bend straps taken from certification tests, was fabricated for a welding inspector. The design and fabrication were labor-intensive, but the cost of materials was minimal.

Project Description

Make a template to use with an optical tracer torch, and use a milling machine to fabricate Mark 3-C. Cut the four pieces required for Mark 3-A, using the optical tracer torch. After cutting the diesel wrist pins with the chop saw for Mark 3-B, weld M3-A to M3-B, using a stainless alloy stick electrode (WPS #3), and weld M3-A to M3-C using an E71T-1 gas shielded flux core wire (WPS #1 and #4). This completes the guided bend brackets.

To support the guided bend brackets use 11-gauge, 1-3/4" square tubing. One piece of tubing can be used to support a pivot point (M4-A), and the other for height adjustment purposes (M5-A). For the actual pivot point, drill M4-A to accommodate M4-B, a piece of 1" round stock drilled through with a 5/8" drill bit on a machine lathe. Make the height adjustment assembly from the same square tubing and a piece of 1/2" plate ground down to size to make an inset cap for the tubing (M5-B). Join the tubing and plate with a 1/8" beveled groove weld (WPS #4) and grind it flush, then tap for the 5/8" height adjustment bolt.

For a base to mount the tubing on, use a section of C6 x 10.5" channel iron (M1-A). Cap the ends of the channel with a 5/16" beveled groove weld and grind it flush (WPS #4). For extra support, add a section of 5/16" plate to span from one end cap to the other (M1-B) (WPS #1). Next, center the bottle jack on the piece of channel iron and tap holes for fastening the jack to the base.

Before welding the sections of tubing to the base, first lay out and scribe all of the necessary dimensions to ensure that the tubing will be in the correct positions. Recheck all dimensions for accuracy. Now, tack weld the pieces of tubing in place. Make sure that the tubing is perpendicular to the base, and that each section of tubing is parallel to the other, before welding. To join the square tubing to the base, use a 1/4" fillet weld with no joint preparation (WPS #2). To keep the pieces of tubing parallel after repeated use of the bend tester, add a section of 5/16" plate between the two pieces of tubing on either side of the jack (WPS #2).

Finally, fabricate the plungers that provide the appropriate bend radius for the straps pulled from the certification test. Because the brittleness of steel increases with higher carbon content, a larger bend radius must be provided when testing higher grades of steel. This means that for the device to be versatile enough to test different grades of steel, it needs three separate plungers as shown by Mark 2 in the Bill of Materials. This also explains why the bend brackets must be able to adjust in and out from the jack to provide adequate clearance for the 3/8" straps. The thickness of the straps, as well as the clearance needed for testing, will not change, only the bend radius will change.

When fabrication is complete, sand and blast all surfaces to prepare for painting. Apply primer and finish paint to the completed guided bend tester.

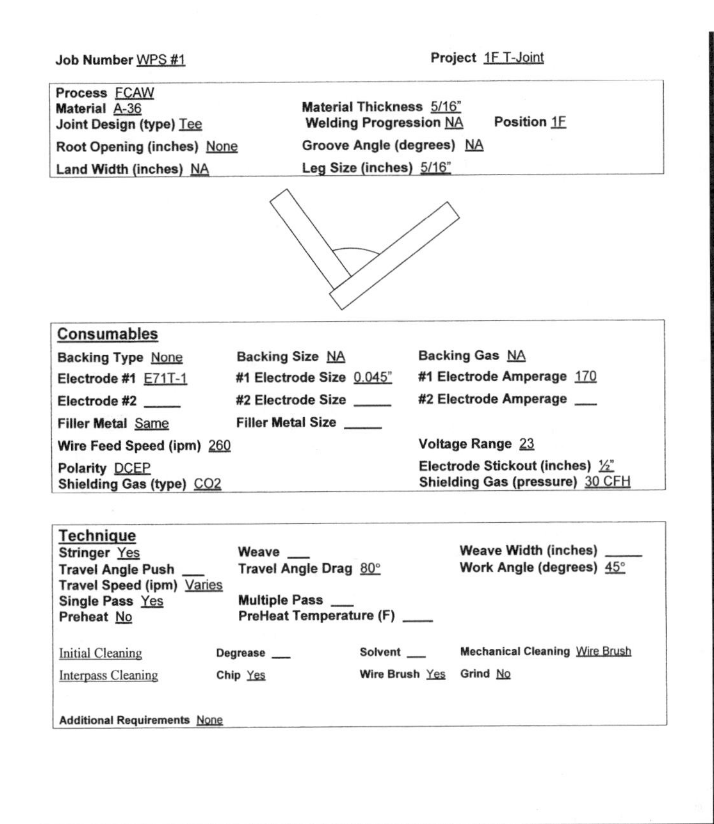

Job Number WPS #1 — Project 1F T-Joint

Process FCAW
Material A-36 — Material Thickness 5/16"
Joint Design (type) Tee — Welding Progression NA — Position 1F
Root Opening (inches) None — Groove Angle (degrees) NA
Land Width (inches) NA — Leg Size (inches) 5/16"

Consumables
Backing Type None — Backing Size NA — Backing Gas NA
Electrode #1 E71T-1 — #1 Electrode Size 0.045" — #1 Electrode Amperage 170
Electrode #2 ___ — #2 Electrode Size ___ — #2 Electrode Amperage ___
Filler Metal Same — Filler Metal Size ___
Wire Feed Speed (ipm) 260 — Voltage Range 23
Polarity DCEP — Electrode Stickout (inches) ½"
Shielding Gas (type) CO2 — Shielding Gas (pressure) 30 CFH

Technique
Stringer Yes — Weave ___ — Weave Width (inches) ___
Travel Angle Push ___ — Travel Angle Drag 80° — Work Angle (degrees) 45°
Travel Speed (ipm) Varies
Single Pass Yes — Multiple Pass ___
Preheat No — PreHeat Temperature (F) ___

Initial Cleaning — Degrease ___ — Solvent ___ — Mechanical Cleaning Wire Brush
Interpass Cleaning — Chip Yes — Wire Brush Yes — Grind No

Additional Requirements None

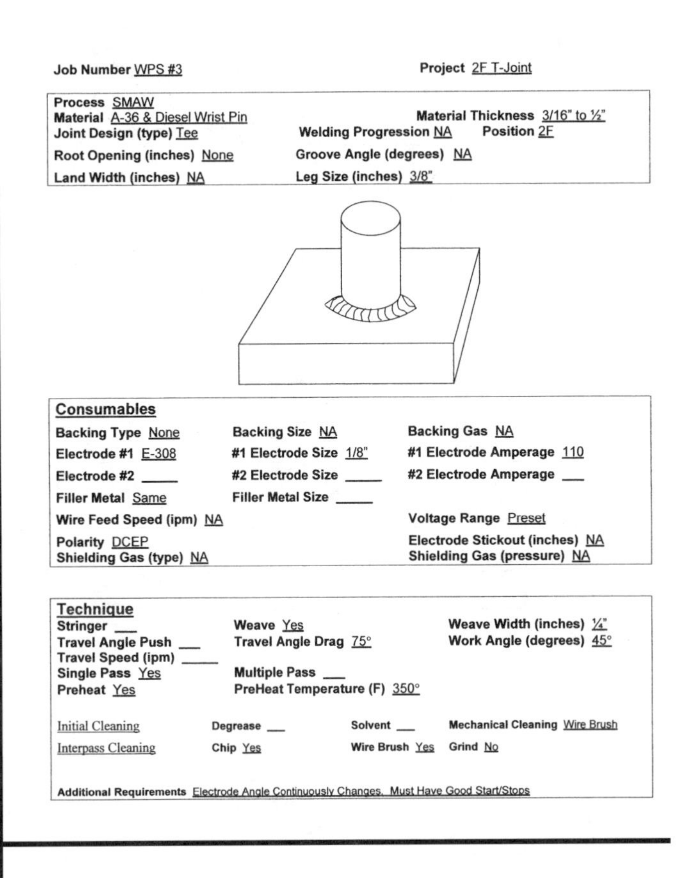

Job Number WPS #3 — Project 2F T-Joint

Process SMAW
Material A-36 & Diesel Wrist Pin — Material Thickness 3/16" to ½"
Joint Design (type) Tee — Welding Progression NA — Position 2F
Root Opening (inches) None — Groove Angle (degrees) NA
Land Width (inches) NA — Leg Size (inches) 3/8"

Consumables
Backing Type None — Backing Size NA — Backing Gas NA
Electrode #1 E-308 — #1 Electrode Size 1/8" — #1 Electrode Amperage 110
Electrode #2 ___ — #2 Electrode Size ___ — #2 Electrode Amperage ___
Filler Metal Same — Filler Metal Size ___
Wire Feed Speed (ipm) NA — Voltage Range Preset
Polarity DCEP — Electrode Stickout (inches) NA
Shielding Gas (type) NA — Shielding Gas (pressure) NA

Technique
Stringer ___ — Weave Yes — Weave Width (inches) ¼"
Travel Angle Push ___ — Travel Angle Drag 75° — Work Angle (degrees) 45°
Travel Speed (ipm) ___
Single Pass Yes — Multiple Pass ___
Preheat Yes — PreHeat Temperature (F) 350°

Initial Cleaning — Degrease ___ — Solvent ___ — Mechanical Cleaning Wire Brush
Interpass Cleaning — Chip Yes — Wire Brush Yes — Grind No

Additional Requirements Electrode Angle Continuously Changes. Must Have Good Start/Stops

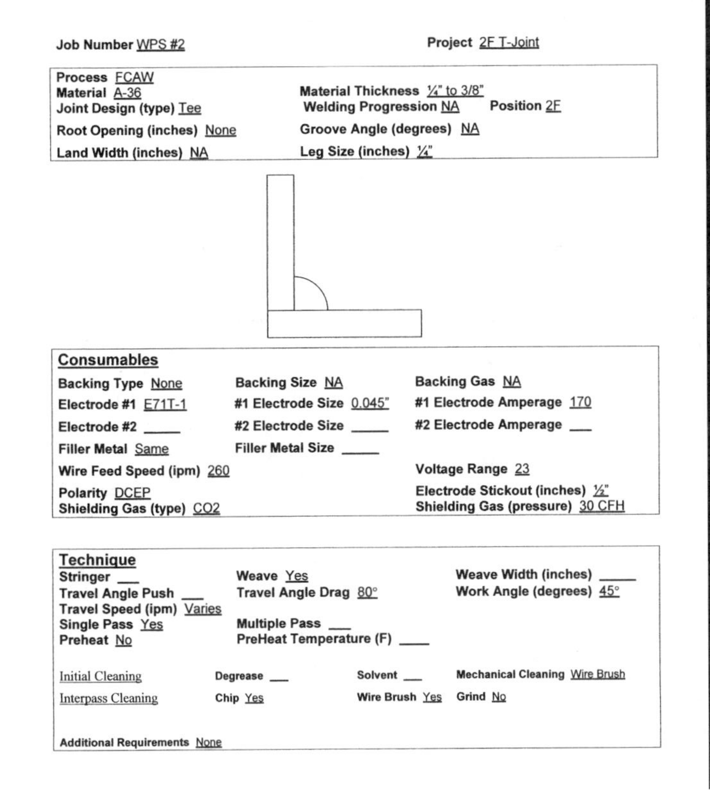

Job Number WPS #2 — Project 2F T-Joint

Process FCAW
Material A-36 — Material Thickness ¼" to 3/8"
Joint Design (type) Tee — Welding Progression NA — Position 2F
Root Opening (inches) None — Groove Angle (degrees) NA
Land Width (inches) NA — Leg Size (inches) ¼"

Consumables
Backing Type None — Backing Size NA — Backing Gas NA
Electrode #1 E71T-1 — #1 Electrode Size 0.045" — #1 Electrode Amperage 170
Electrode #2 ___ — #2 Electrode Size ___ — #2 Electrode Amperage ___
Filler Metal Same — Filler Metal Size ___
Wire Feed Speed (ipm) 260 — Voltage Range 23
Polarity DCEP — Electrode Stickout (inches) ½"
Shielding Gas (type) CO2 — Shielding Gas (pressure) 30 CFH

Technique
Stringer ___ — Weave Yes — Weave Width (inches) ___
Travel Angle Push ___ — Travel Angle Drag 80° — Work Angle (degrees) 45°
Travel Speed (ipm) Varies
Single Pass Yes — Multiple Pass ___
Preheat No — PreHeat Temperature (F) ___

Initial Cleaning — Degrease ___ — Solvent ___ — Mechanical Cleaning Wire Brush
Interpass Cleaning — Chip Yes — Wire Brush Yes — Grind No

Additional Requirements None

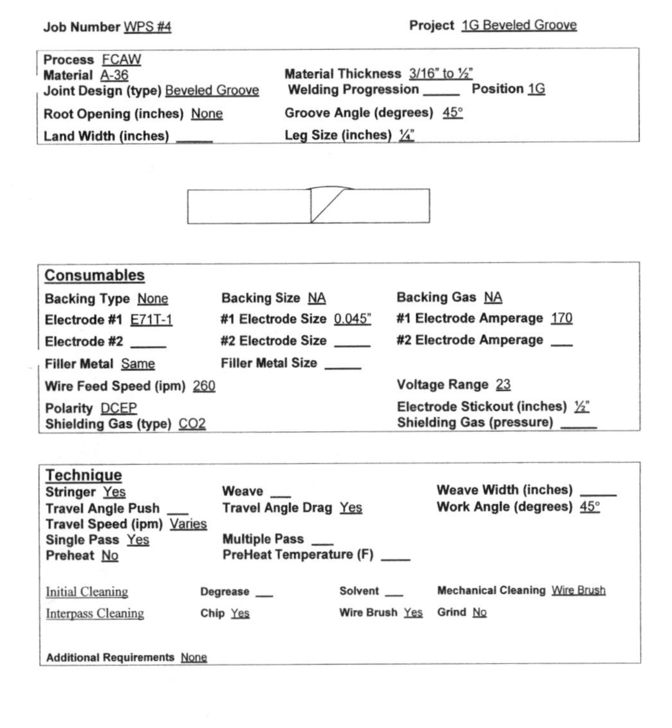

Job Number WPS #4 — Project 1G Beveled Groove

Process FCAW
Material A-36 — Material Thickness 3/16" to ½"
Joint Design (type) Beveled Groove — Welding Progression ___ — Position 1G
Root Opening (inches) None — Groove Angle (degrees) 45°
Land Width (inches) ___ — Leg Size (inches) ¼"

Consumables
Backing Type None — Backing Size NA — Backing Gas NA
Electrode #1 E71T-1 — #1 Electrode Size 0.045" — #1 Electrode Amperage 170
Electrode #2 ___ — #2 Electrode Size ___ — #2 Electrode Amperage ___
Filler Metal Same — Filler Metal Size ___
Wire Feed Speed (ipm) 260 — Voltage Range 23
Polarity DCEP — Electrode Stickout (inches) ½"
Shielding Gas (type) CO2 — Shielding Gas (pressure) ___

Technique
Stringer Yes — Weave ___ — Weave Width (inches) ___
Travel Angle Push ___ — Travel Angle Drag Yes — Work Angle (degrees) 45°
Travel Speed (ipm) Varies
Single Pass Yes — Multiple Pass ___
Preheat No — PreHeat Temperature (F) ___

Initial Cleaning — Degrease ___ — Solvent ___ — Mechanical Cleaning Wire Brush
Interpass Cleaning — Chip Yes — Wire Brush Yes — Grind No

Additional Requirements None

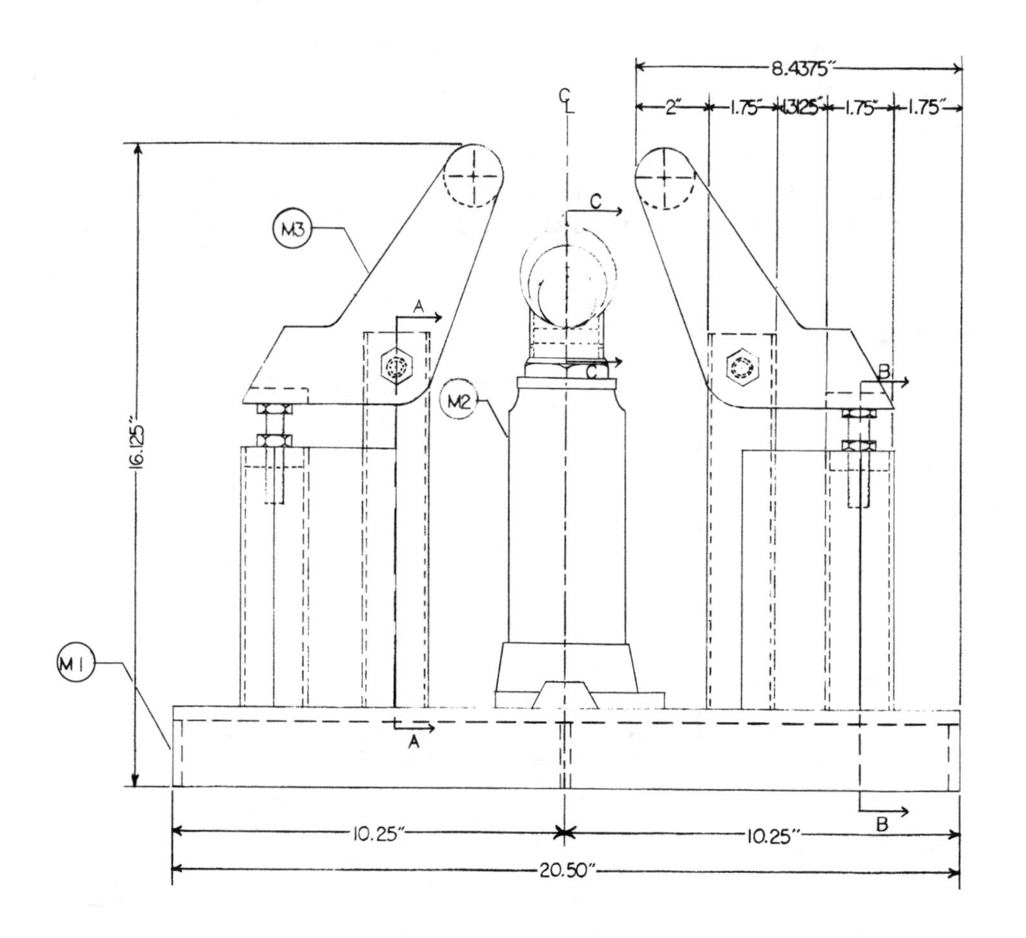
8.4375"
2"
1.75"
.3125"
1.75"
1.75"
CL
C
M3
A
M2
B
16.125"
M1
A
B
10.25"
10.25"
20.50"

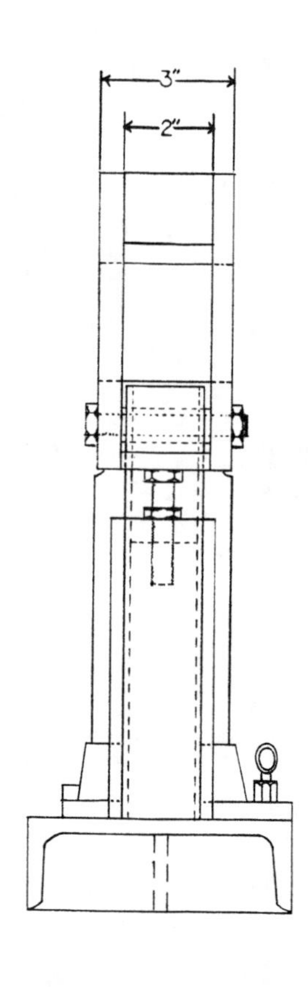
3"
2"

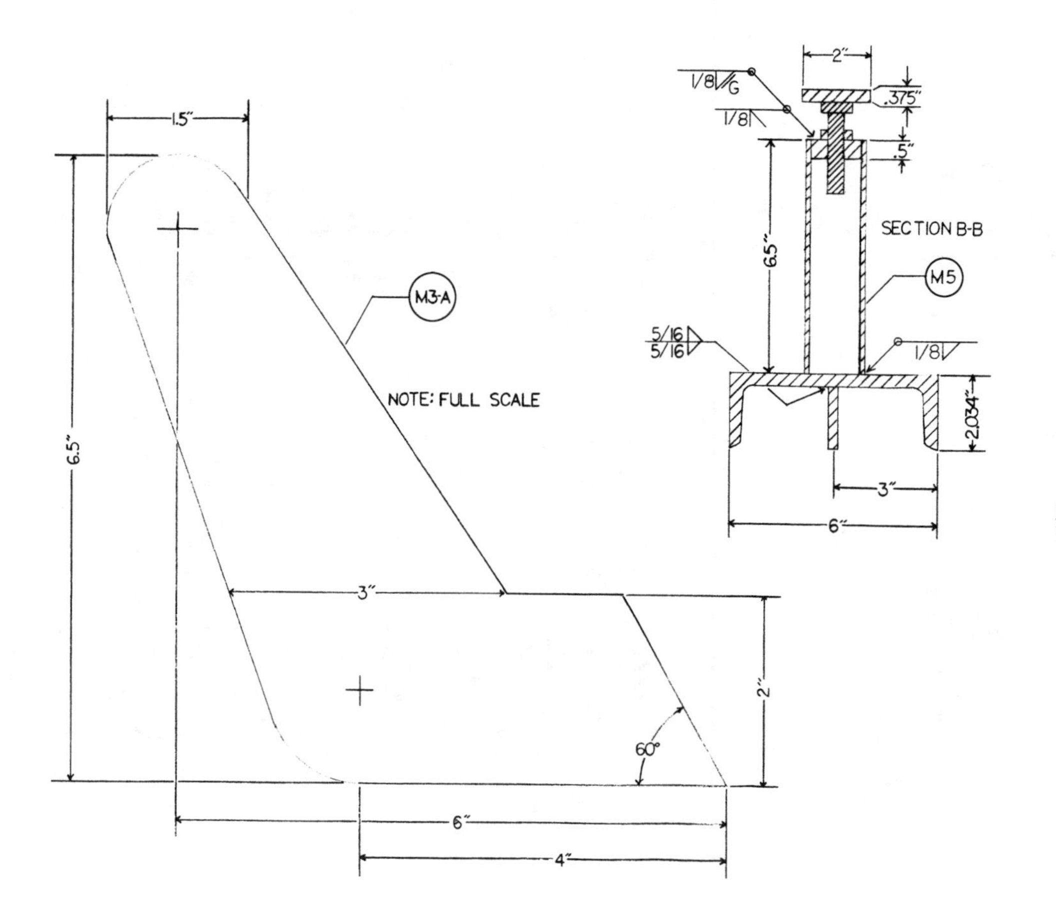
1.5"
M3-A
NOTE: FULL SCALE
6.5"
3"
2"
60°
6"
4"
2"
.375"
1/8 G
1/8
.5"
6.5"
SECTION B-B
M5
5/16
5/16
1/8
2.034"
3"
6"

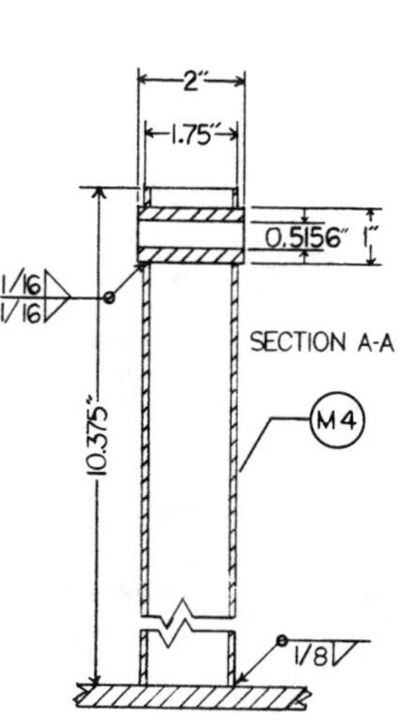
2"
1.75"
0.5156"
1"
1/16
1/16
SECTION A-A
10.375"
M4
1/8

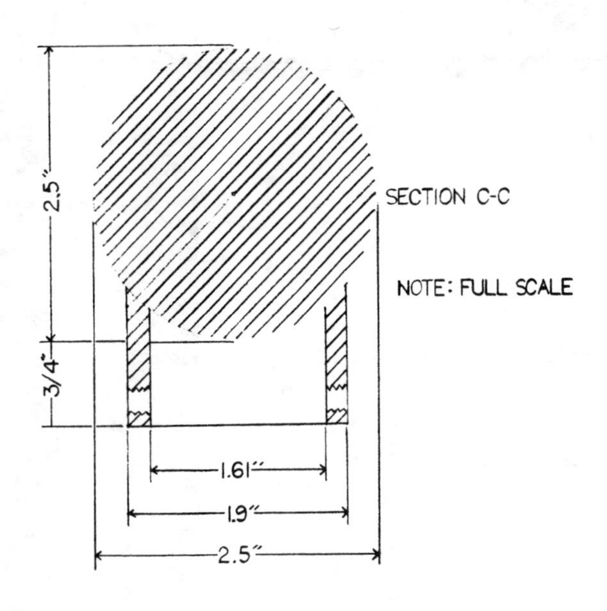
2.5"
3/4"
SECTION C-C
NOTE: FULL SCALE
1.61"
1.9"
2.5"

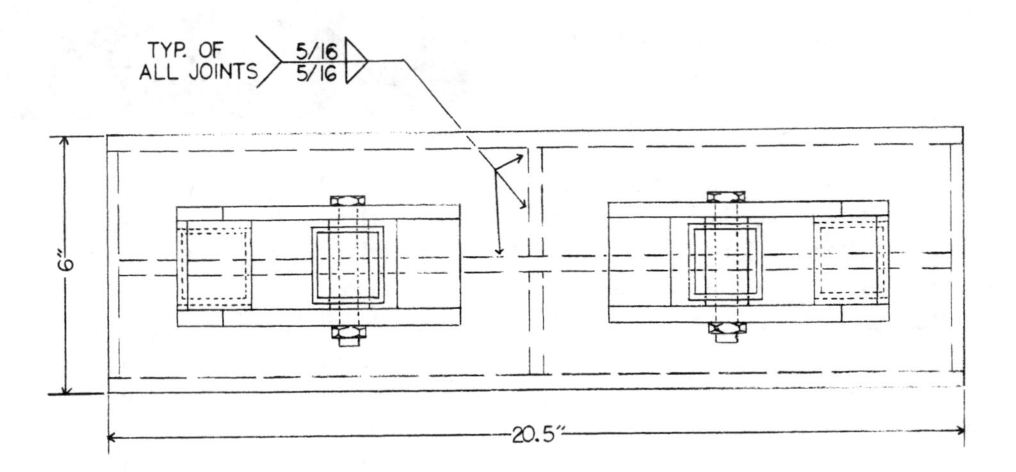
TYP. OF
ALL JOINTS
5/16
5/16
6"
20.5"

Oil Filter Crusher

Author: Jared Schmid
Instructor: Dan Saunders
School: Monroe High School
City & State: Monroe, WI

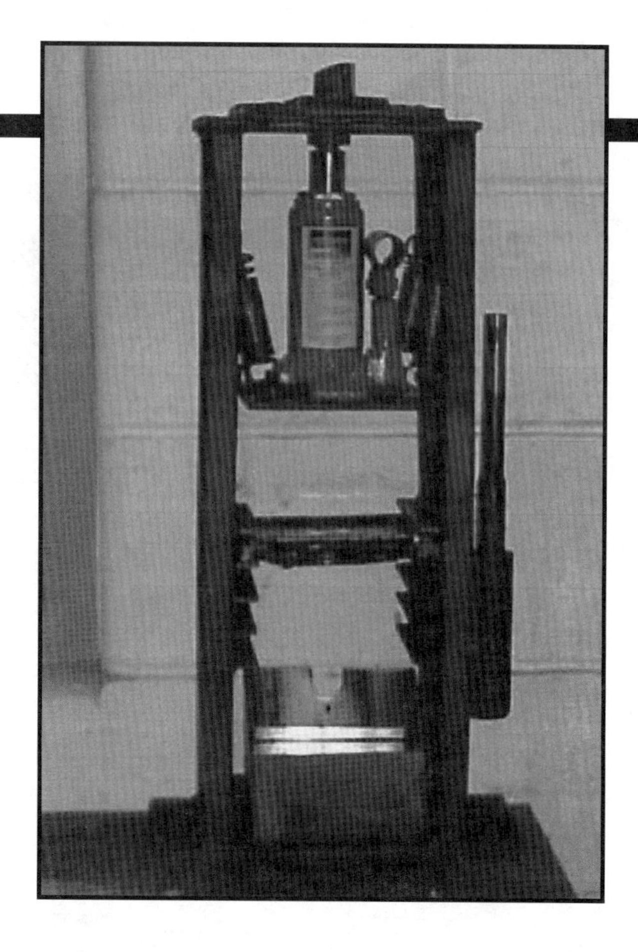

Introduction

In an automotive shop, used oil filters can take up a lot of space in garbage receptacles. An oil filter crusher works like a heavy duty pop can crusher. It aids the environment by removing the excess oil trapped inside the filters, and by saving landfill space. This crusher reduces filters to about one quarter of their original size.

Project Description

To fabricate the sides of the crusher, begin by cutting four pieces of 1" square tubing to 22" long. Next, cut ten pieces of 8" x 1" x 3/16" angle iron. Notch 1" out of each end so that they will fit between the square tubes. This also gives extra places to add welds for strength.

Weld the first piece on 6" from the bottom on both sides. Stack the rest on above that one, which will leave a 4" gap between the top piece and the bottom of the jack plate. This will allow crushing of the smallest filters on the top level; moving the plate all the way down to the bottom level will allow room for very large filters.

Make the base from a 12" x 8" x 1/4" piece of plate, which can be cut using a plasma cutter. Drill six holes, three on each end, to allow the base to be secured to the workbench. Weld the sides to the base, 8" apart.

To make the top, begin by making an 8-1/2" frame from two 8-1/2" pieces and two 5-1/2" pieces of flat bar, 1-1/2" wide. Cut an 8" x 8" x 1/4" piece of plate. On top of that plate, add two pieces of 1-1/2" x 3/16" angle iron for added strength.

For the plate used to crush the filters, cut a 6" x 6" x 1/2" piece of plate, using the plasma cutter. Using a 1/4" drill bit, drill nine holes in the plate, 1" apart, to allow the oil to drain out properly. On the top side of this plate, weld 1/4" steel rods all the way around the edge to keep the oil from spilling over the sides.

Underneath the plate, weld the same 1/4" rods in a circle with a 4" diameter to keep the oil from running to the sides of the plate and missing the drip tray.

Make the drip tray from 10-gauge steel. Measuring 5-1/2" wide and 7" long, it will be 3" high on one end and 4-1/2" high on the other end. On the tall end, create a place to allow the oil to be easily poured out. This can be done by punching a 1" hole on the ironworker, 1/2" from the edge in the center. Use a grinder to grind the edges of the hole smooth.

Use a 4-ton bottle jack as the crushing power. Attach springs to the jack so it can be raised back up after crushing the filters. Cut four springs, one for each corner. If springs long enough to cover the full height of the jack are not available, 3-1/2" long springs can be used.

To make up for the shortness of the springs, use chains to cover the extra distance. An advantage to using the chains is the flexibility they provide. If the springs become stretched out a little, they can be be hooked to a higher link and they will stay tight.

To attach the springs to the jack plate and the chains to the top, cut eight 1-1/4" pieces of 7018 welding rods and bend them into a U shape. Hook the chains into them and weld each side of the U. To make the jack plate, cut a 6" piece of 6" x 1/4" flat bar. Next, cut three narrow pieces of 1" angle iron and weld them onto it so that they cradle the jack and keep it in place. To keep the jack centered against the top plate, cut a 3/4" piece of 1" steel pipe and weld it right in the center of the top plate. The top of the jack will fit into it perfectly.

For the finishing touches, cut a 4" piece of 1" pipe and weld a cap around the bottom. Weld it to the outside of the frame to have a place to put the jack handle when it is not in use. Grind any weld spatters, any sharp edges, and clean up the whole unit. Prime it and paint it for a good appearance and rust protection.

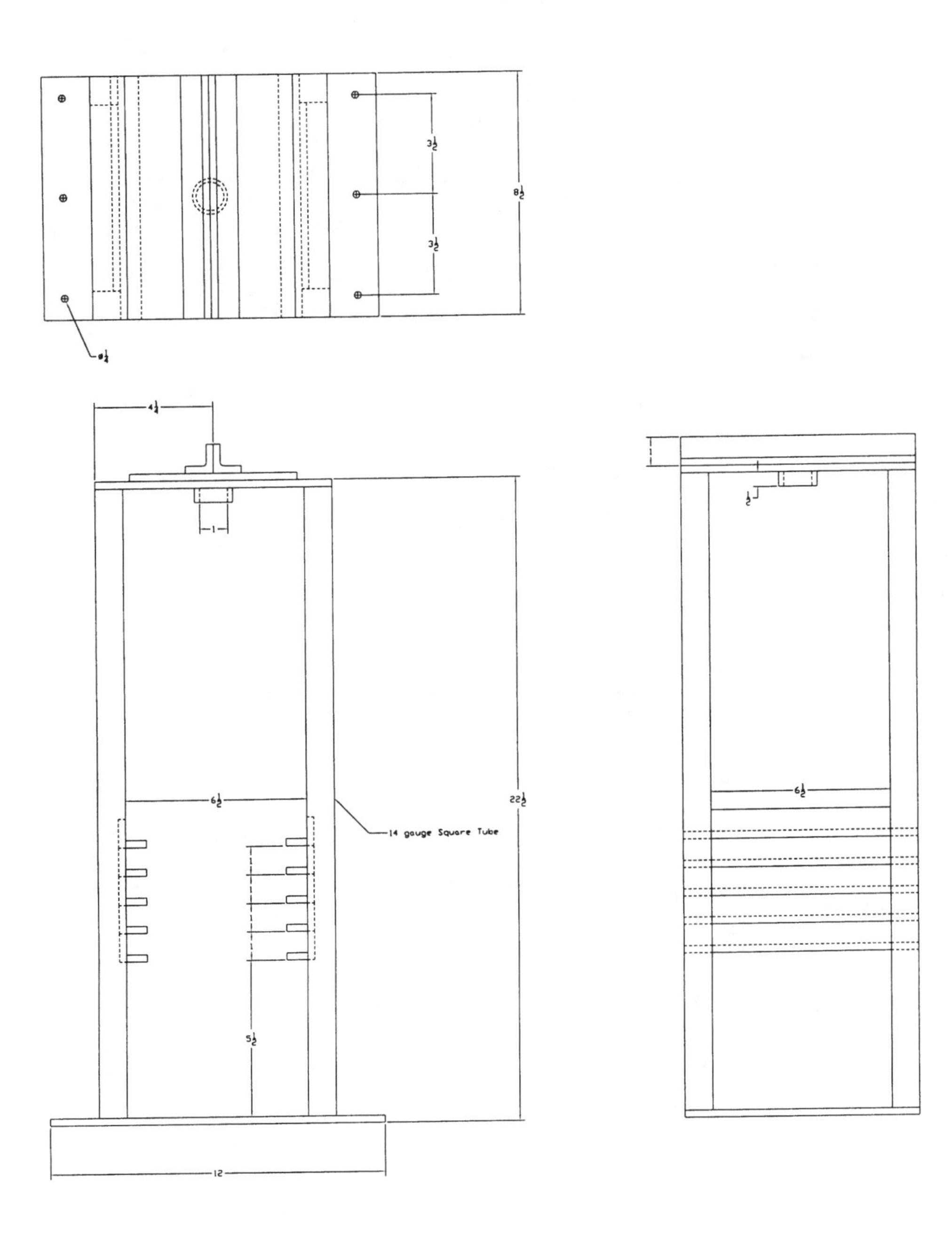

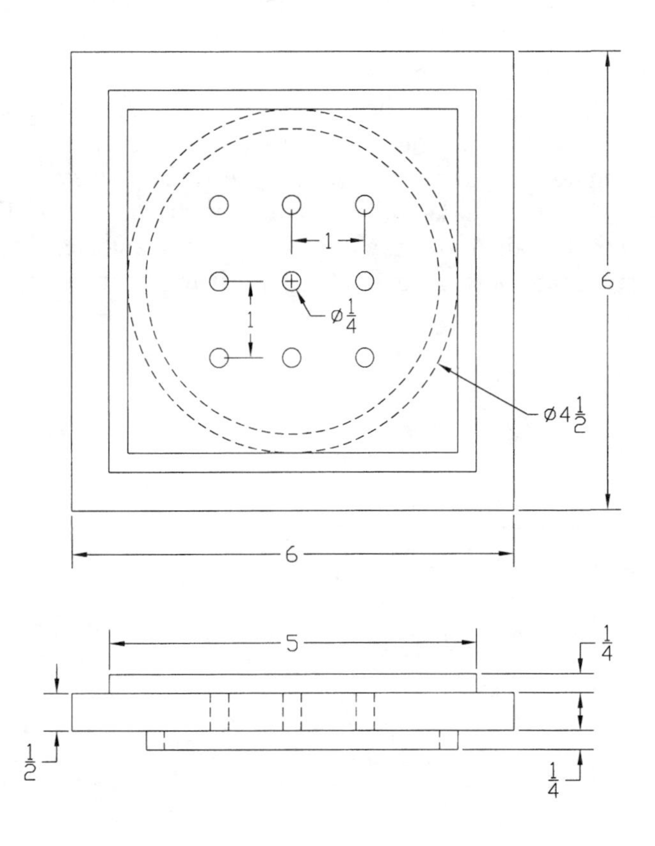
1
1
Ø1/4
Ø4 1/2
6
6
5
1/4
1/2
1/4

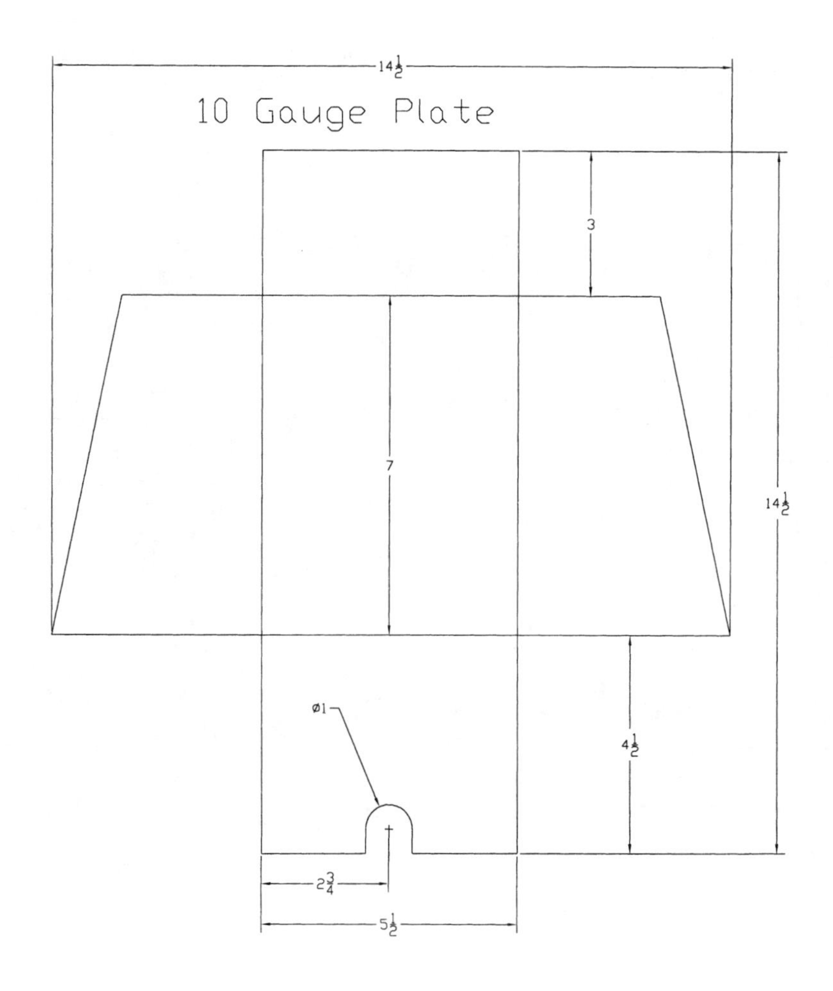
14 1/2
10 Gauge Plate
3
7
14 1/2
Ø1
4 1/2
2 3/4
5 1/2

Arc Table

Author: Timothy Horn
Instructor: Nicholas Regets
School: William D. Ford Vo-Tech Center
City & State: Westland, MI

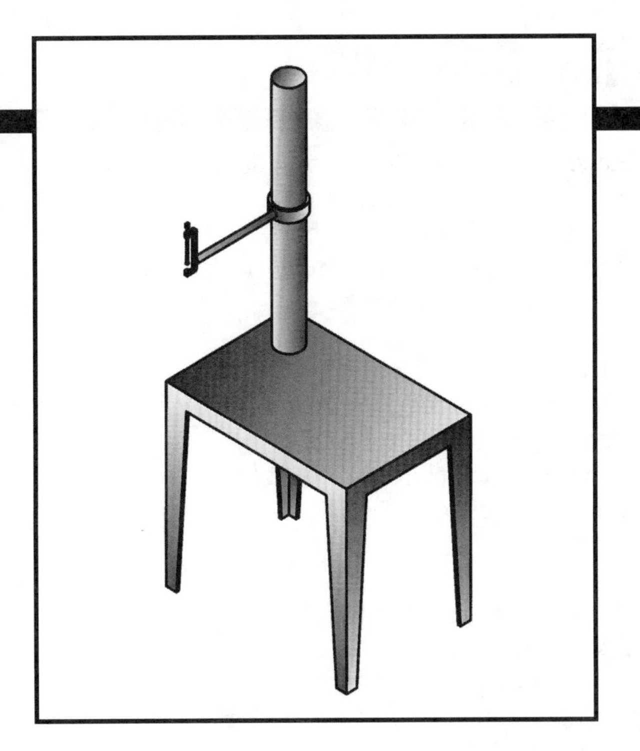

Materials List

Qty	*Dimensions*	*Description*
1	5″ x 10′ x 1/8″	steel plate
1	3″ x 10′ x 1/8″	steel plate
1	3″ x 3′8″ x 1/8″	pipe
1	2′x 3′ x 1/4″	steel plate
1	4″	clamp
1	1″ x 1′2″	pipe
1	5/8″ x 1′3″	pipe
1	5″ x 1′x 1/8″	steel plate
3		tee handles
3		tee handle nuts

Cut to Size

Qty	*Dimensions*	*Description*
4	32″ x 5″	Legs
2	2′ x 3″	Frame/Sides
2	3′ x 3″	Frame/Front & Back
1	2′ x 3′	Top
1	3′8″x 3″	Pipe
1	1″ x 1′2″	Pipe
1	5/8″ x 1′3″	Pipe (Arm)
1	3′ x 1″	Angle Iron (Inner Form)
2	11-3/4″ x 1″	Angle Iron (Inner Form)

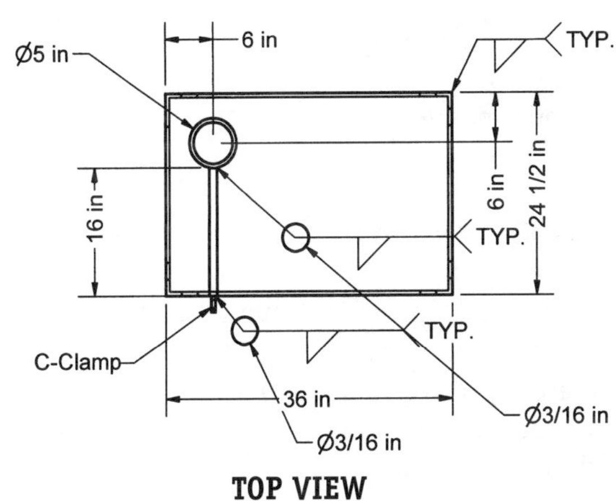

TOP VIEW

Procedure

1. Lay out metal and materials.
2. Cut and size pieces individually.
3. Cut and size inner frame pieces and outer frame pieces.
4. Deburr all edges.
5. Square all sides and tack frame together.
6. Tack in inner frame bars.
7. Square and tack on all legs.
8. Size top and tack on.
9. Weld in solid 1″ pitch welds throughout table.
10. Weld on pole and clamp with arm to top of table.
11. Grind down all welds and deburr all edges.
12. DA the table.
13. Paint black and clear coat.

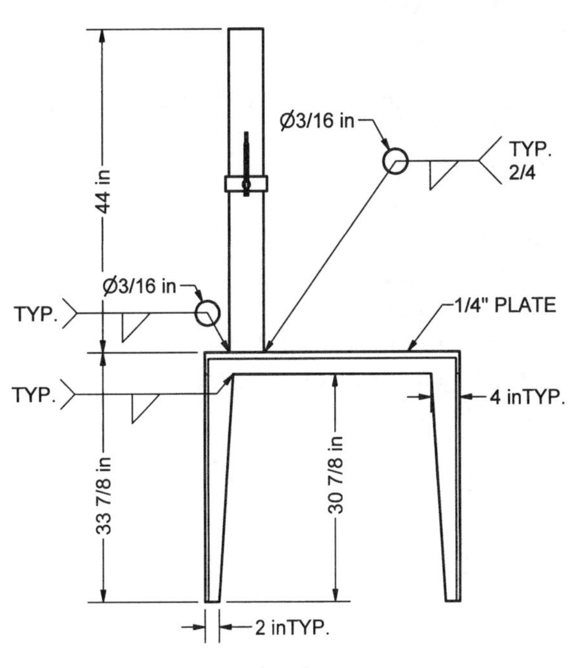

SIDE VIEW

Shop Cabinet

Authors: Brian K. Longton, John Jones, and Charles Asch
Instructor: Nicholas Regets
School: William D. Ford Vo-Tech Center
City & State: Westland, MI

Materials List

For the 36 drawers, all 14 ga. mild steel:
72 side plates, 2-1/4″ x 6-1/2″
72 front and back plates, 4″ x 6-1/2″
36 bottom plates, 4″ x 6-1/2″
36 handles, 1-1/4″ x 2″

For the Cabinet, all 12 ga. mild steel:
1 back sheet, 2 ft. 6-1/4″ x 2 ft. 1/8″
2 side sheets, 2ft. 7″ x 2 ft. 1/8″
Top and bottom sheets, 7″ x 2 ft. 1/8″
30 pieces, 3-15/16″ x 6-15/16″
5 pieces, 29-1/4″ x 7″

Black spray paint

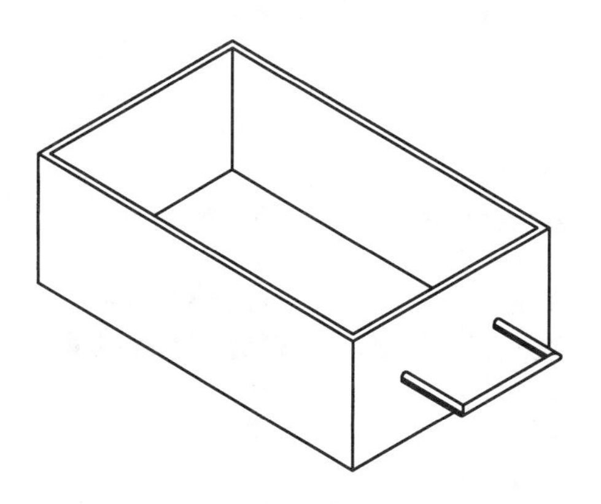

Step-by-Step Procedure

For the Drawers:

1. For the first drawer, cut metal on shear.
2. Hammer metal flat.
3. Create jig for tacking drawers.
4. Tack drawer.
5. Stitch-weld drawer.
6. Bend and cut out handle.
7. Weld handle on front of drawer.
8. Smooth out drawer with file and grinder.
9. Spray paint drawer black.
10. Repeat process except # 3 for the other 35 drawers.

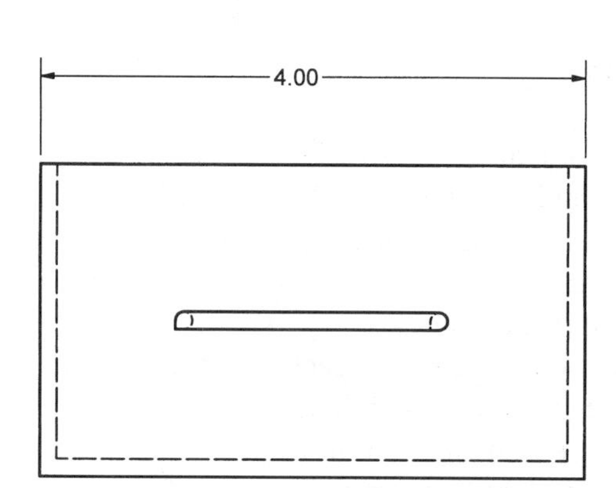

For the Cabinet:

1. Cut metal on shear.
2. Hammer metal flat.
3. Tack cabinet together.
4. Stitch-weld cabinet.
5. Grind and file cabinet smooth.
6. Spray paint cabinet black.
7. Slide drawers in.

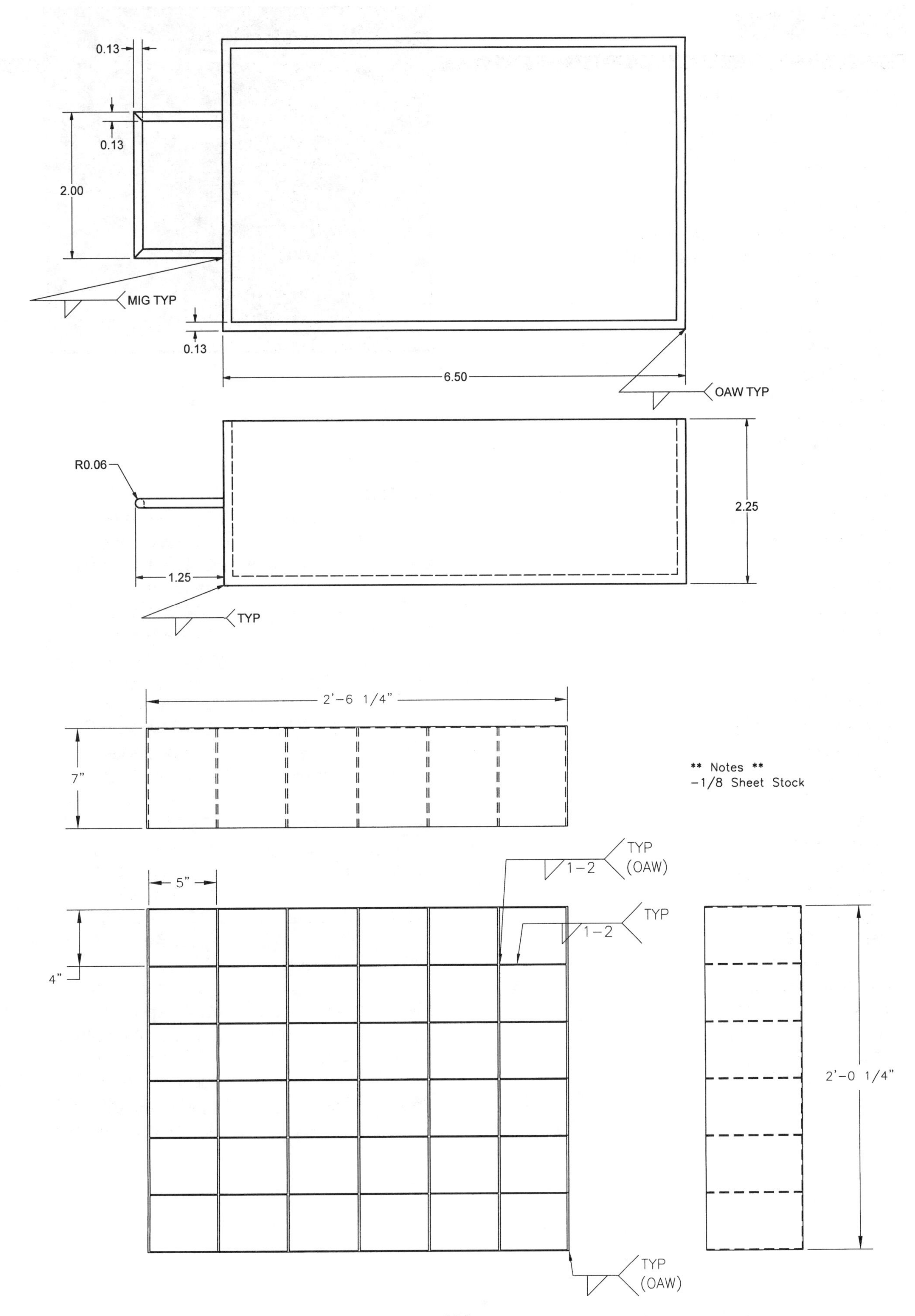
0.13
0.13
2.00
MIG TYP
0.13
6.50
OAW TYP
R0.06
2.25
1.25
TYP
2'-6 1/4"
7"
** Notes **
-1/8 Sheet Stock
TYP
(OAW)
1-2
5"
TYP
1-2
4"
2'-0 1/4"
TYP
(OAW)

Band Saw

Author: Amanda L. Furr
Instructor: Dan Froneberger
School: Winona High School
City & State: Winona, TX

Bill of Materials

- 1/4″ Diamond Plate (one sheet)
- 3/16″ Flat Plate (one-half sheet)
- Guide Blocks
- Adjustment Brackets
- Blade Wheels (2)
- Hex Jam Nut
- Clexis Block
- Descent Cylinder
- Switch Acluster
- On/Off Switch
- Clamp Lock
- Clamp Bar
- Movable Vice Jaw
- Stationary Jaw
- Nut
- Slide Plate
- Lock Plate
- Spring
- Rod
- Hand Wheel
- Motor Mount
- Compression Spring
- Gear Reducer
- Reducer Pulley
- Motor Pulley
- Wheel Axle
- Bearing Bolt
- Hex Nut
- Blade Guides (2)
- Angled Leg
- Bearings Wide (4)
- Bearings Narrow (2)
- Eccentric Bolts (2)
- Blade Guide Carriage
- Spa Care Sunk Pump
- Utility Table
- Safety Doors (2)

Cut List

Cut 8	4″ pieces	3/4″ flat stock for the corners
Cut 4	12-1/4″ pieces	3/4″ flat stock for the top and bottom sides
Cut 4	8″ pieces	3/4″ flat stock for the top and bottom back and front
Cut 2	10″ pieces	3/4″ flat stock for the supports
Cut 2	11-1/2″ pieces	3/4″ flat stock for the bracket
Cut 2	6″ pieces	3/4″ flat stock for handle supports
Cut 2	15″ pieces	3/4″ flat stock for wheels
Cut 1	8″ piece	3/8″ round stock for bar handle
Cut 10	2″ pieces	3/8″ round stock for spokes
Cut 4	4″ pieces	3/4″ flat stock for bottom supports
Shear	8x12	1/4″ expandable metal for the bottom
Shear	4x12	1/4″ expandable metal for the sides
Shear	4x8	1/4″ expandable metal for the front and back

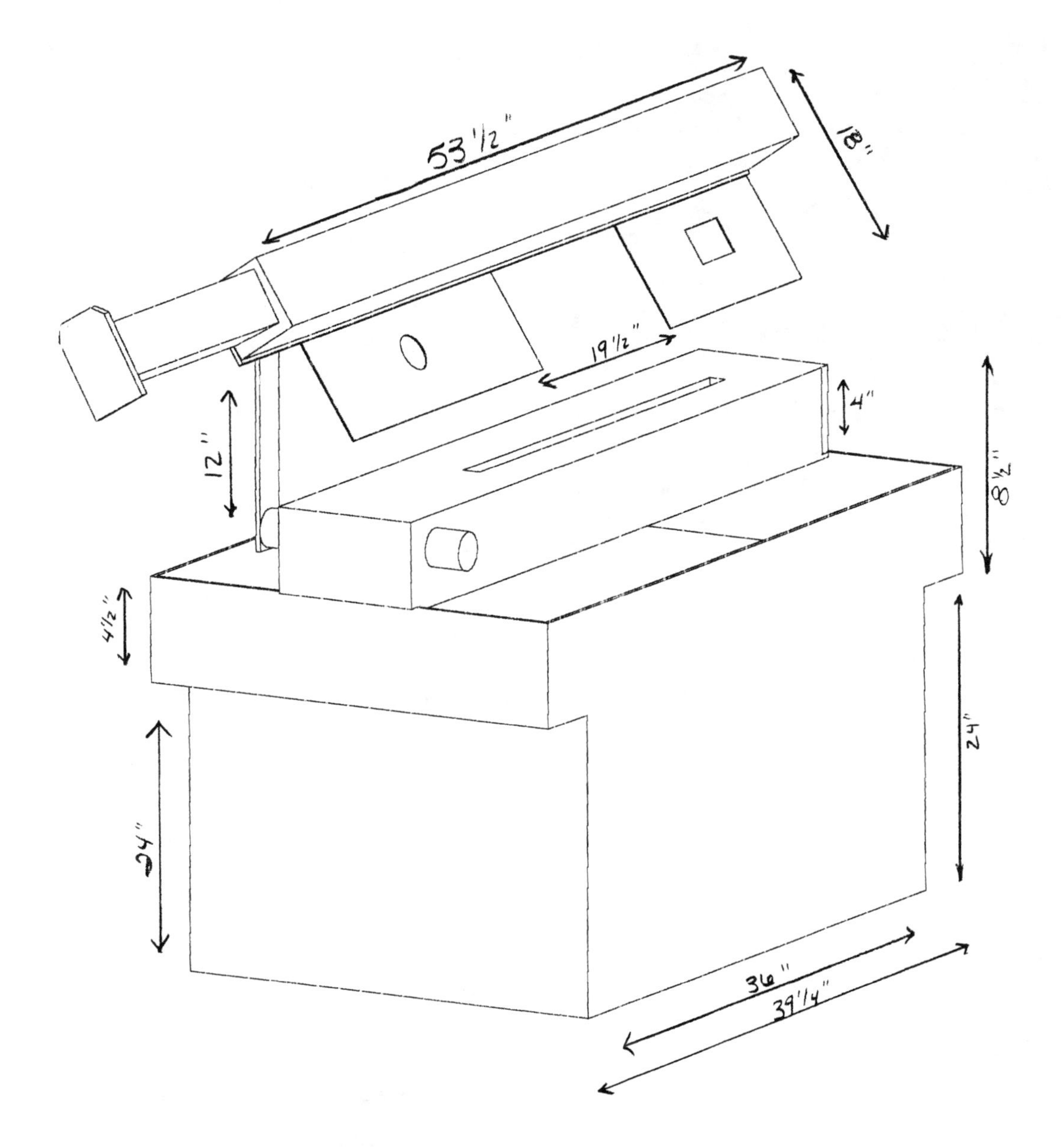

Procedure

1. Weld the base together, tacking the pan together and then welding the pan to the rectangular base.

2. Cut the metal that will be used to hold up the cutting base from the pan.

3. Weld the metal that will be used to hold up the cutting base (and used as a side to the cutting base) to the cutting base.

4. Grind the welds down to make the edge flat, straight and clean.

5. Following the grinding, place the metal evenly into the pan almost 5 inches from each side. Tack it into place so it won't move.

6. Weld the cutting base into the cutting pan.

7. Cut 1/4" plate, one 16" x 20" and the other 14" x 16", which will be used to hold the wheels.

8. Pad-grind down the edges, and then cut a circle out of the square.

9. Weld the angle iron on the arm bolted to the base.

10. Punch holes for the roller (blade support).

11. Cut holes for the top pulley and the bottom pump.

12. Build a foot on the cutting arm, for support, and for the arm to rest in an upright position.

13. Drill holes and weld bolts into the arm to allow tightening the foot.

14. Bolt on the braces to hold the metal in place.

15. Drill holes and add a stopper to prevent the arm from falling all the way down and hitting the base.

16. Add the cutting support (brace) to the cutting pan.

17. Drill a hole for the shaft on the side.

18. Put the turning shaft under the cutting pan and connect the wheel to turn the shaft.

19. Drill holes for automatic cut-off switch.

20. Make the lever for the switch to hit the automatic switch to stop the machine.

21. Mount the Blade wheels.

22. Measure and cut out 2 pieces of metal for the spring support.

23. Weld on supports for the spring; one to the arm, and the other to the base.

24. Add the braces for the motor onto the arm.

25. Mount motor to the braces and then support it.

26. Take a bolt and cut off the head.

27. Taking the two pieces of metal, make a "T". Weld them together to make a handle for the blade tension.

28. Take another bolt and cut off the head.

29. Taking the 2 pieces of metal, make a "T" by welding them together to make a handle for the rollers (width of metal).

30. Put on pulleys and belt on motor and pump.

31. Drill holes through wheel and rotating shaft to move braces that support the metal.

32. Cut open the end of the arm.

33. Add scrap metal to the end, adding weight.

34. Measure the water tube holder for the blade.

35. Make the safety doors.

36. Make and add blade guides: one with a hole for the water tubing (to keep blade cool).

37. Pad grind the edges and brush grind everything down.

38. Disassemble the band saw arm.

39. String up the parts of the band saw to be painted.

40. Paint the band saw with primer.

41. Paint the band saw parts (wheels, braces, and other little parts that connect on the arm and pan).

42. After the paint dries, reassemble the band saw.

43. Place the pump in the bottom of the water pan, hook up the 3/8″ clear water tubing to the connectors and pump.

44. Place the blade on the saw and the stickers on the arm.

45. Connect the protective doors over the wheels.

46. Put foam on the edges of the door to prevent scraping the paint.

47. Connect the wheel. Put the pin through it to hold it on.

48. Draw the plans for the vertical cutting table.

49. Cut it out with the plasma cutter.

50. Weld it to the stand.

51. Brace it in the arm connector.

52. The Band Saw is finished and ready to operate.

Home & Recreation

Table of Contents

Relief Art Mural

Author: James Cuhel
Instructor: David H. Murray
School: Ferris State University
City & State: Big Rapids, MI

Selecting the Material and Welding Process

The art mural is fabricated from 6061-T6 mill finished aluminum, 0.125″ thick. The material was chosen because of its welding properties and its ability to be polished. The GTAW process was chosen because the welder is able to control the weld puddle with more precision than using other semi automatic methods such as GMAW. Another welding process used is cross wire welding. The resistance welder is used to weld the 1/8″ braze rods together that are found on the bridge. To join the 1/16″ braze rod used for spokes on the bicycle, traditional oxy/acetylene and a fluxed braze rod is used.

Starting the Relief Art Mural

After the design was drawn by hand using paper and pencil, a detailed drawing was made using AutoCAD. Using AutoCAD, all the items were drawn to scale and dimensioned. After all the items were drawn using AutoCAD, the file was saved to a DXF format so the computer-controlled plasma cutter could read the AutoCAD file. The file was then transferred to the plasma cutter. Once the plasma cutter was able to read the AutoCAD file and the plasma cutter was set up, the desired parts were then ready to be cut out.

Cutting Out the Parts

To cut the 0.125″ 6061 T6 aluminum, the plasma cutter was set up with H35 shielding gas. H35 represents 35% hydrogen and the balance is argon. The plasma gas used is pure nitrogen. The pressure required for the shielding gas is 60 psi. The pressure for the plasma is 36 psi. To get the desired cut quality, the travel speed of the plasma cutter was turned down from the indicated speed. The travel speed used to cut out the pieces is 110 ipm at 100 amps.

Instead of cutting the parts out all at once, it was decided to have them cut out one at a time. This was done to help accommodate some of the smaller, more intricate parts. The large backdrop was originally going to be sheared. However, due to the size limit on the squaring shear, the backdrop had to be cut using the plasma cutting table. The backdrop measured 4′ x 5′ with a 5″ radius on the corners.

Cleaning the Parts

When all the parts were cut out, they were taken to a belt sander. All the burrs and sharp edges are taken down. A small die grinder fixed with a scotch bright pad was used on the smaller parts. This helped to reach in the tight corners. In order to have all the individual parts look similar, the edges should not be sanded. It would be impossible to reach inside some of the smaller parts and have them look similar to the other parts.

Locating Parts on the Backdrop

All the parts are placed on the backdrop after all the sharp edges and dross have been removed. This helps to see that the individual parts have been cut properly. It also helped to visualize where they would go. After placing all the parts on the backdrop, the next step was to connect the parts together. Duct tape was used to temporarily join the parts together. Temporary standoffs were made using scrap material left over from the plasma cutter. A hand brake is used to bend the small pieces of aluminum into various-sized pieces of channel. Next, the items plus the standoffs are attached to the backdrop using duct tape.

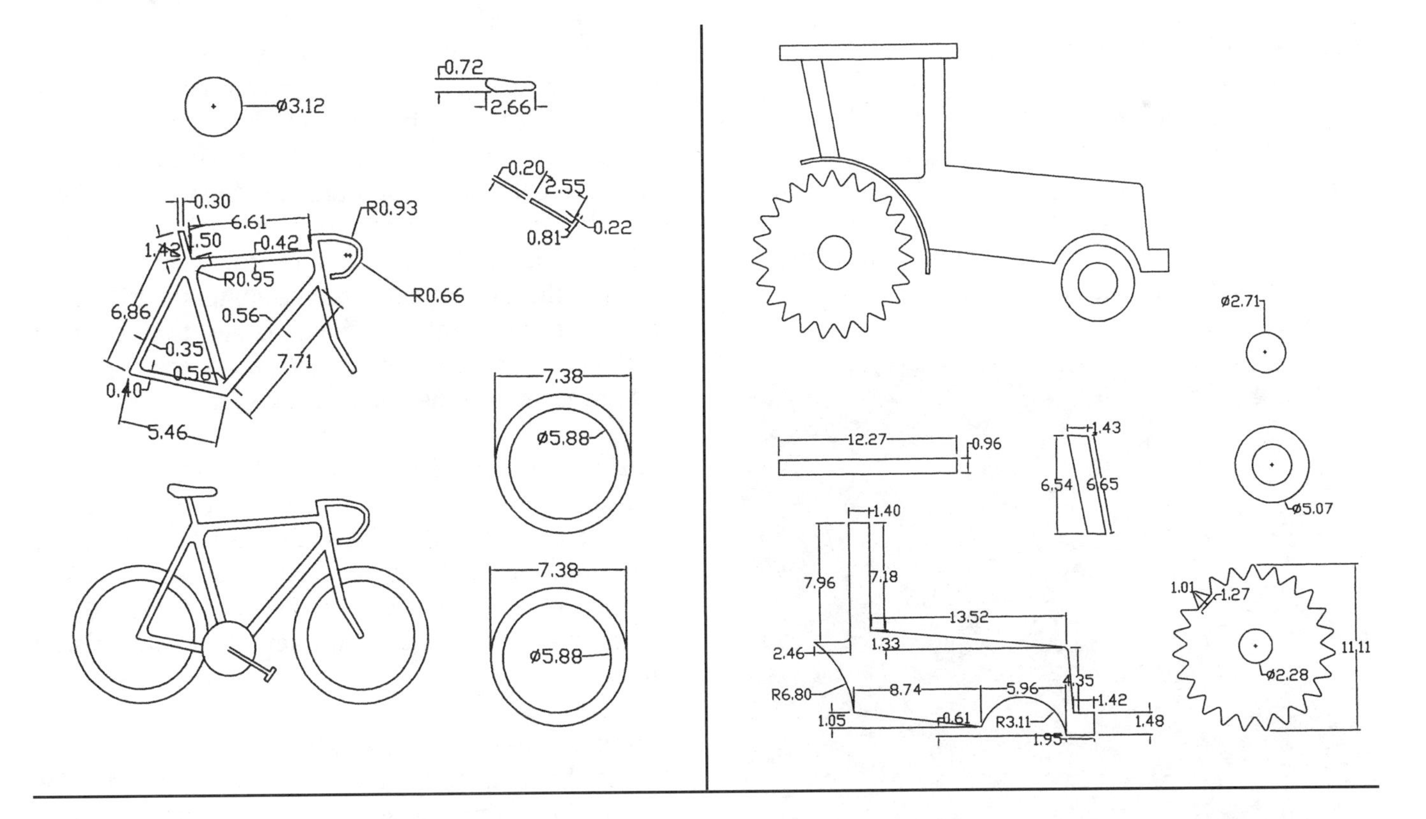

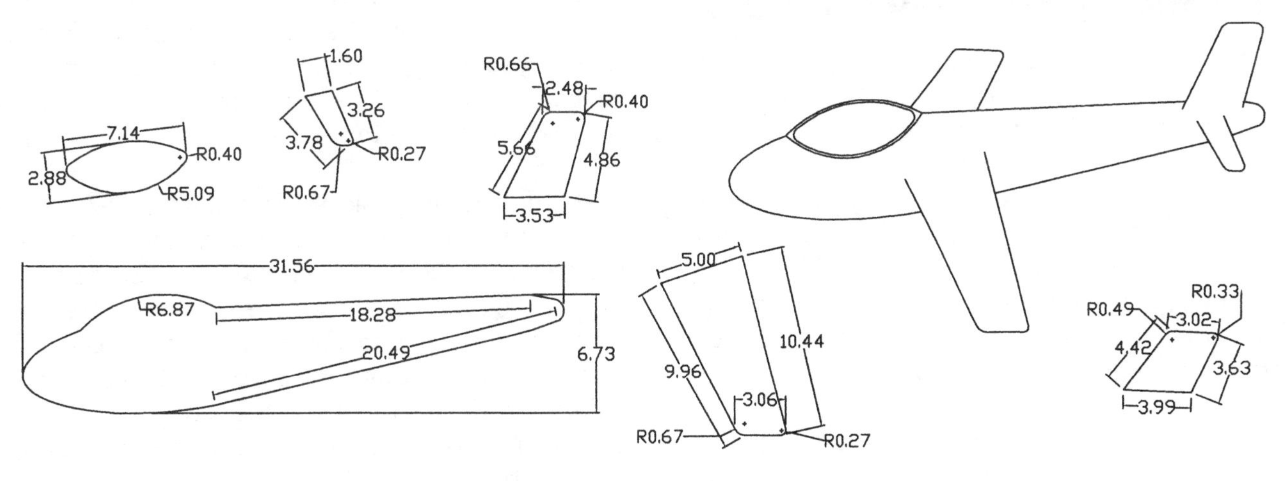

Most of the changes from the initial drawing were made when the items were temporarily attached to the backdrop. Some of the major changes included the use of brazing rods on a couple of items. The brazing rods are used on the bicycle, bride and tractor. This was done to add color. Also, adding water around the ship helped separate the ship from the tractor.

Welding the Items Together

The GTAW process is used to join the aluminum parts together. The welder used is a constant current AC/DC power source equipped with high frequency starting capabilities. The torch is a water-cooled 300 amp TIG torch equipped with a 1/8″ collet and collet body. The ceramic nozzle used is a size 6. The tungsten is a 1/8″ pure tungsten. To get the desired arc characteristics, the tungsten is sharpened to a point using a diamond grinding wheel. The angle of the sharpened tungsten is 20 degrees.

Compressed argon with a flow rate of approximately 20 SCFH is used as the shielding gas. In preparing the items to be welded, the first step is to remove the oxide on the surface of the material using a stainless steel brush. To weld the members together, ER5356 filler rod in 3/32″ size is used. The welder is set up with a max output of 230 amps. The welding machine is equipped with a foot control, allowing the operator to control the arc characteristics and limit the amount of heat put into the part.

Joining the Brazing Rods

After the smaller aluminum parts are welded, the next step is to join the brazing rods together. The spokes on the bicycle wheel are 1/16″ thick. The traditional oxy-acetylene method is used to braze the rods together. A number 0 welding tip with 5 lb of oxygen and 5 lb of acetylene is used to join them. The brazing rod used to join the spokes together is 3/32″ and coated in flux.

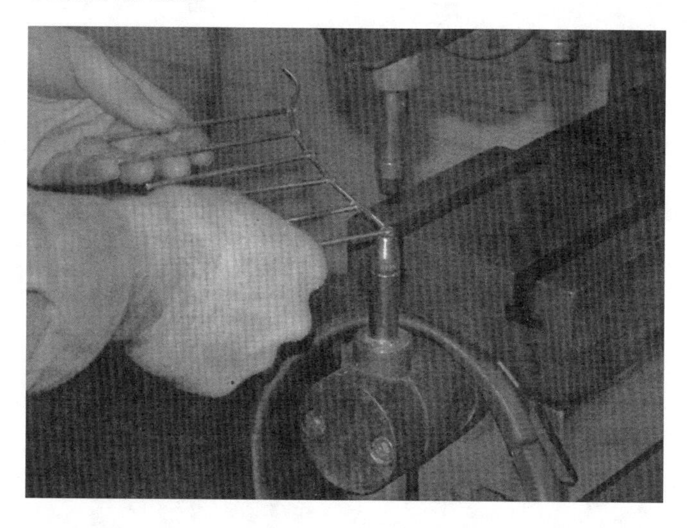

The process is called cross wire welding. A press series resistance welder is used to create the heat and pressure required to join the two rods together. The desired length of the braze rods was marked out with a pen and cut on a Beverly shear. Once the braze rods are cut to size, they are then ready to be welded. The two pieces of rod are laid on each other. Using this process saved a lot of time over the traditional oxy/acetylene method. The cross wire welding method does not require any preheating or the use of flux.

Because of all the small parts associated with the bridge, a separate table is used to construct it. This helped to see where all the small parts should go and where the best places to weld them together would be.

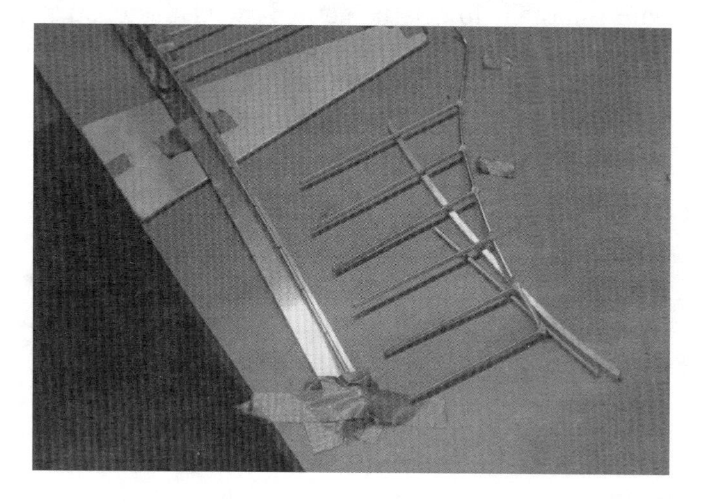

To connect the braze rods with the aluminum, a unique method is used. Pieces of 1/8″ aluminum filler rod are welded to the back of the items. The aluminum filler rod is cut so that they are about 3/4″ long. Then the pieces of aluminum filler rod are bent over the braze rod to secure it. Not all of the braze rods are joined at equal distance from each other. As a result, the spacing between the aluminum filler rods is not in identical positions. Using this type of method allowed for greater tolerances when joining the braze rods to the aluminum items.

Putting a Special Finish on the Items

Each item has its own unique surface finish. To create such finishes, Scotch Brite pads fixed to a small die grinder are used. By manipulating the speed of the die grinder and the speed of the die grinder moving across the item's surface, different types of finishes are possible.

Welding Items to the Backdrop

To join the items to the backdrop, the same method is used as in joining the smaller parts together to form the items. However, this time the backdrop has the slots drilled as opposed to the parts having the slots punched. To locate the exact location of the slots, the standoffs are welded to the back side of the items first.

Next, the backdrop had to have a finish put on the surface prior to welding. This is done because some of the items would be welded close to the backdrop. There would not be enough space between the backdrop and the items to allow for the dual action finisher. The biggest problem encountered while putting a finish on the backdrop was the water and oil in the air line. The water and oil would get on the backdrop and create smudges. It was very hard to get the smudges out of the finish. Another problem was getting the entire backdrop to look similar. Focusing the dual action finisher in one spot for too long cause the overall look to suffer.

In order for the items to be perfectly level with the surface of the backdrop, pieces of 2″ x 2″ 16 ga. square tubing were cut in different lengths. The square tubing was cut in four lengths. The lengths are 1″, 3″x 3-1/2″, and 4″. The backdrop must be laid down so that the items are facing the ceiling. This is the best way of positioning the items up to be tack welded. Unfortunately, it requires that the items be tack welded in a very difficult overhead position. The backdrop was approximately 24″ off the ground and resting on two chairs. The GTAW process is used

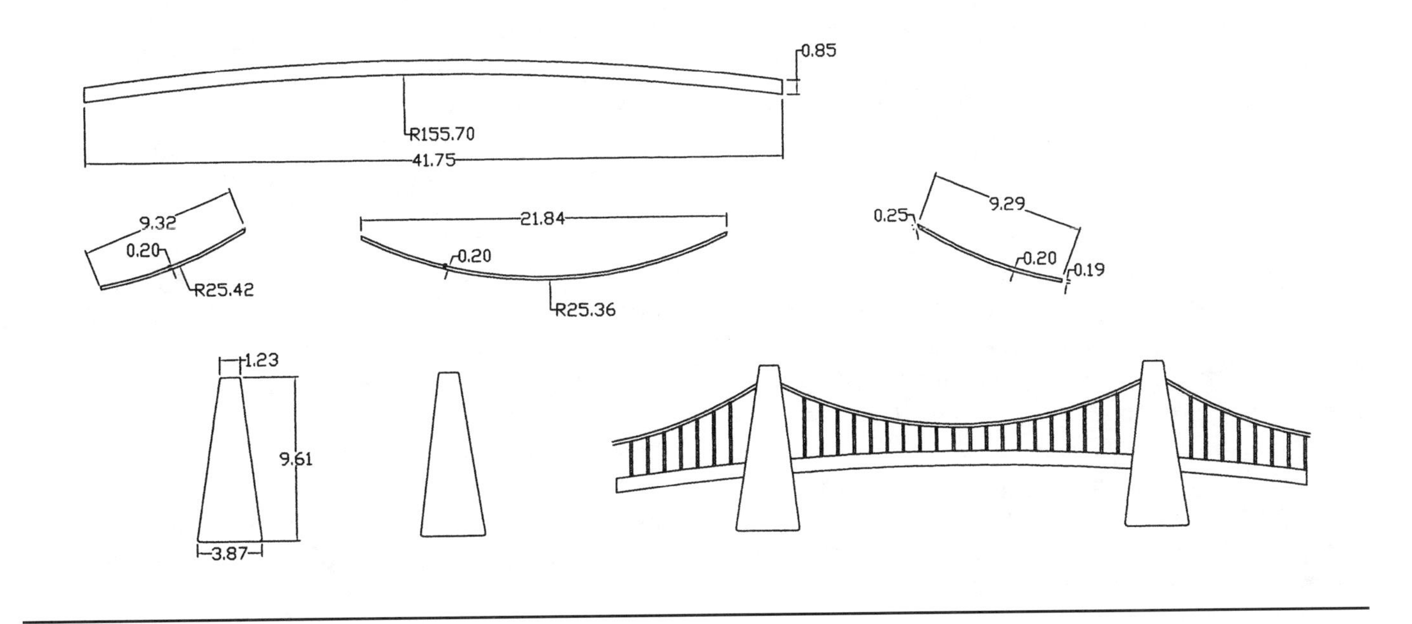
0.85
R155.70
41.75
9.32
0.20
R25.42
21.84
0.20
R25.36
0.25
9.29
0.20
0.19
1.23
9.61
3.87

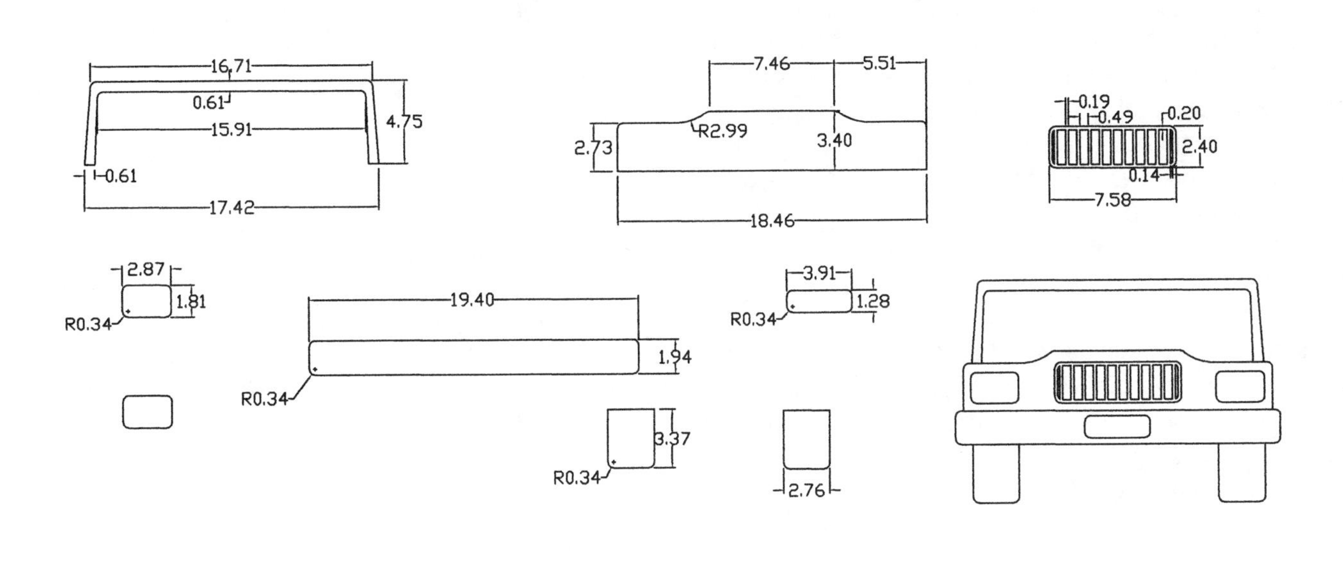
16.71
0.61
15.91
4.75
0.61
17.42
7.46
5.51
2.73
R2.99
3.40
18.46
0.19
0.49
0.20
2.40
0.14
7.58
2.87
1.81
R0.34
19.40
1.94
R0.34
3.91
1.28
R0.34
3.37
R0.34
2.76

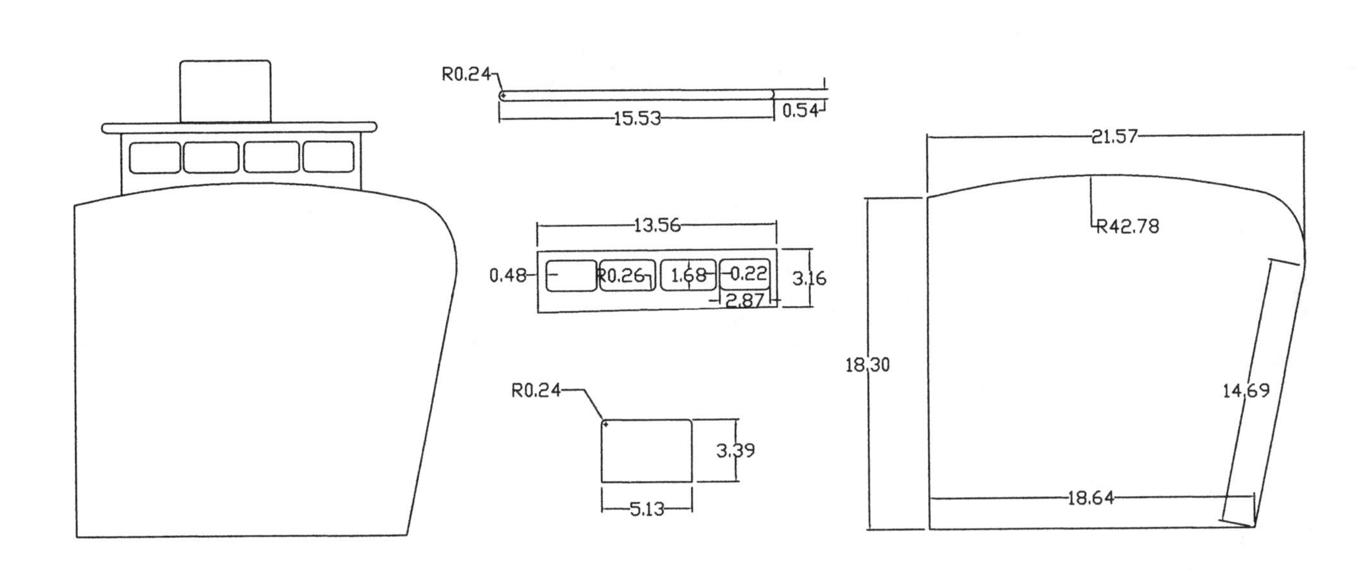
R0.24
15.53
0.54
21.57
R42.78
13.56
0.48
R0.26
1.68
0.22
3.16
2.87
18.30
14.69
R0.24
3.39
5.13
18.64

to join the items to the backdrop. The same welding procedure is used as when welding the aluminum parts together. Also, the same procedure is used to remove the oxide from the standoffs and the back of the backdrop.

The first item to be tack welded on was the car. It is in the top left hand corner. After the first item was tack welded, slight warpage was noticed on the backdrop. As a result, the remaining standoffs were tack welded one item at a time by skipping around. A tack would be placed on one item, then the next tack would be placed on the opposite side of the art mural. The reason for skipping around was to limit the amount of warpage caused by the heat in a concentrated area of the backdrop. There was still some distortion but the majority of it is not noticeable.

The next step is to remove the temporary metal square tubing from under the items. All of the items on the backdrop have different lengths. This was done to give the art project a sense of depth and shadows. After tack welding, the backdrop is carefully picked up off the chairs and leaned vertically up against a table. The welds are made in the horizontal position. Once the welds on the sides are complete, the backdrop is flipped over and the opposite sides of the joints are welded. To limit distortion, the smallest possible welds are made. After all the items had been welded, the remaining standoffs that stuck out past the welds are broken off and ground flush with the backdrop. A 4″ right angle grinder is used to take down the remaining welds.

Finishing the Art Mural

The finished art mural must have some kind of protective coating put on it because of the oxidation characteristics of the aluminum. A clear coat type of spray paint can be used.

General Outline of Fabrication Steps

Design using AutoCAD.

Convert AutoCAD file to plasma cutting table software.

Place a 60″ x 120″ sheet of 6061 mill finished sheet of 1/8″ aluminum on the cutting table.

Cut out parts and backdrop using the plasma cutter.

Deburr parts using the belt sander.

Braze rod for the spokes and bridge supports are cut out using diagonal cutting pliers or Beverly shear.

Clean the ends of the spokes and bridge supports with a file or belt sander.

Organize parts on the backdrop.

Shear out leftover aluminum for standoffs.

Duct tape the set offs to the parts to visually confirm correct placement of parts on backdrop.

Tack weld standoffs to back of parts.

Put a finish on the front of the items using a 4″ right hand grinder set up with finishing pads.

Weld joining parts together.

Put a finish on the backdrop.

Weld items to the backdrop.

Spray paint with clear coat.

Square Fire Pit

Author: Dom Coscia
Instructor: Nicholas Regets
School: William D. Ford Vo-Tech Center
City & State: Westland, MI

Materials List
1/4″ wire fence
1-1/2″ x 1-1/3″ x 45′ angle iron
1 – 4′ x 8′ sheet 11 ga.

Procedures
Cut four angle iron 2′-1/8″.

Grind edges to perfect 24″.

Tack bottom frame, tack top frame, and use four angle iron to join frames.

Cut 11 gauge metal into two sheets, 24″ x 24″.

Tack bottom base to frame with one 24″ x 24″ piece.

Cut four holes with torch into top part of angle iron base.

Tack the 1″ metal locks onto the lid.

Tack on handle for lid.

One series of tacks, tack four 23″ x 23″ wire fence.

Cut six square bars and cut a 45 degree angle.

Tack all bars together and to fire pit.

Completely weld all tack welds.

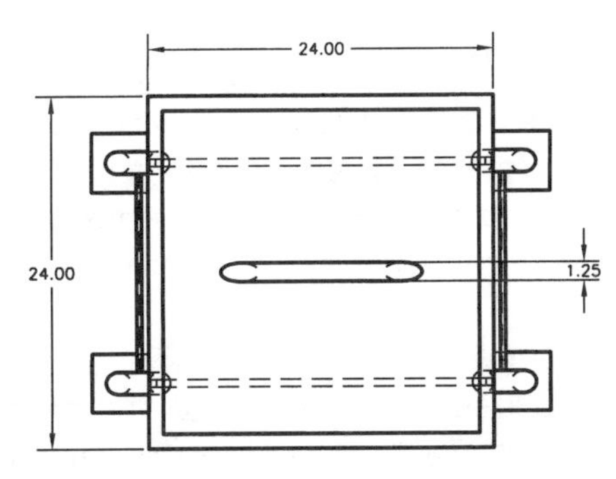

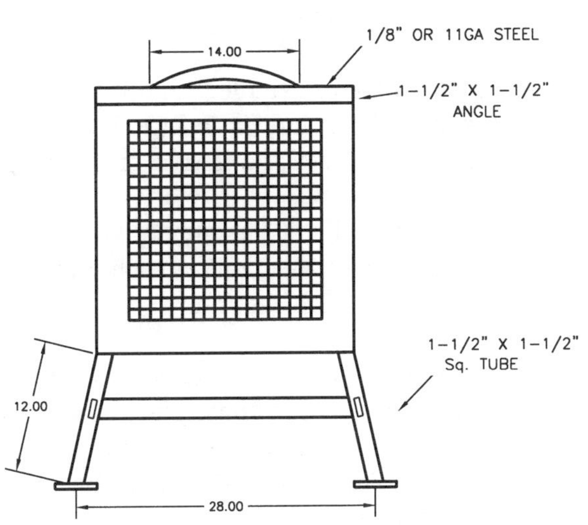

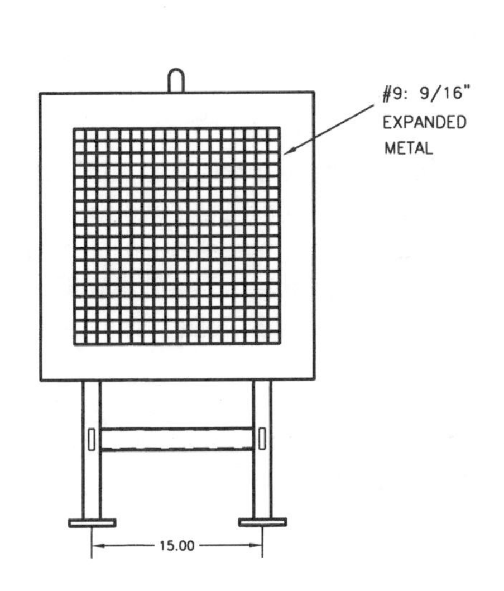

Iron Furniture

Author: Jadyn Fisher
Instructor: Robert Shook
School: University of Montana
College of Technology
City & State: Missoula, MT

Bill of Materials

20′	3/8″ Hot rolled round
20′	5/8″ Hot rolled round
20′	1/2″ Hot rolled round
20′	1/4″ Hot rolled round
4′ x 8′	1/2″ Expanded raised steel

Introduction

These tables and stools were made for the owners of a small restaurant. They are sturdy and decorative. They would also be suitable for a private home or garden.

A great feature of this project is that once a simple jig is made to fabricate the stools and tables, the frames can be made within hours. Cutting the pieces for the vines requires no measuring. They all look consistently the same and yet each one is an original.

Tables

Cut all the desired lengths for the circles for the tables and chairs.

Using a slip form roller, make each of the straight pieces into circles of desired dimensions. Be sure to have an assistant to ensure it does not roll out of the circle's plane.

Cut all legs for tables and chairs. Then, using a power slip form roller, form the legs into an eye-pleasing arc. Be sure to keep the arcs consistent.

Make a fixture out of 4″ x 4″ square tubing, 5/8″ pipe nipples, and couplings.

Square the tubing and weld to a flat and level surface. Measure to where Pc. D and Pc. F would reside if the tables and chairs were upside down, find center of tubing and mark this area. Level couplings at the measured points, then tack weld, and repeat for all eight couplings.

Screw the pipe nipples into the couplings and ensure they are level. This is the completed jig.

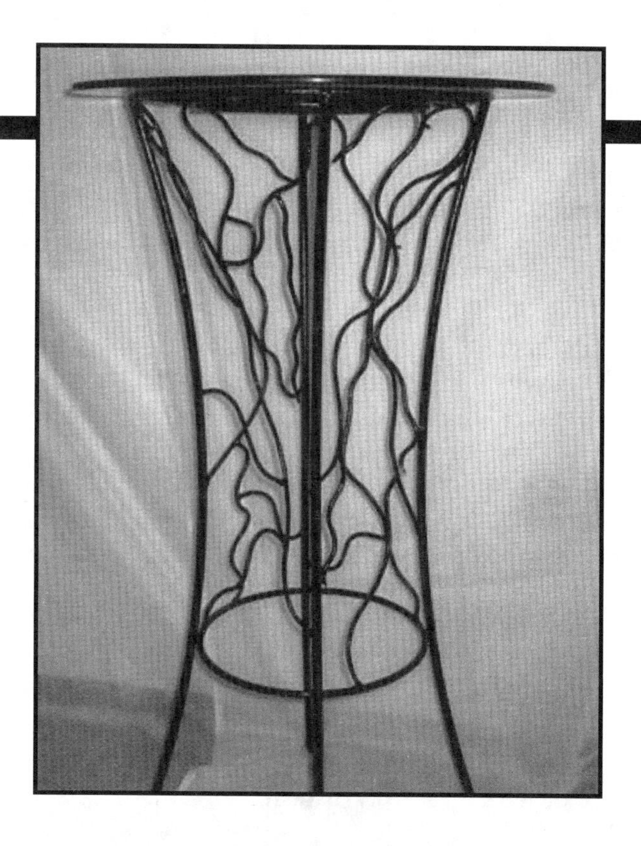

Clamp Pc. B to a level table, put a piece of 1/4″ plate 1′ x 1′ square in the center of the circle, then center Pc. A into Pc. B and clamp down.

Cut random lengths 5″ to 11″ long of the 1/4″ round stock. Cut four longer pieces and six shorter pieces for each tabletop.

Bend pieces into "S" shapes and other random curves. Place bent pieces in space between circles. Ensure nothing is cluttered. Repeat steps seven through nine for the next two tabletops.

Find the four equal points of circles Pc. B and Pc. F. Place fabricated tabletop around the jig and center, then clamp. Place Pc. F on the pipe nipples and move until all points match up.

Take the prefabricated legs and match up with points marked on Pc. B and Pc. F. Tack weld. Stand back and ensure that everything is true and then weld.

Cut 1/4″ round stock randomly keeping a variable of 5″ to 12″ and cut enough for each section. There need to be four for each section.

Cut twelve pieces of 1/4″ round stock to the length of 3′6″, heat one end of these pieces until red hot, hammer on anvil and draw out ends to a point. Bend each piece into pleasing curves, using the pointed ends to your advantage.

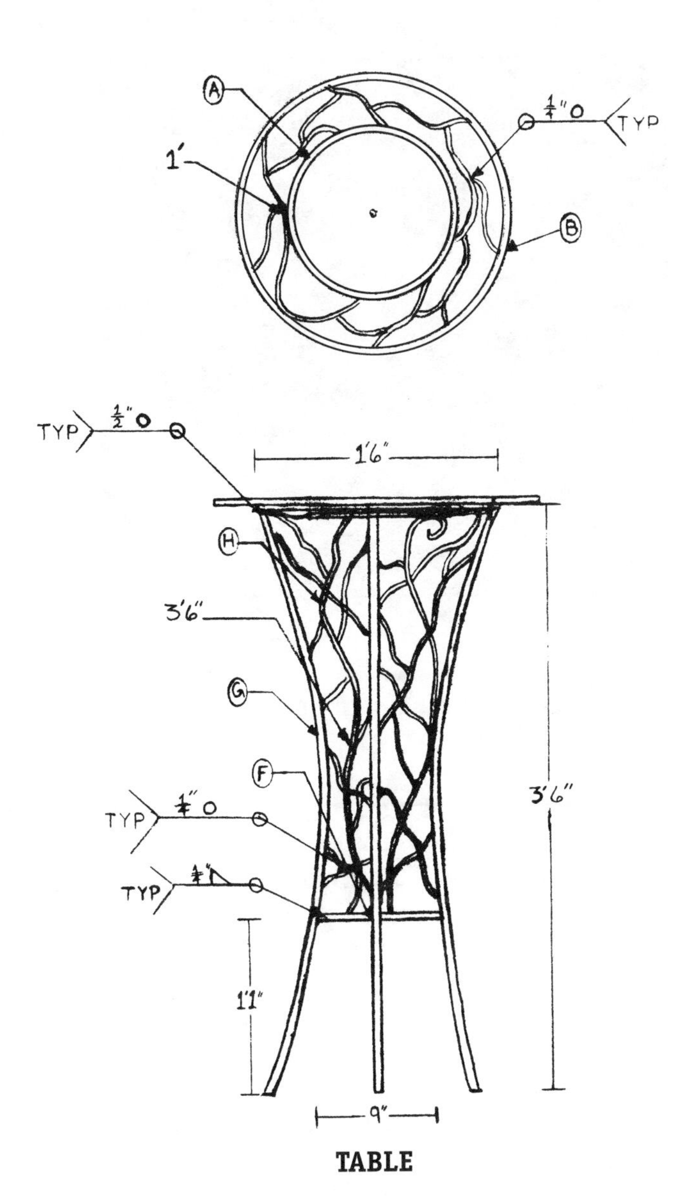

TABLE

Take these bent pieces and place Pc. H in the desired place first, tack and weld. Then rotate the sections one by one and weld smaller pieces in place. Be sure to move around the work to prevent distortion. Clean up any excess spatter and grind any unsightly welds.

Stools

Move pipe nipples to the lower couplings. Find four equal points of circle Pc. E and Pc. D. Center Pc. E on the jig and center Pc. D on the pipe nipples, match up points and clamp down. Tack weld each leg onto circles and then weld. Take out of the fixture. Repeat five more times.

Cut ninety-six pieces of 1/4″ round stock at random lengths anywhere from 5″ to 9″ and then bend to curve.

Cut twenty-four random pieces from 1/4″ round stock at 2′5″ and then bend at random. Weld Pc. I onto stool frame first.

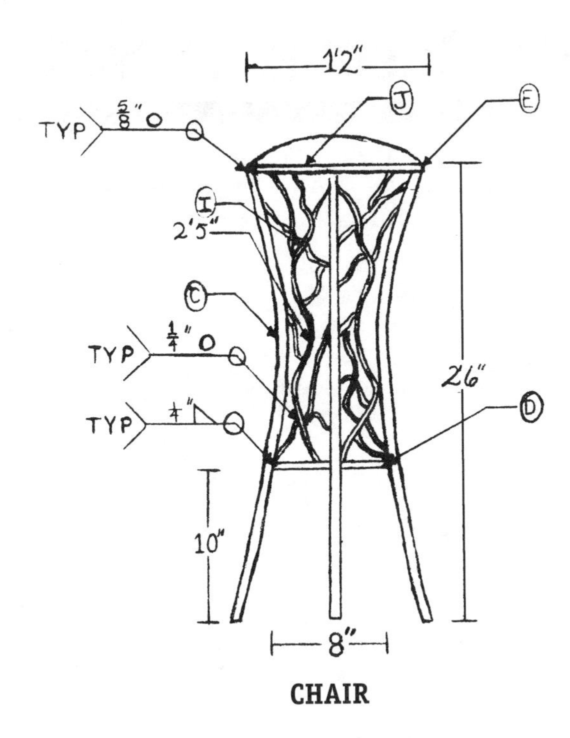

CHAIR

Piece by piece, weld smaller bent pieces to the frame while rotating the stools to prevent distortion. Clean up spatter and any other imperfections.

Trace seat circles onto expanded steel square 1′2″ by 1′2″ and cut out with ironworker. Place Pc. J on the stool frame and spot weld in at least four places. Grind down all these welds to ensure the cushions will fit nicely onto the frames.

Send off all pieces to be cleaned and powder coated.

Upholstery

Cut six 1′2″ circles out of plywood and drill a 1/4″ hole that would also enter the appropriate place into the expanded steel.

Buy six 1-1/4″, 1/4″ carriage bolts and 1/4″ 20 wing nuts. Put the carriage bolt into predrilled holes.

Cut 2″ high-density foam into 1′2″ circles. Also cut 2′4″ circles out of vinyl. Center the plywood with a bolt onto the foam circle. Center the foam onto the vinyl.

Kneel onto the plywood and pull the vinyl tight then, using 1/2″ staples, staple the wood. Continue this while rotating the seat until the opposite side is uniform and wrinkle-free.

Place upholstered cushion onto the stool frame and attach with wing nut.

Dutch Miser Wood Stove

Authors: Sean Lucier and Jonathan Osterman
Instructors: Mr. Guyette and Mr. Viner
School: Blackstone Valley Regional Vo-Tech High School
City & State: Upton, MA

Materials List

2	22″ x 27-1/2″	14 Gauge
1	47-1/2″ x 17″	14 Gauge
1	28-1/2″ x 19″	14 Gauge
4	21″ x 3-1/2″	14 Gauge
2	24″ x 16″	11 Gauge
1	12″ x 16″	11 Gauge
1	18″ x 22″	11 Gauge
2	12″ x 24″	11 Gauge
4	1-1/2″ x 1-1/2″ x 1/8″	Iron
1	10″ x 12″	6 Gauge
1	18″ x 11-3/4″	11 Gauge
1	4″ x 18.87″	14 Gauge
1	1-1/2″ x 25.12″	14 Gauge
1	48″ x 96″	14 Gage
1	48″ x 96″	11 Gage
1	48″ x 96″	3/16 Gage
1	20′	1-1/2″ x 1-1/2″

Procedure

The first step is to cut all the pieces for the inside firebox. Next, align the pieces to prepare for assembly. After the measurements are correct, tack and weld the inside firebox.

Next, cut and weld four 21″ x 3-1/2″ pieces of 14 gauge sheet metal with a 90 degree angle. These pieces are used as support for the outer covering.

Then, cut a 21″ x 17″ piece of 11 gauge steel. Plasma cut an 11″ square in the center and MIG weld it to the front on the fire box.

Weld a smoke stake 6″ in diameter and 4″ high on the top of the fire box. Then put a 28-1/5″ x 19″ piece of 14 gauge sheet metal on top.

The cover has a 6″ in diameter hole and an 8″ in diameter hole on the top. Weld it in place.

Next, make the door. First, cut a 10″ x 12″ piece of 6 gauge. To construct the holes for the damper, use a CNC Bridgeport and machined a blot pattern. Also drill a hole for the handle. Also using the CNC Bridgeport, machine the door hinges and pins, and weld them to the door and firebox.

The next thing to do is to manufacture the outside covering and top out of 14 gauge mild steel. The bottom and back piece are made from one long piece, split with a 90 degree angle. The two sides and top were single pieces with 90 degree 1/2″ bends to secure and weld in place

Attach all the outside pieces with magnets to prepare for welding. Next, drill a hole in the front of the stove and put angle iron on the inside and weld it in place.

Then cut a piece of 18″ x 11-1/4″ 11 gauge and drill a hole near the top. Bend a 1/4″ piece of round stock and put it through the hole in the front of the box and inside to the sheet of metal. This makes a damper.

Finally, attach four 10″ x 1-1/2″ x 1-1/2″ legs to the bottom. This completes the Dutch Miser Wood Stove.

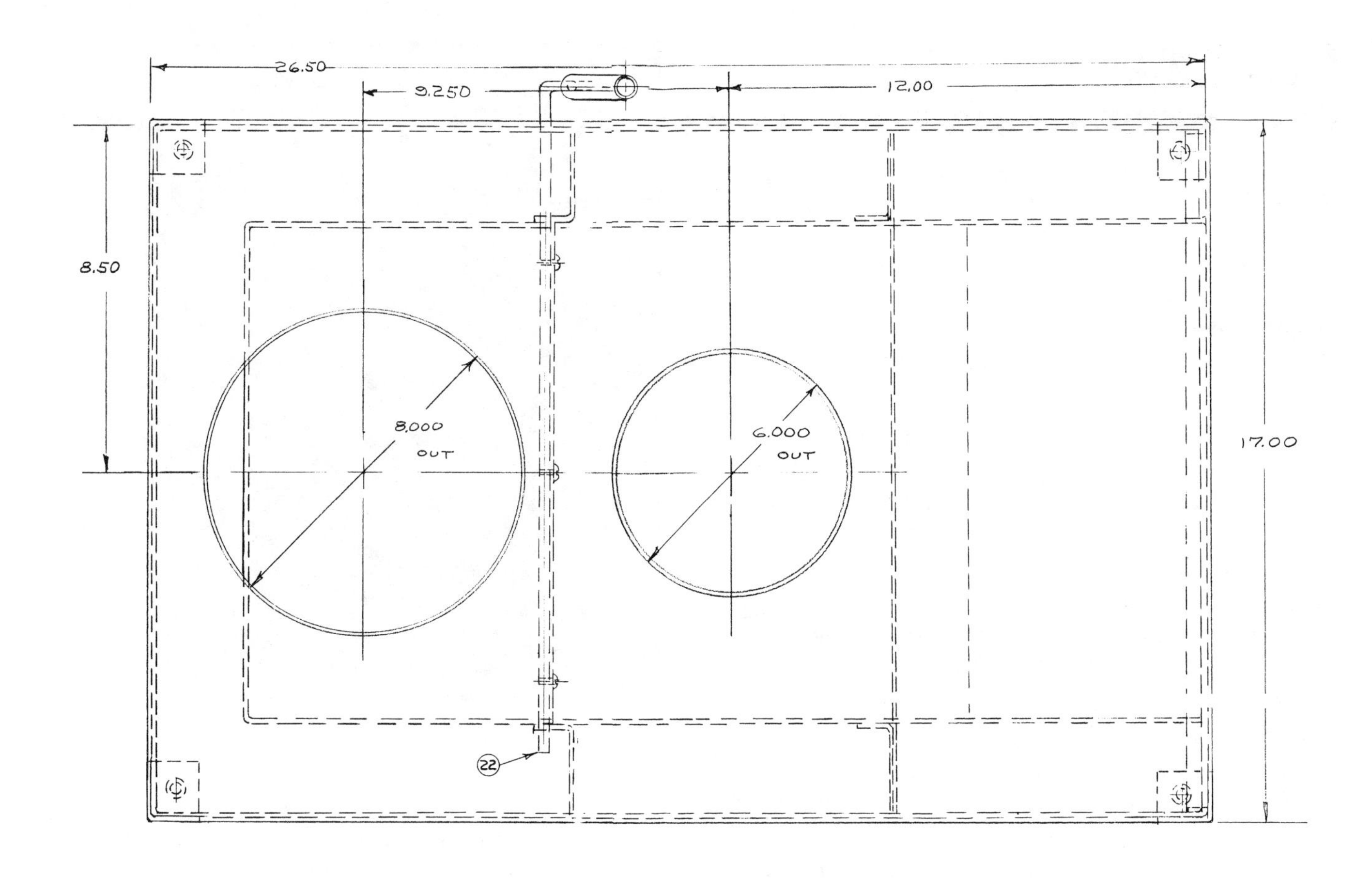
26.50
9.250
12.00
8.50
8.000 OUT
6.000 OUT
17.00
22

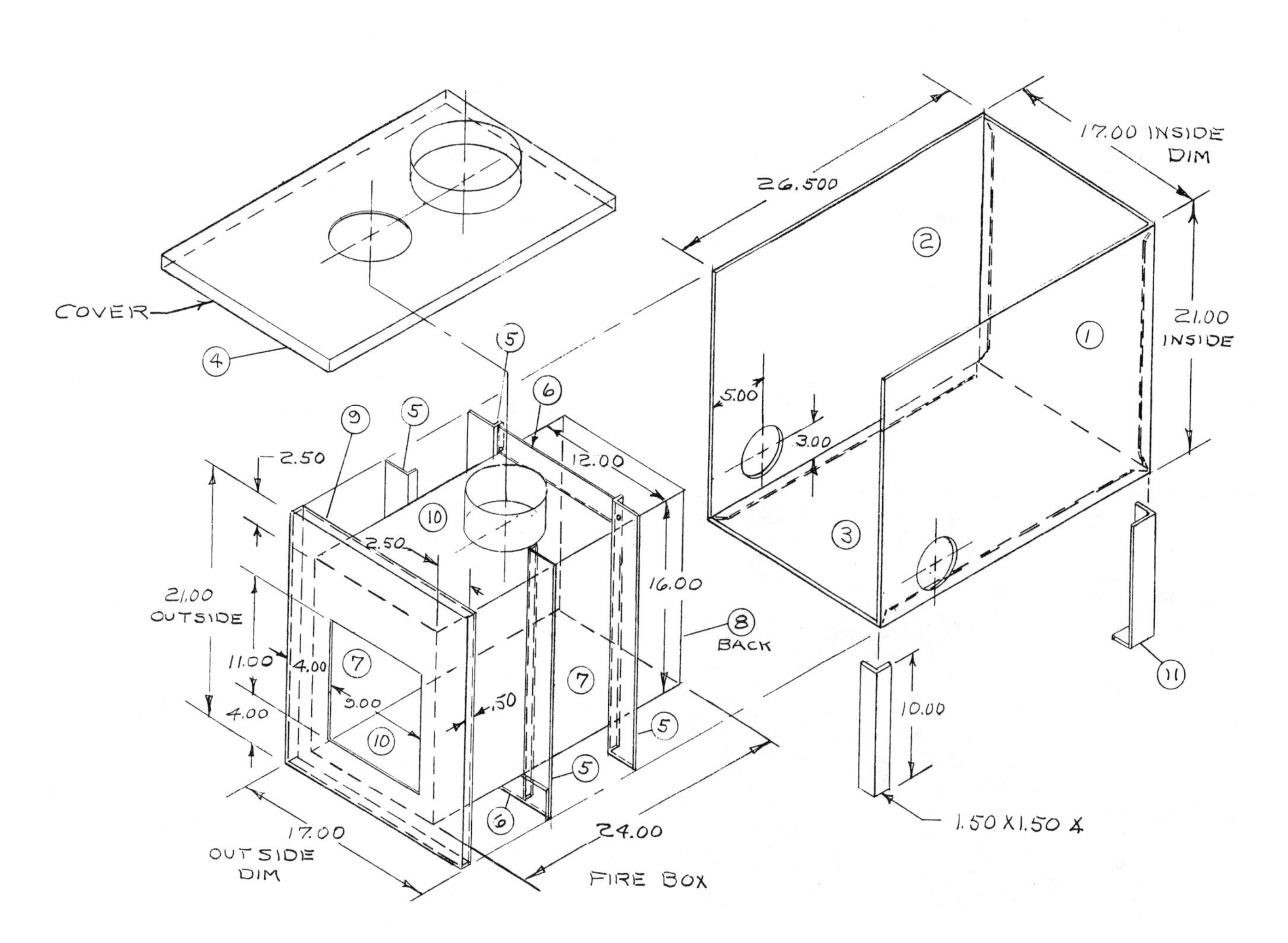
COVER
4
26.500
17.00 INSIDE DIM
2
21.00 INSIDE
1
5.00
3.00
3
5
6
9
5
2.50
12.00
10
2.50
21.00 OUTSIDE
16.00
8 BACK
11.00
4.00
7
7
3.00
.50
4.00
10
5
5
6
10.00
11
17.00 OUTSIDE DIM
24.00
FIRE BOX
1.50 X 1.50 ∡

Gun Safe

Author: Matthew Smith
Instructor: Garry Wilfong
School: Tulia High School
City & State: Tulia, TX

Building the Base
First, take an 8′ x 8′ x 1/4″ steel plate. Use a shear to cut it at 5′ tall by 61″ wide. Measure 14″ from each end of the 5′ side and mark each measurement. Using a metal break, bend the plate to 90 degrees at each mark. After it is bent, it should look like a horseshoe.

Top and Bottom
Next, use the metal shear to split a 64″ x 28″ x 1/4″ plate in half. After the metal is sheared, there should be two pieces, each 32″ x 14″. These pieces will be the top and bottom. Round off the corners that will go against the U shape. Using the wire welder, tack one piece on at a time. Make sure the pieces are inset and flush with the edge all the way around. Once each end is tacked, make sure everything is where it needs to be. If so, weld it up. Skip weld on the inside and weld the outside solid with the wire welder.

Doorframe
For the doorframe, form zee shapes out of 1/8″ x 4-1/4″ flat strap. Use the metal break to make the first lip 2″, the second lip at 1-1/4″, and the third lip 1″. Repeat these steps for four pieces (two at 31-3/4″ for the top and bottom, and two at 60″ for the sides). Set the pieces like a picture frame, with the 1″ edge on the inside. Next, cut 45 degree angles in the corners with the chop saw. After they are cut, tack them into place onto the base with the wire welder. If they do not fit exactly, use the grinder to make modifications until they do. After all four pieces are tacked into place, weld them the same way as the top and the bottom. Skip welds on the inside and a solid weld on the outside. Next, use a grinder to knock down every weld on the outside. Use a die grinder to grind the corners because a hand grinder will not fit. Then use the tiger paw to smooth off the welds.

Installing the Levelers
First, choose the side that is going to be the bottom. Tack a leveler at each corner. Put three tacks on each one. Make sure they are square, then weld all the way around them.

Building the Door
Use 1/4″ plate for the door. Shear the plate at 28″ x 56-1/4″. Next, use 1″ square tubing for the door jam. Using the chop saw, cut two pieces of square tubing at 26-1/2″ and two pieces at 56-1/4″. After the pieces are cut at these lengths, cut 22-1/2 degree angles on each end of each piece. Next, make triangles out of the tubing by cutting 22-1/2 degree angles on each side of a 1″ piece of tubing. Make four triangles - one for each corner. Set up the tubing like a picture frame with the tubing in the corners. Tack the pieces in place, then weld them up with the wire welder. After they are welded, grind the welds off smooth. Next, set the tubing flat on the door. Make sure it is setting at least 1/8″ from the edge all the way around and tack it in place. Put the tacks on the inside of the tubing. Once everything is in place where it needs to be, put skip welds around the inside of the tubing.

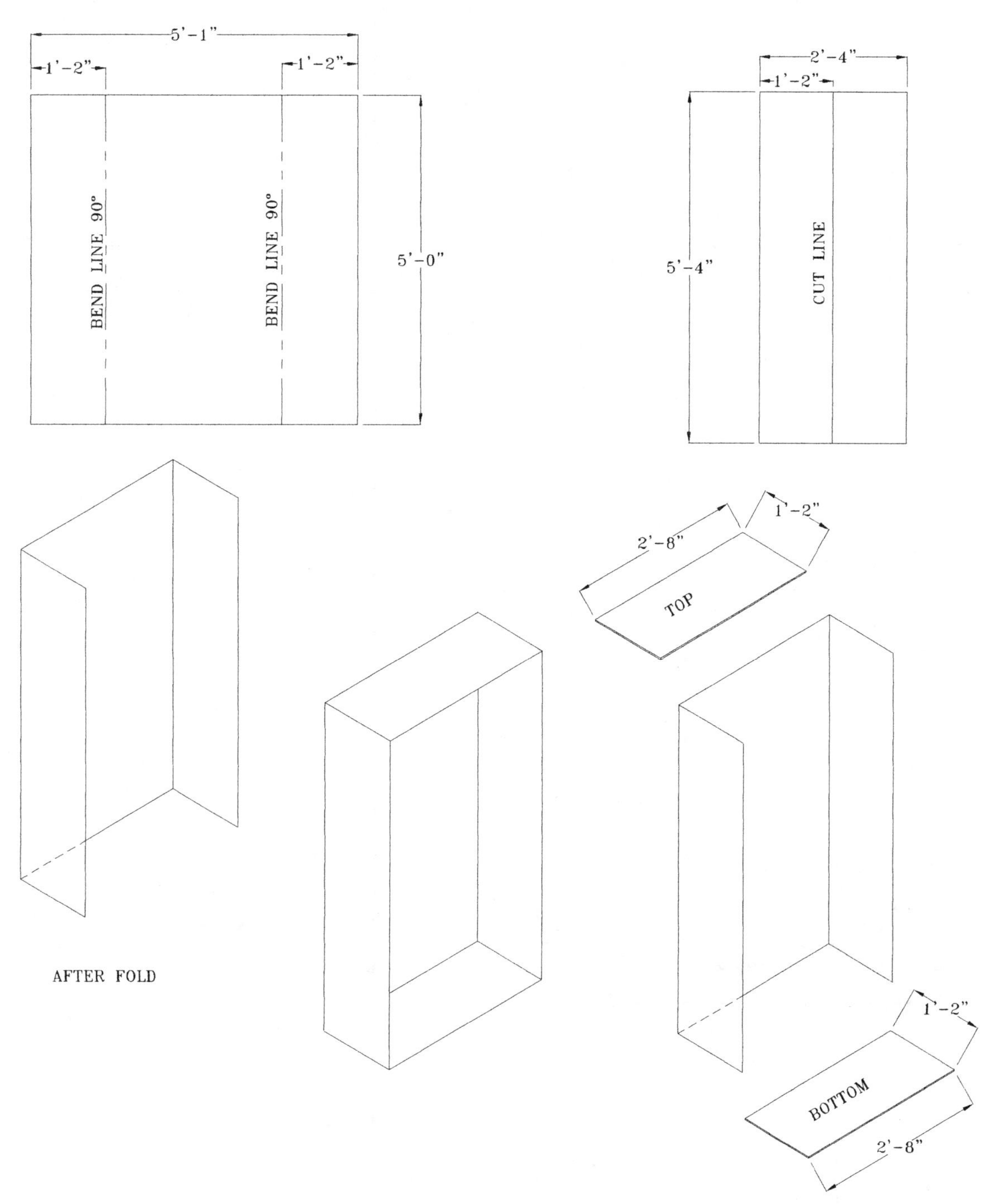

BASE

Installing the Hinges

Lay the base on the ground with the open end up. Next, set the door on top of the base with the door jam inset into the doorframe. On the left side, put a hinge 5″ from the top, 22-1/2″ from the top, 5″ from the bottom, and 22-1/2″ from the bottom. To install a hinge, put a weld on one side of the top of the hinge, then put another weld on the bottom of the hinge on the opposite side of the top. Be sure to weld at least one of the hinges different from the rest, so the door does not fall off when you stand it up.

T-Handle

Drill a 7/8″ hole 20″ from the top of the door and 3″ from the right side of the door. Next, insert the T-handle into the 7/8″ hole; use the handle as a pattern for the mounting holes. Use a piece of soapstone small enough to fit inside hole to mark where the hole needs to be. Drill a 1/4″ hole at these marks. Bolt the handle on the door.

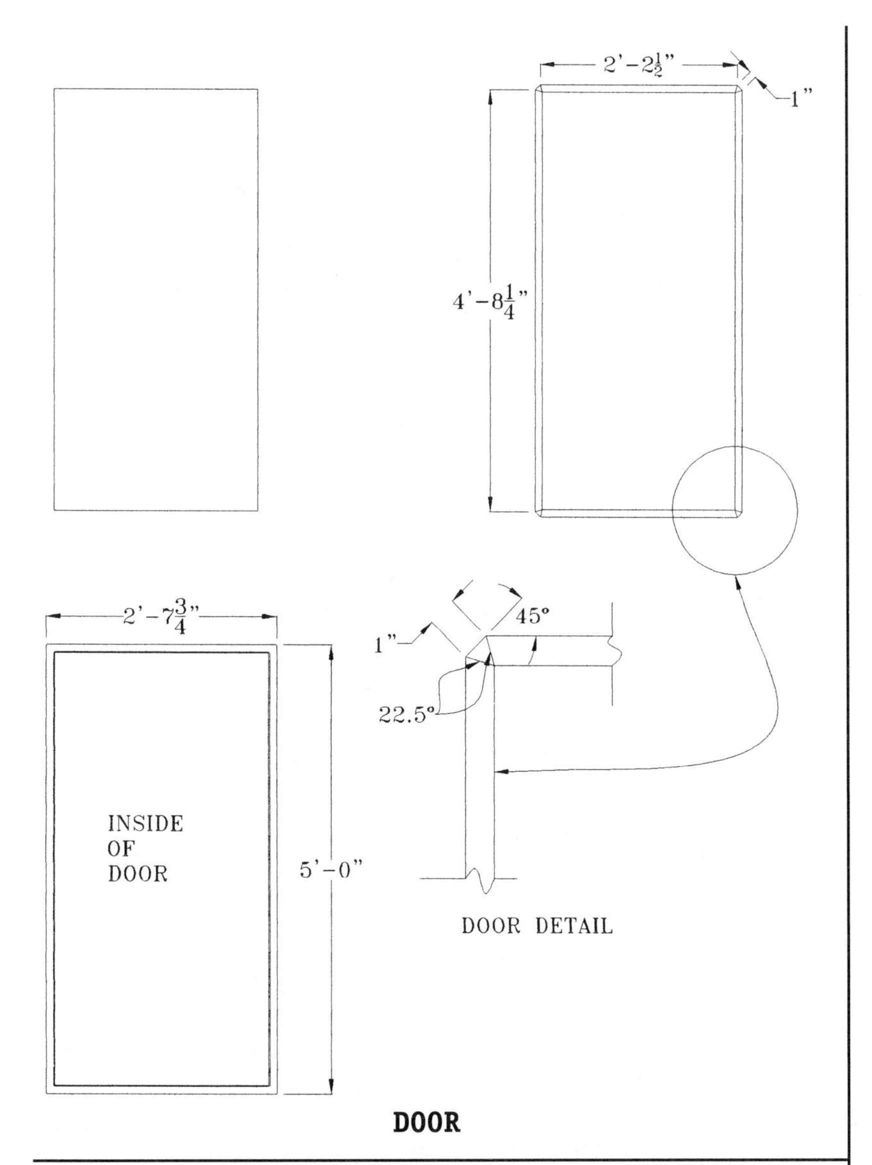

DOOR

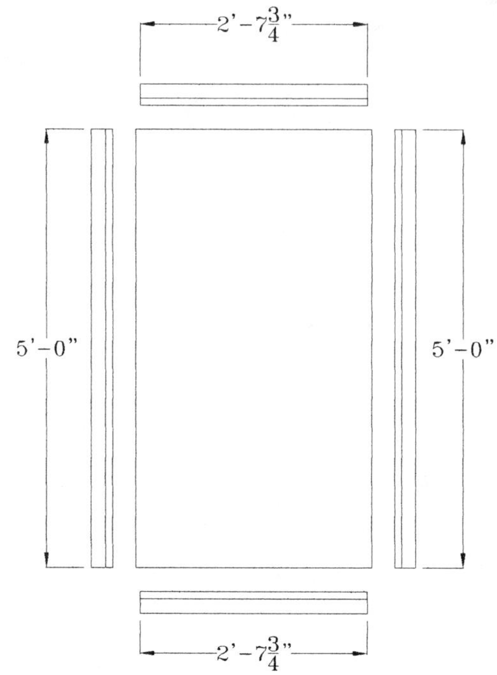

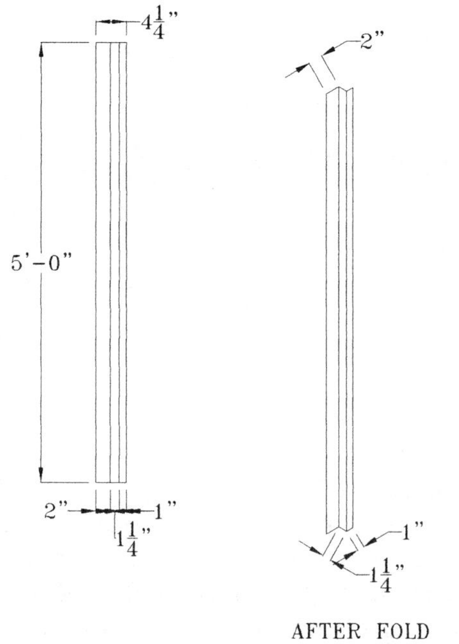

DOOR FRAME

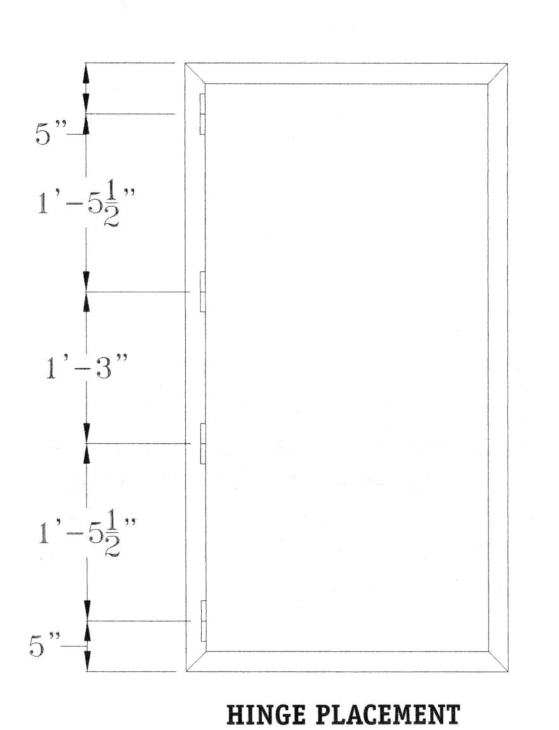

HINGE PLACEMENT

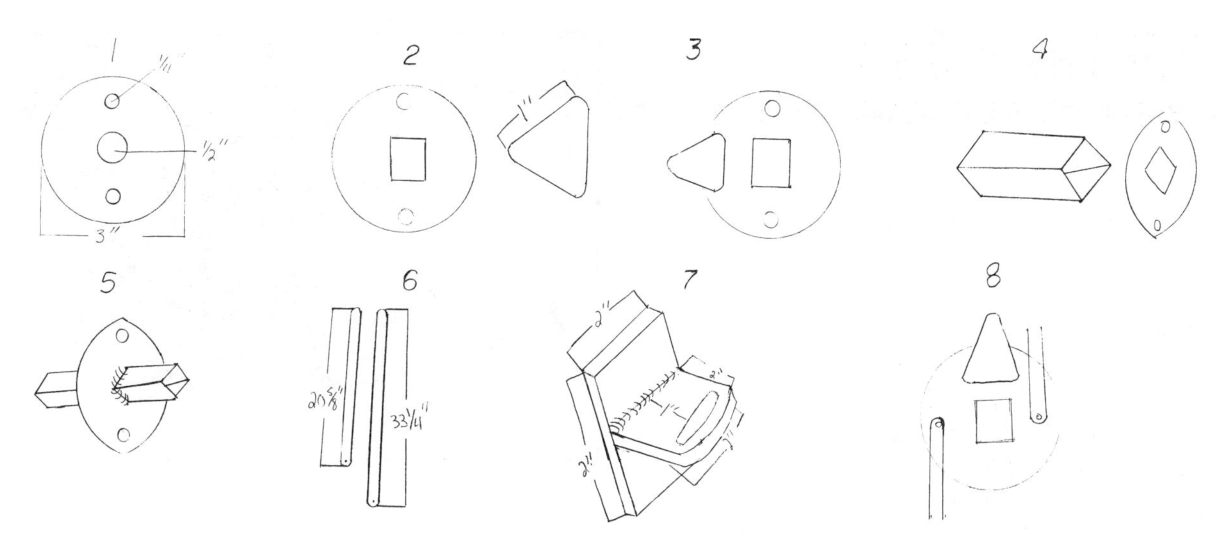

LOCKING MECHANISM

Building the Locking Mechanism

Use the pattern torch to cut out a 3″ circle out of 1/4″ sheet. Next, drill a 1/2″ hole in the center and a 1/4″ hole on each side of the 1/2″ hole, 1/4″ from the outside edge. Use a square punch to punch a square in the center hole. After that, cut a 3-1/4″ piece of square tubing. Insert the tubing into the square of the circle. The circle should be in the center of the tubing. Make sure it is square and weld all the way around both sides. Next, cut a 1″ triangle out of 1/4″ flat strap. Round the edges with the grinder, then clamp it to one side of the circle. Clamp it on one between the two 1/4″ holes with one of the points sticking out 1/4″. When the triangle is setting equal, weld around the front side. After that, cut two pieces of 1/2″ x 1/4″ flat strap for the locking straps. Cut one 20-5/8″ and the other piece 33-1/4″ long. The short piece will be for the top. The long piece will be for the bottom.

Round both ends of each piece. Then drill a 1/4″ hole in one end of each piece. Bolt them on where they are setting on the same side as the triangle. After that, build tracks for the locking straps out of 1/4″ flat. Cut four pieces 2″ x 2″. Then weld them together squarely to make two T shapes. Use the cutting torch to make a slit in the side that is sticking out. Make the slit big enough that the locking strap has enough room to move up and down and side to side. When the locking straps are turned to the lock position, they should latch behind the lip of the doorframe, so the slit should be 1″ from the edge of the track. Weld one track directly above the T-handle and the other one directly below the T-handle. The one that is on top should be 3-1/2″ from the top and 3-1/2″ from the side. The one on bottom should be 4″ from the bottom and 3-1/2″ from the side. Put a weld on the top and bottom of each track to hold them on securely.

Assembling the Lock

Assemble the locking mechanism by sliding the square tubing over the backside of the T-handle. Make sure the handle is horizontal and pointing to the right, the triangle should be facing up, the short sliding strap is on the right, and the long sliding strap is on the left. Insert the sliding straps into the tracks and bolt them onto the circle. When the handle is turned vertical facing down, the safe is locked. When the handle is turned to the right, the handle is unlocked.

Texas Emblem

Using a plasma cutter, cut the Texas shape 13″ x 13″ out of 3/8″ metal. Drill a 1/4″ hole on each end and on each side of the backside of the Texas. Tap these holes with a tap and die set. Center the Texas on the door, then drill holes to match in the door. Make sure the holes match, then paint it. Once the paint is dry, bolt it on.

Painting and Carpeting

To prepare for the painting and carpeting, sand the entire safe, inside and out, with 320 grit sandpaper. After that, wipe the entire safe down with mineral spirits to remove any grease that might be on the metal. Paint the inside of the door black, then close the door and paint the entire outside of the door with black bedliner. Once everything is dry, carpet the inside. Cut four pieces of carpet; one at 32″ x 62″ for the shelf, one at 51-1/2″ x 61″ for below the shelf, one at 9″ x 61″ for above the shelf, and two pieces at 14″ x 31″ for the bottom and the top. Next, spray glue on one area at a time and put the carpet in. Once all the carpet is in, hand press it all over so there will not be any wrinkles. After that, trim off all the edges and flatten any spot that might not be sticking well.

E-Z Tilt Lawn Cart

Author: Mitchell Austin
Instructor: Dan Saunders
School: Monroe High School
City & State: Monroe, WI

Bill of Materials

Qty	*Item*	*Material*
2	side frame	3/16″ x 2″ x 2″ x 48″ angle iron
2	front-rear frame	3/16″ x 2″ x 2″ x 36″ angle iron
2	center support	3/26″ x 1″ x 2″ x 35-1/2″ channel
4	pivot supports	3/16″ x 1″ x 2″ x 14″ channel
2	pivot	1″ I.D. x 6″ Pipe
1	axle bracket	1″ I.D. x 20″ Pipe
2	gusset	1/4″ x 3″ x 3″ HR plate
2	hitch	1/4″ x 2″ x 7″ HR flat
1	tongue	1/8″ x 2″ x 2″ x 48″ Square tube
1	center support	3/16″ x 1″ x 2″ x 22″ channel
4	corner support	3/16″ x 2″ x 2″ x 16-1/2″ angle iron
4	side support	3/16″ 2″ x 2″ x 16-1/2″ angle iron
1	axle	1″ DIA. X 3″ CRS rod
2	tail gate guide	1/8″ x 1″ x 1″ x 10″ long angle iron
1	tilt lock	1/2″ rod 6″ long
2	lock guide	1/4″ x 2″ x 3″ hot rolled flat
2	side	1″ x 8″ boards
1	front	1″ x 8″ boards
1	bottom	3′ x 4′ 12 ga. Steel plate
1	tail gate	1/2″ plywood 2-2 x 4-32″ long
3	top trim	1/4″ x 2″ x 2″ flat bar x 11′

Start by making the frames of the lawn cart out of 3/16″ x 2″ x 2″ angle iron. Notch the ends out to accommodate a corner weld. Then after the frame is welded together, put the center supports in. They are made of 3/16″ x 1″ x 2″ channel iron. The pivot supports run down from the center supports. They meet together at the end. Next, lay a 3/4″ inside diameter pipe on top of the pivot supports. The axle is 3/4″ cold rolled steel. Thread one end on the lathe to 3/4″ pipe and put two nuts on the other end. Then weld a large washer on the other end.

Between the pivot supports, measure and cut another piece for the axle bracket. The tongue is 2″ x 2″ x 1/8″ square tubing, 48″ long. In one end of the tongue, drill a 3/16″ hole 2″ in. Slip the axle bracket through the hole and weld it. After the tongue is on, use 1/4″ bar stock for a gusset support. Then run a piece of 3/16″ x 1″ x 2″ channel from the frame down to the tongue. Before welding it on, punch holes in it for the locking mechanism and the sideboards.

The corner and side supports are 1/8″ x 2″ x 2″ angle iron, 16-1/2″ long. On top of the corner and side supports, put 1/4″ bar stock for a trim. Instead of putting in a plywood floor, use 12-ga. steel plate. For sideboards, use 2-1″ x 8″ stacked on top of each other. Paint the boards gloss black and attach them with 1/4″ carriage bolts.

For the tongue, cut a piece of 2″ x 3″ rectangular tubing to build it up so it will sit level when hooked to a lawn tractor. On top of the rectangular tube, use 1/4″ bar stock for a hitch. The latching mechanism is two pieces of flat bar four inches apart. They have 1/2″ holes drilled in them with two springs between.

A cotter pin is at the trailer end so it pulls against the springs. Add a two-inch ring for a handle. After grinding the burrs off, paint the cart and put two clear coats of sealer on it. Undercoat the bottom of the trailer with rubberized tar spray.

Electric Start Log Splitter

Author: William S. Chemin
Instructor: David H. Murray
School: Ferris State University
City & State: Big Rapids, MI

Bill of Materials

Structural steel

1	3/8 x 8 x 8 x 42 Square tubing
1	3/8 x 6-1/2 x 8 x 75 H-beam
1	3/16 x 2 x 3 x 71 Rectangular tubing
1	1/4 x 2 x 2 x 74 Angle iron

Mild steel

1	1/4 x 48 x 96 Plate
1	3/8 x 12 x 12 Plate
1	1/2 x 12 x 12 Plate
1	1 x 18 x 18 Plate

Miscellaneous

1	Axle
1	1-1/2″ o.d. x 1″ i.d. x 3″ Bushing
1	Tool steel
4	1″ Dia. X 3″ Black pipe coupler
2	1 Black pipe plug
1	2″ Housing with 2″ ball opening hitch
2	3/8″ x 4″ Zinc bolts with nuts
1	12 Horsepower engine
1	16 gallons per minute two stage pump
1	Love connector
1	5″ Dia. X 26″ stroke Cylinder
1	Filter
1	Control valve
1	12 volt Battery

Introduction

This custom log splitter features an electric start 12 HP engine; a 16 gallon per minute, two-stage pump for fast cycle time; and the ability to split logs horizontally or vertically. The splitter has a 4″ diameter by 30″ long cylinder and is capable of splitting a log up to 26″. The cylinder is mounted on a heavy duty 6-1/2″ x 8″ H-beam for maximum strength. For convenience, a built-in log cradle was made to assist the user. The cradle protects the user from a rolling log. The log splitter also contains heavy-duty fenders to support the operator.

Procedure

Fabrication was started by preparing the previously used material: the axle and the square tubing that was used for the hydraulic oil reservoir. First, all of the unnecessary brake lines and brackets were cut off of the axle with the oxy-acetylene torch. Then, using an electric grinder with a wire wheel attachment, the axle was cleaned. The same method of cleaning was used to prepare the outside of the hydraulic tank. For the inside of the tubing, a 7″ sanding pad attached to a pneumatic sander was utilized to remove the oxidizing rust.

After cleaning, the 8″ square tubing (Part C) was ready to be customized. Added to the tubing were four 1″ diameter, black pipe couplers (Parts D). Before welding the couplers, four 1″ holes were drilled into the tubing with a drill and a 1″ diameter hole saw bit. This is to enable the flow of hydraulic oil into and out of the tank. Two of the couplers were tacked directly over the two holes on the top surface of the tank and fully welded (see Welding Procedure 1). The other two couplers were welded using the same procedure; one at the bottom of the tank, used for a drain, and the second one was located at the lower side of the tank to feed the pump.

Welding Procedure 1

Process	Polarity	Amperage	Voltage
GTAW[1]	DCEN[2]	90 A	12 V
Electrode	**Electrode Size**	**Filler Material**	**Filler Material Size**
2% Thoriated Tungsten	1/8 in.	R70S-3	1/16 in.
Electrode Angle	**Joint Type**	**Position**	**Travel Speed**
60°	Tee Joint	2F (horizontal)	8 in./min.
Shielding Gas	**Gas Flow Rate**		
100% Argon	15 CFH		

In addition to preparing the tubing, end caps (Parts K) were tacked and then welded to each end of the tube to close the reservoir (see Welding Procedure 2). Next, four pegs were tacked, and then flux core arc welded to the bottom of the tank to properly attach the reservoir to the axle. The prepared tank was then mounted by welding the other end of the pegs to the axle. The last two steps also use Welding Procedure 2.

Welding Procedure 2

Process	Polarity	Amperage	Voltage
FCAW[3]	DCEP[4]	350 A	28 V
Wire Feed Speed	**Filler Material**	**Filler Material Size**	**Shielding Gas**
200 in./min.	E70T-1	.045 in.	100% CO_2
Joint Type	**Position**	**Travel Speed**	**Gas Flow Rate**
Tee Joint	2F (horizontal)	36 in./min.	35 CFH
Electrical Stickout			
1 in.			

[1] Gas Tungsten Arc Welding
[2] Direct Current Electrode Positive
[3] Flux Core Arc Welding
[4] Direct Current Electrode Positive

The sealed tank should now be ready for the other add-ons; this includes the tongue, H-beam mounting, motor mount, and others. This is an appropriate time to attach the tongue. Offsetting the tongue is more of a preference, but for easier hauling, it is advised to assist in distributing the weight. The tongue, Part I was first clamped to the reservoir, and then with a measuring tape and framing square, it was aligned perpendicular to the reservoir. Using the GMAW (Gas Metal Arc Welding) process, the tongue was tacked in the proper position. Since the tongue is made up of only 1/8th in. thick material, gas metal arc welding was used to fully weld the tongue (see Welding Procedure 3).

Welding Procedure 3

Process	Polarity	Amperage	Voltage
GMAW[5]	DCEP	175A	22 V
Wire Feed Speed	**Filler Material**	**Filler Material Size**	**Shielding Gas**
200 in./min.	ER70S-3	.035 in.	75%/25% Ar, CO_2
Joint Type	**Position**	**Travel Speed**	**Gas Flow Rate**
Tee Joint	2F (horizontal)	20 in./min.	35 CFH
Electrical Stickout			
1/2 in.			

[5] Gas Metal Arc Welding

Next, the mounting bracket for the H-beam was fabricated. This bracket consists of Parts F, G, and H. Parts F were first beveled to a forty-five degree angle with a disc grinder in order to properly fit up against the welded tongue. After beveling, they were clamped to their designed location, one on each side of the tongue, and tack welded with the GMAW process. Using the parameters listed in table two, the brackets were fully welded with the FCAW process. Another plate, Part H, was placed on top of the tongue and the side brackets; this was added for support of the hinge bracket. The GMAW process was selected for this weldment because of its visual appearance.

Now the hinge was ready to be welded. A 3" bushing (Part G) with a 7/8" inside diameter was welded to the top surface of the support bracket. Two additional tabs, Parts L, with 7/8" diameter holes drilled through them were welded to the underside of the H-beam to complete the hinge assembly. The tabs were welded with Welding Procedure 2, after rotating the H-beam to the underside of the horizontal position. Then a support pit (smaller than a 7/8" diameter) slides through one of the tabs attached to a beam through the bushing and through the other mounting tab to create the pivot of the beam.

Now the durable H-beam was ready to be mounted, but, before mounting, modifications were added to the beam. The first justification was a base plate, Part K, to rest the log against. A 1" plate was used for this application to ensure maximum strength. The plate was welded to the end of the H-beam using the variables described in Welding Procedure 2.

Next, using the same procedure previously described, angle iron (Parts J) was flux cored arc welded to the side of the beam for the wedge tracking device. To ensure proper positioning, a 5/8" spacer was made to fit under the angle iron. Then the angle iron was tacked using the GMAW process. Another bracket (Part Q) was placed on the underside of the beam. This was to level the beam onto the rectangular tubing used for the tongue. To do this, mount the beam with the hinge assembly and then level the beam using a four foot level. Make an H-shaped bracket to accommodate the space between the tubing and the bottom of the beam and weld the bracket to the beam (Welding Procedure 2). To lock the beam to the horizontal position, a hole was drilled into the bracket for a pin to slide through to lock the beam in place.

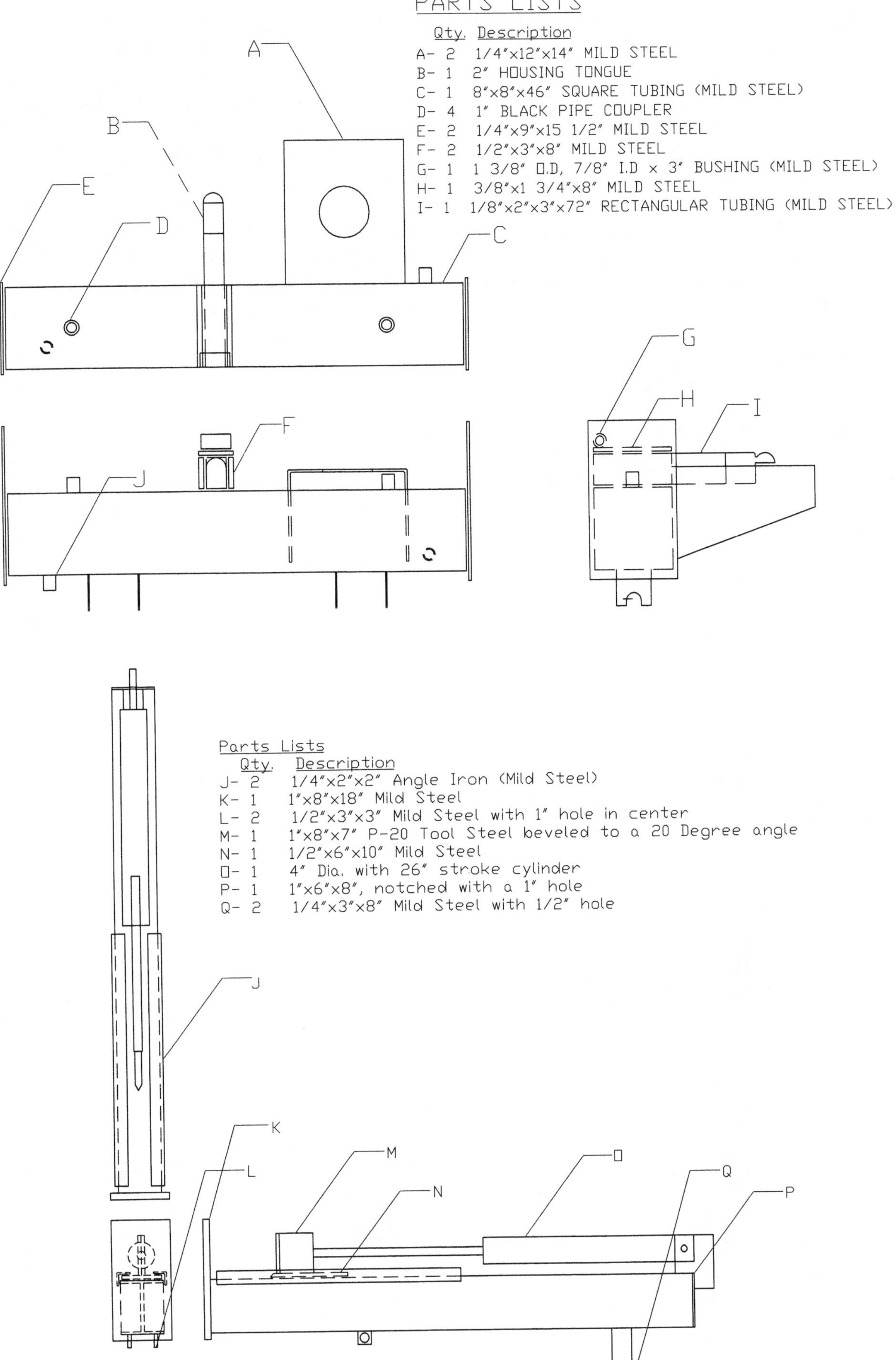
PARTS LISTS
Qty. Description
A- 2 1/4"x12"x14" MILD STEEL
B- 1 2" HOUSING TONGUE
C- 1 8"x8"x46" SQUARE TUBING (MILD STEEL)
D- 4 1" BLACK PIPE COUPLER
E- 2 1/4"x9"x15 1/2" MILD STEEL
F- 2 1/2"x3"x8" MILD STEEL
G- 1 1 3/8" O.D, 7/8" I.D x 3" BUSHING (MILD STEEL)
H- 1 3/8"x1 3/4"x8" MILD STEEL
I- 1 1/8"x2"x3"x72" RECTANGULAR TUBING (MILD STEEL)
A
B
E
D
C
G
H
I
F
J
Parts Lists
Qty. Description
J- 2 1/4"x2"x2" Angle Iron (Mild Steel)
K- 1 1"x8"x18" Mild Steel
L- 2 1/2"x3"x3" Mild Steel with 1" hole in center
M- 1 1"x8"x7" P-20 Tool Steel beveled to a 20 Degree angle
N- 1 1/2"x6"x10" Mild Steel
O- 1 4" Dia. with 26" stroke cylinder
P- 1 1"x6"x8", notched with a 1" hole
Q- 2 1/4"x3"x8" Mild Steel with 1/2" hole
J
K
M
O
Q
L
N
P

Now the engine could be mounted. A custom mount (Part A) was cut out using 1/4" material with the plasma cutter. After cutting, the plate was bent 90° at the two ends. A press break was used to bend the ends of the plate. Next, the holes for the bolt pattern needed to be drilled. For locations of these holes, contact the manufacturer or make your own template. Then drill the holes with the drill press in the same locations as the template. The mount was now welded to the tank using the following procedure. Before mounting the engine, the two stage pump was mounted to the engine using love couplers and a bracket that the pump bolts to.

Welding Procedure 4

Process	Polarity	Amperage	Voltage
GMAW	DCEP	175A	22 V
Wire Feed Speed	**Filler Material**	**Filler Material Size**	**Shielding Gas**
200 in./min.	ER70S-3	.035 in.	75%/25% Ar, CO_2
Joint Type	**Position**	**Travel Speed**	**Gas Flow Rate**
Tee Joint	3F (v-down)	20 in./min.	35 CFH
Electrical Stickout			
1/2 in.			

The cylinder, Part O, was now ready to be mounted to the top surface of the H-beam. Using the horizontal band saw, 1" thick parts were cut and then drilled using the drill press. With Welding Procedure 2, these parts were welded onto the end of the cylinder using the FCAW process. Another 1" thick piece was cut, drilled, and welded to the beam so the cylinder can be properly mounted.

Attached to the other end of the cylinder is the wedge, Part M. This is made out of a P-20 tool steel beveled at one end. Additional brackets were made to reinforce the wedge while splitting. The wedge was mounted on a piece that had been cut and drilled to bolt the wedge up to using a 1" hardened bolt.

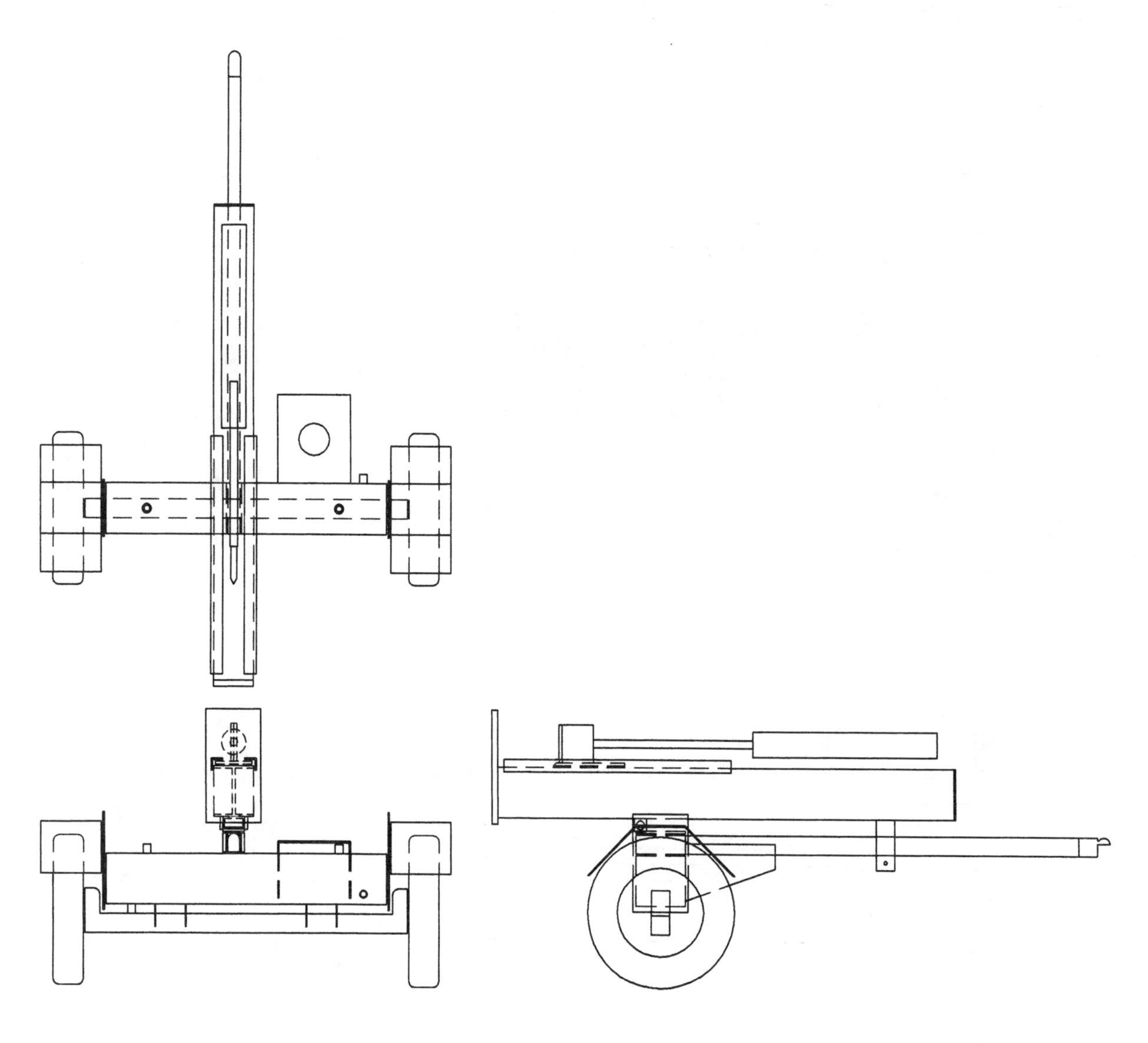

Articulating Boat Dock

Author: Jacob Miller
Instructor: John Mulcahy
School: Lassen Community College
City & State: Susanville, CA

Introduction

This boat dock is designed to be easy to take apart and remove from the water in the winter so that it will not freeze into the lake. It is also relatively light so it can be transported easily from place to place. "Articulating" means that the wings that come off of both sides have a pivot point so they hinge. The dock has a 15′ PVC pipe under it to keep it from sinking. It is covered with 2 x 12 rough-cut wood so that one can walk on it. As designed, it can accommodate four boats; it can be extended if necessary.

Materials and Processes

The basic structure was fabricated of 5″ steel channel, along with some 2″ angle iron and some pieces of 1/2″ plate. It was TIG and MIG welded using primarily 1/8″ 6011 and 1/8″ 7018 electrode. The 6011 was chosen because of its penetration, digging action, and its all-position capabilities. The 7018 was used in the areas where fit-up had to be excellent and also because of its high tensile strength.

Procedure

When this project was started, the first thing that had to be done was the main perimeter of the main part. This was a very critical part of the project. The dimensions had to be perfect. To determine if it was square, the diagonals were checked repeatedly and often. The overall length had to be 27′ 3″. The metal that was purchased was only 20′ 1-3/8″, so a 7′ 1-5/8″ piece had to be cut. Then a 4′ x 7′ piece of 3/8″ thick metal was used to patch the inside where the two pieces met, and a weld was made all the way around the plate. The outside of the juncture was then welded. 6011 was used for all of this welding.

Next, 45-degree angles were cut on the ends using a cutting torch. Then they were ground down using a 4-1/2″ angle grinder. Then two pieces were cut at 5′1″ out of 5″ channel using a band saw and the ends were cut at 45-degree angles using the same method as the long ones. The 45′s were fitted and diagonals were measured. After some persuasion, they finally lined up and were tacked on top and bottom using 6011. After all the corners were tacked, the top and bottom of the corners were then welded fully using 6011.

The cross members came next. Five of them were cut 5′5″ using the band saw. When welding in the patches to make the overall length, the metal bowed because of heat distortion. The cross members did not fit as they should have, but this was okay because bar clamps were used to either pull it or forc it out. After all of them had been tacked into place, they were welded using 7018.

After this, holes had to be drilled. There were two sizes of holes, so all of the 7/16″ holes were drilled first, then the 9/16″ holes were drilled. Now the tabs had to be machined. So using the Piranha, six 3″ x 3″ tabs and four 4″ x 4″ diagonal tabs were cut. The holes were punched using the punched on the Piranha. On the square tabs, the holes were punched 1″ up on the tab and 1-1/2″ over. On the corner diagonal tabs, the holes were made 1″ up and 1″ over from the corner. Then they were lined up and tacked in. After all of the measurements were taken and checked, they were welded using 6011.

Next came the areas where the two 20′ wings would hinge. These were a challenge because the dimensions were critical and there was no room for mistakes. First, two pieces were cut at 1′ 1-1/4″. Four more were cut 1′ and then laid out according to the prints. The diagonals were checked and the width was checked. After they were all perfect, they were tacked into place. Just as a precaution, all of the measurements were checked again.

Finally, the project was ready for the leg braces. There were two of them that were both 2′ tall and 5′5″ wide. Four pieces of channel, 2′ each, were cut using a band saw. Then a Piranha was used to cut two 5′ 4″ pieces of angle iron. After that, the 2′ pieces were tacked into place using a framing square. Then, using vise grips, the 5′5″ piece was attached to the uprights. The legs were hammered into where they needed to be.

After checking the measurements, it was welded into place using 7018 (where the channel met the mid-frame) and 6011 (where the angle met the channel).

All of this construction was done indoors, but to finish the dock, it had to be moved outside. A forklift was borrowed and five people helped to flip it over. After it was flipped over, all of the topside was welded. The corners of the main part were MIG welded using 0.035 - ER70S-2 wire. The cross members were welded with 7018 and the rest was done with 6011. Two 1-1/4" holes were blown in the end for a pipe to be inserted for articulation. Then, using a flatbed trailer, the dock was transported to the lake where all the floatation was put on.

The next parts that had to be built were the two wings that hinge onto the main part. They are identical so only one will be explained. For the mainframe, the angles were cut using a cutting torch. This time, the band saw was used. Since the longest length on this one was only 20', no patching of the extension was required. After the band saw was set, the two 20' pieces were cut. After that, the 4' 10-3/4" pieces were cut.

Then came the tedious task of getting all the dimensions right. It was a little easier this time because the angles were more accurate. After it was set up, the diagonals were checked to determine if the frame was square. After that, the corners were tack welded like the main frame with 6011. After double-checking, the corners were welded fully except for the very outside corners. The cross members were also cut with the band saw after it was set back at 90 degrees. They were cut at 4'11". Unlike last time when the cross members were welded then drilled, the holes were first punched on the Piranha. Doing it this way was faster and did not require sharpening the drill bits all the time.

At last, it was time to put the tabs on. This time there were only four center tabs and four corner tabs. They were all cut and with holes punched in the same fashion as the previous ones on the main section. They were welded on with 6011. Then it was time to put on the leg braces. They were identical to the ones on the main part except that they were 4'11", so the same method was used — cutting the channel on the band saw, chopping the angle on the Piranha, tacking them in, taking measurements, then welding them in fully.

After all that was done, the wing was ready to be dragged outside and flipped. The forklift was borrowed again, and with some man and machine power, the wing was moved outside and flipped. After it was flipped, the topside was ready to be welded. The tabs and leg braces were welded with 6011. Unlike before (on the main part), the corners could be welded with 6011 because there was hardly any gap. The cross members were welded with 7018. Then 1-1/4" holes were blown on both sides for the hinge points. After all that, it was transported to the lake in the same way as the others.

A couple of different methods of welding and cutting were used on this project. One of them was a whip 'n pause using the 6011 rod. This is where an arc is struck, then the electrode is whipped away from the puddle, letting it cool, then more filler is added. The same method was used on all of the vertical welding. Another type of welding that was used was the drag motion. It is done with a 7018 rod and the motion is exactly how it sounds — the rod is just dragged across the metal. It is important to keep an arc gap about 1 times the rod, though. When cutting with the torch, an 00 tip was used for everything. Those are all the procedures of welding and cutting that were used on this project.

Cooking Pan

Author: Ruiz Quinonez
Instructor: Dan Crookham
School: Exeter High School
City & State: Exeter, CA

Materials

36″ length, 1″ x 1/8″ thick Flat Stock
18″ length, 3/4″ Round Stock
Tractor Disc

Procedure

The first step in building the cooking pan is to fill in the 2″ x 2″ gap in the very center of the tractor disc, which will be the center of the cooking pan. Using a MIG welder set at 18 volts and 150 wire speed, weld it up. Then take a Makita grinder set with a wire brush plate and remove all the rust around the interior and exterior of the pan.

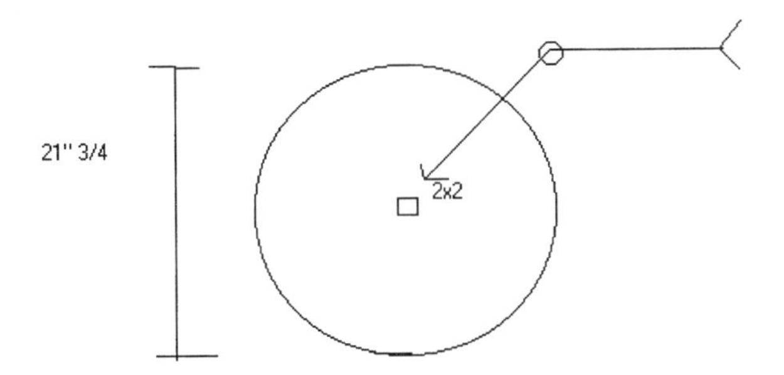

The next step is to bend the round stock in order to make the handles. Cut 6″ out of a 1-1/2″ long piece of round stock using an abrasive chop saw. The bend requires using a Shop Outfitter Compact Bender to bend the stock to 90 degree angles. Fist heat up the very middle of the 6″ piece of round stock, then take it to the bender and apply the right amount of pressure to get the correct angle. Grind down the ends to make them smooth.

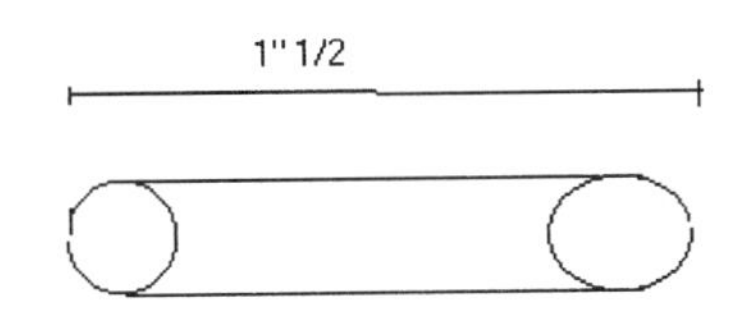

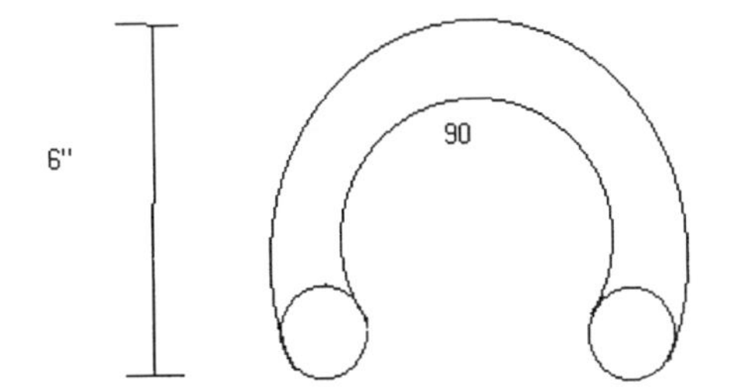

After bending the round stock handles, use a MIG welder set at 20 volts and 200 ipm wire speed to tack each end, making sure they are square and even across the center. Then weld the inside and exterior of the handles.

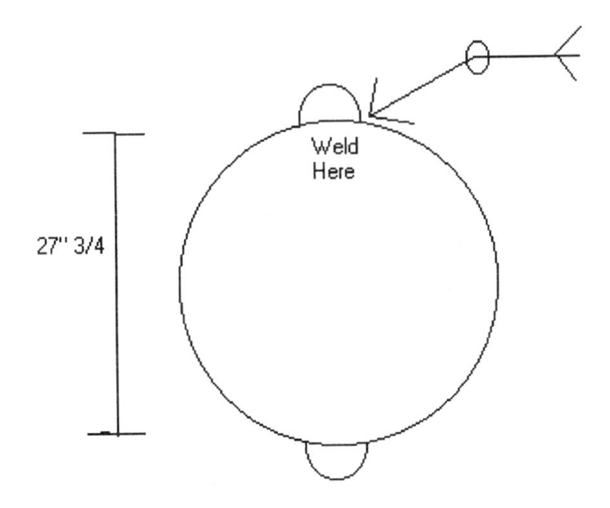

Using 36″ of 1″ x 1/8″ flat, carefully tack the lip around the pan. After all the tacks are complete, skip-weld a 2-1/2″ gap of weld each time for strength and sturdiness.

Recreational Park Grill

Authors: Steven O. Alexander and Terry L. Williams
Instructor: Charles Bell
School: Central Carolina Community College
City & State: Sanford, NC

Bill of Materials

Qty	*Material*	*Size*
2	Flat Bar	1/4″ x 8″ x 14″
1	Flat Bar	1/4″ x 8″ x 20″
1	Flat Bar	1/4″ x 14″ x 20″
2	Angle Iron	1″ x 1″ x 3/16″ x 13-1/4″
4	Angle Iron	1″ x 1″ x 3/16″ x 11-1/4″
1	Flat Bar	1/4″ x 1″ x 18″
1	Round Bar	5/8″ x 40″ long
17	Round Bar	3/8″ x 14.5″ long
2	Round Bar	1/2″ x 14.5″ long
1	Round Bar	5/8″ x 19.5″ long
2	Pipe	3/4″ Sch. 40 x 6″ long
1	Pipe	3″ Sch. 40 x 4″ long
2	Flat Bar	1/4″ x 1″ x 8-1/2″
1	Post/Pipe	2-1/2″ Sch. 40 x 48″ long

Step 1
Select a 21′ piece of 1/4″ flat bar and a 1″ x 1″ x 3/16″ angle iron. Using a plasma cutter, cut out two 8″ x 14″ flat bar for the left and right side of the grill. Next, cut out one 8″ x 20″ flat bar for the grill back plate, and one 14″ x 20″ flat bar for the grill bottom plate. Smooth the edges using a side grinder and a sander.

Step 2
Cut two 1″ x 1″ x 3/16″ x 13-1/4″ angle iron and four 1″ x 1″ x 3/16″ x 11-1/4″ angle iron. Using a tri-square, draw a line down the back end of the left and right side plates. Draw the line a 1/4″ off the back end. Since the angle iron will be attached to the left and right side plates with a one inch space between the angle iron from the top and in between, cut two extra angle iron pieces 2″ long and 1″ wide to use as spacers. This will save having to measure out each individual line.

Step 3
On the right side plate, place a one-inch spacer at the top. Then put the 1″ x 1″ x 3/16″ x 11-1/4″ angle iron one quarter inch from the back edge of the plate under the spacer. Clamp each end of the angle iron down. Now, put another spacer under that angle iron and place another 1″ x 1″ x 3/16″ x 11-1/4″ angle iron under that spacer and clamp it down. Place another spacer under that angle iron, put a 1″ x 1″ x 3/16″ x 13-1/4″ angle iron under that spacer and clamp it down.

Step 4
Tack weld each end of the angle iron, top and bottom, using the MIG process with wire feed speed (WFS) of 275 ipm and 19.5 volts with 0.035 wire size.

Step 5
Repeat Step 3 with the left side plate. MIG weld a 3/16″ staggered, intermittent fillet weld on the angle iron that was previously tack welded on the left and right side plate, after removing all of the clamps.

Step 6
Take the back plate and measure out a one-inch dimension to be cut with a tri-square and cut them both off using a Piranha Ironworker.

Step 7
Lightly grind the outer edges, front and reverse sides of the left, right, bottom and back plates.

Step 8
Line up the back end of the left and right side plate to the front side of each end of the back plate (from above, it looks like a U-shape). Ninety degree shaped magnets can be used to hold the plates in place. Make sure to form two outside corner joints on each end of the back plate.

Step 9
Tack weld the left, right and back plate together. Now, turn the weldment upside-down and tack weld the bottom plate to the first assembly, making sure to form an outside corner joint. Use a hammer to flatten any gaps. MIG weld all outside corner joints with a fillet weld using a WFS of 320 ipm and 19.5 volts, with 0.035 wire size.

Step 10
Grind all fillet welds flush on the outside corner joints. Grind the mill scale off the front end of the grill (bottom plate) about an inch back from the edge. Repeat the process on the opposite side, so that a 1/4" x 1" x 18" flat bar can be attached to it. Clamp the flat bar to the table and grind the mill scale off of both grill sides. Using a sander, round the two edges on the front end of the left and right side plates (for safety) and smooth the front ends of the six angle iron pieces.

Step 11
Switching the welding machine over to the TIG process, position the flat bar at 30 degrees. Tack weld the flat bar (with mill scale removed) to the front end of the bottom plate, making sure there is an inch of space from each end of the flat bar to the end of the bottom plate. This flat bar prevents charcoal from rolling out of the grill bottom onto the ground.

Step 12
Cut one 5/8" x 40" round bar, seventeen 3/8" x 14.5" round bar, two 1/2" x 14.5" round bar, and one 5/8" x 19.5" round bar, as noted on the materials list.

Step 13
Using a tape measure, mark off an 8" dimension and a 31" dimension starting from the end of 5/8" x 40" round bar. Place the round bar in a jig to be bent. Using a torch with a rosebud tip, heat the bar until it is red hot, then bend it to a 90 degree angle (making sure not to overheat the metal, which will lead to metal deficiencies). Using a tri-square, check to make sure the bent angle is square. If it is not, keep re-bending it until it is. Then submerge the bent end into a container of water to cool. Place the round bar back in the jig, clamp it down and measure out another eight inches to bend, starting from the 31" mark. Repeat the procedure done to the previous bend. While the round bar is still in the jig, use a pocket-sized magnetic level to make sure that both bends are level. Cut the end off of the longer bent end to make both bent ends an even 8".

Step 14
Place the 5/8" x 19.5" round bar on the table. Place both the 1/2" x 14.5" round bars, one at each end of the 5/8" x 19.5 round bar to form a U-shape. Change the welding machine back over to the MIG process, using the same settings as earlier. Tack weld the U-shape together, then check it with a tri-square to make sure the corners are square.

Step 15
Place the round bar on the table with the two bent ends facing forward. Take the U-shaped weldment (Step 14), turn it upside-down, and place it in the middle of the round bar with two bent ends (Step 13), making sure the U-shape is facing you. Check to make sure there is equal spacing on both sides of the U-shape and between the bent ends. Now tack weld both ends of the U-shaped weldment to the bent ends.

Step 16
Place two 7/8" spacers on the inside of the U-shaped vertical bars on each end. Place a 1/2" x 14.5" round bar up against the spacers on the left and right sides, making sure the bars rest on the table. Tack weld both bars to the frame. Next, remove the 7/8" spacers and put two 1/2" spacers against the two 1/2" x 14.5" round bars. Now, place a 3/8" x 14.5" long round bar on the left and right side against the 1/2" spacers. Tack weld the 3/8" round bar to the frame. Repeat this process until all 3/8" round bars have been tack welded to the frame. Flip the grate over and finish welding the bars to the frame. This completes the grate for the grill.

Fabricator's Note: *While adding the 3/8" bars to the frame, I noticed while trying to fit the bars to the frame that they became too tight to install. To eliminate the problem, I took about six bars to the bench grinder and shaved a little bit off the ends. Also, I found that taking one of the 3/8" bars and placing it in the middle of grate helped cut down on the grate pulling together, which was the reason for the tight fit. After flipping the grate over, I kept a slight tension on the bars to keep the bar from pulling away while I finished welding.*

Step 17
Cut two, 3/4" Sch. 40 x 6" long pipe and grind the burrs off the end of the pipe. Place the grate in a vise, with the two bent ends facing up. Place one of the pipes over the bent ends of the grate. Then put a piece of bent 0.045 MIG wire between the inner pipe wall and the bent end of the grate, so that there will be an even amount of space between the pipe and the

bent end. This will also protect the user from burning his or her hands while moving the grate.

Make sure that the pipe is 1/4″ down from the bent end edge. Place a tack weld on each end of the pipe, remove the MIG wire, and place another tack weld on each end of the pipe. Using the MIG wire to hold the pipe in place while welding it to the bent end, repeat the process for the other bent end. Now, MIG weld both ends of the pipe on both bent ends. (These two pipes serve as handles for the grate).

Step 18
Cut out two 1/4″ x 1″ x 8-1/2″ pieces of flat bar. Measure and mark 1-1/4″ on both pieces. Place one flat bar in a vise with the 1-1/4″ sticking out. Using a torch with a rosebud tip, heat the flat bar until it turns red hot. Hit it with a hammer to bend it to a 90-degree angle. Using a tri-square, make sure the bend is square, then cool the flat bar in a container of water. Repeat this step with the other flat bar.

Step 19
Place both of the flat bars in a vise with the bent end at the top. Using a sander, smooth out a 30-degree bevel angle across the edges of the bent ends. Place one of the flat bars at the front of the grill, 1/4″ back from the front bottom plate edge, with the bent end resting horizontally on the top edge of the left side plate. Place two tack welds inside the bevel groove of the flat bar. Use a magnetic pocket level to make sure it is plumb. Then place two tack welds on the bottom of the flat bar, attaching it to the bottom plate of the grill. MIG weld a 1/4″ fillet weld (arrow side) across the bottom of the flat bar and bevel groove weld (arrow side) across the top. Repeat this process on the right side of the grill. (The flat bars that are welded on the front of the grill serve as a locking mechanism preventing the grate from being removed.)

Step 20
Turn the grill upside down and mark the center. Take a 3″ Sch. 40 x 4″ long pipe, put it in a vise, and grind the galvanized material off an inch back from the edge all the way around. MIG weld the pipe to the center of the grill with a 1/4″ fillet weld (arrow side) all the way around. Cut a piece of 2-1/2″ galvanized pipe 4′ long. Insert the 2-1/2″ pipe inside the 3″ pipe and weld it all the way around. The 2-2/2″ pipe serves as the grill post.

This completes the grill fabrication process.

Step 21
To install the grill, use a post hole digger and dig a hole in the ground 2′ deep. Drive 4 metal stakes in the ground approximately 4′ away from the hole and place the grill in the hole. Next, tie rope to the grill and the stakes to hold the grill level. Pour concrete into the hole and allow plenty of time for it to harden. Remove the ropes and the metal stakes and get ready to grill.

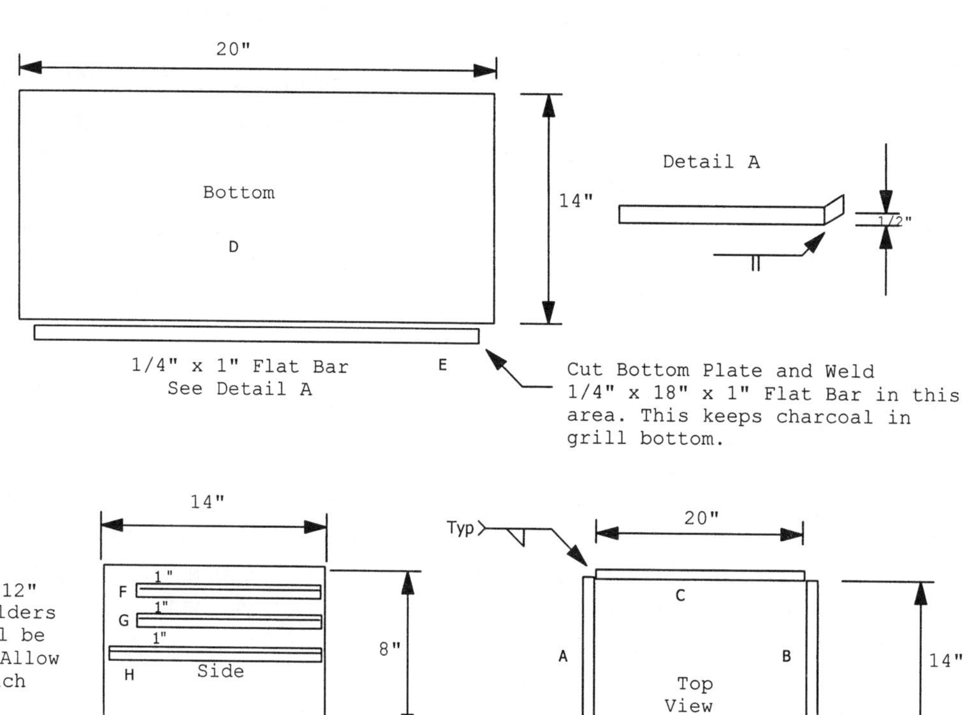

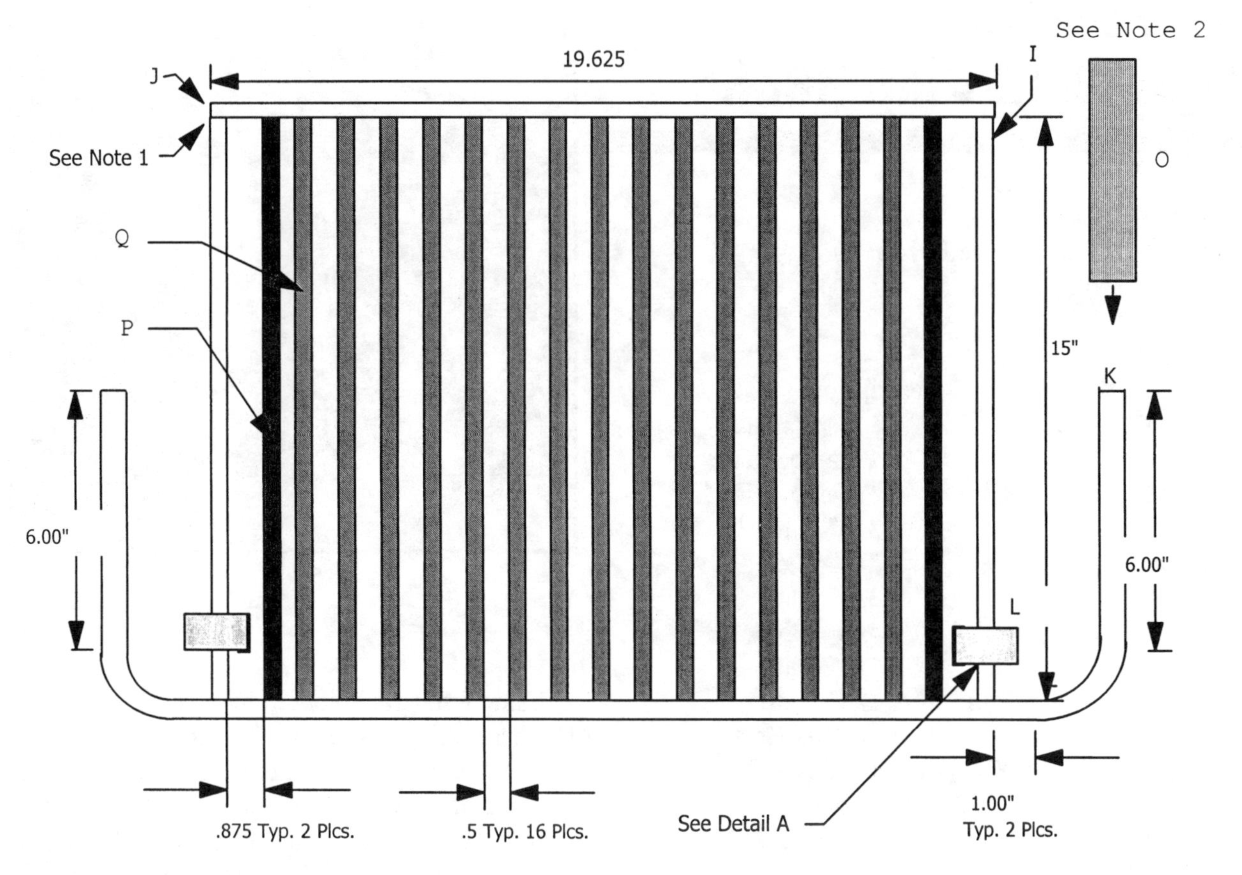

Note 1: Weld all rods on two sides,
Note 2: Weld 3/4" Pipe all the way around 2 sides

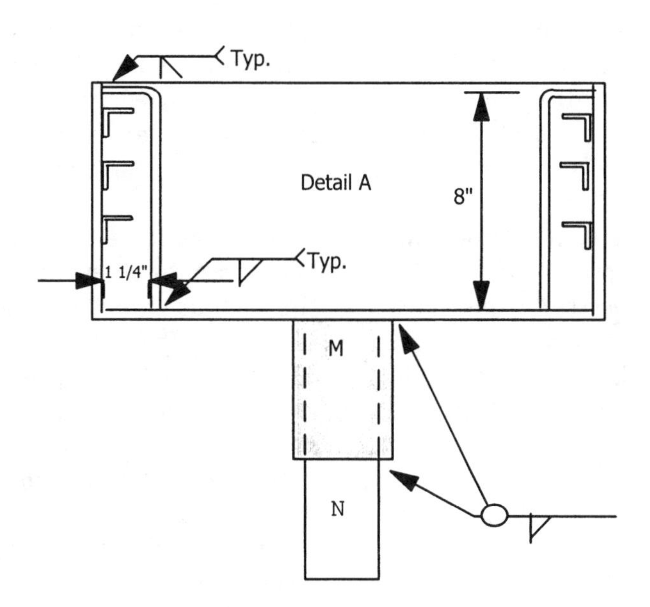

Hitch-Carried Recumbent Bike Rack

Author: Josh Kessler
Instructor: David H. Murray
School: Ferris State University
City & State: Big Rapids, MI

Cut List

Qty	Length	Description
1	76″	2″ x 11ga Sq. tube
1	6″	2″ x 11ga Sq. tube
1	18″	2″ x 11ga Sq. tube
2	4.25″	2″ x 3/8″ Flat
1	40.75″	2″ x 3/8″ Flat
3	6″	2″ x 3/8″ Flat
1	2″	2″ x 3/8″ Flat
14	3″	2″ x 1/4″ Flat
2	11″	2″ x 1/4″ Flat
2	28″	1″OD x 11ga Md tube
4	1-1/2″ x 1-1/2″	1/4″ Sheet
3	1-3/4″ x 1-3/4″	1/16″ Sheet
1	6″	3/4″ x 3/8″ Flat
1	1.125″	3/4″ x 3/8″ Flat
1	20.625″	3/4″ x 3/8″ Flat
1	15.791″	3/4″ x 3/8″ Flat
1	21″	3/4″ x 3/8″ Flat

Introduction

This custom bike rack was designed to mount to the hitch of a Chevrolet pickup truck and carry a specific recumbent bike and also any two other bikes. The original design came from a rack sold by the Draftmaster bike rack company.

The rack is built entirely out of steel. Welding consumables are E7018 for the SMAW process and ER70S-3 for both GMAW and GTAW processes. Three purchased brackets are designed to clamp onto the front fork of the bike. A set of straps is used to help hold the rack stationary while it is carried down the road.

Fabrication Sequence

Drawing A shows the order in which all subassemblies are to be made.

1. Cut all materials to the given sizes and in the given proportions in the cut list.

2. Using a flat table, begin by assembling the parts for the rack supports using Drawing B and Drawing C. Also, while doing this part of the fabrication, the two pieces from the hitch assembly with the holes cut in them will be used. Use a drill press with a 0.385 diameter drill bit to put all of the holes into the pieces requiring them.

When welding the four pieces involved in the pivot, it is important to make sure that the holes line up. This can be done by the use of locating pins. Press the pins through the holes in the two hitch pieces and through the holes of the support pieces with the support sandwiched between them. This particular step should be done first because if the holes do not line up correctly, then the rest of the assembly will fail.

3. Next, fabricate the main pole that the recumbent bike fits onto (Drawing D). The pieces that the rear tire mounts into are at an angle and require some setup to be positioned accurately. To do this, use an adjustable angle clamp set to 110 degrees. Then clamp it to the tire-mounting tab and to the main pole. Position the tab on a pre-scribed line measuring 0.375″. Then tack the tab into place. Next, put the tab directly above the one just positioned. To put this one on, using c-clamps, clamp a piece of any straight material to the already tacked tab and then to the tab yet to be welded. Position it four inches above the bottom of the main pole and tack into place. Do the same for the other two tabs, only they are a mirror image of the first two.

Produce subassemblies in this order:

1. Supports (Drawings B & C)
2. Main Pole (Drawing D)
3. Side Bike Fork Mounts (Drawing E)
4. Side Bike Rear Tire Mounts (Drawing F)
5. Hitch Piece (Drawing G)

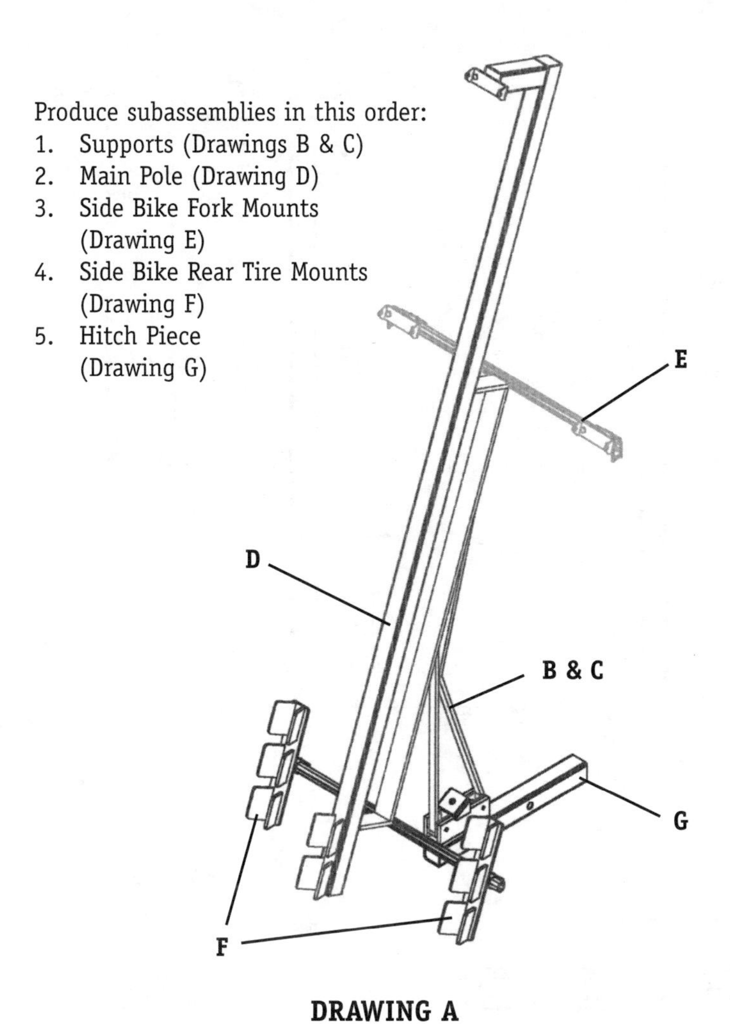

DRAWING A

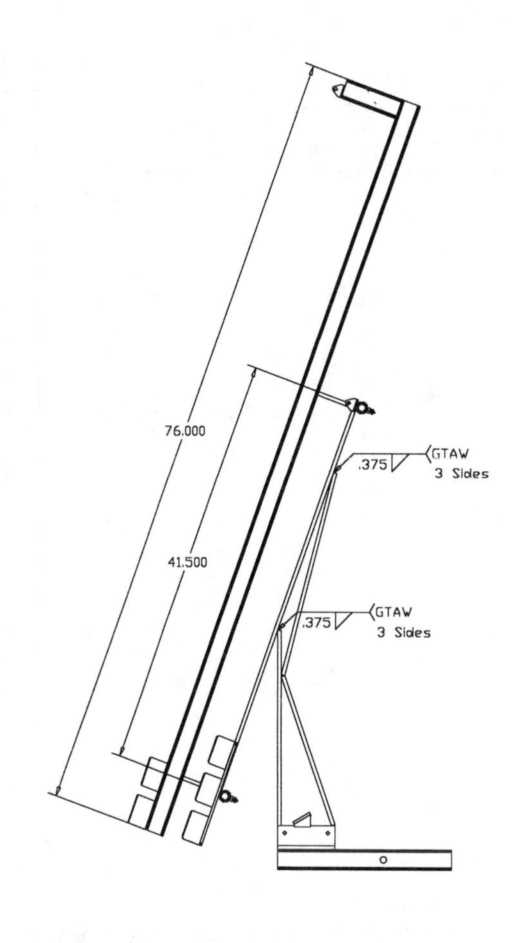

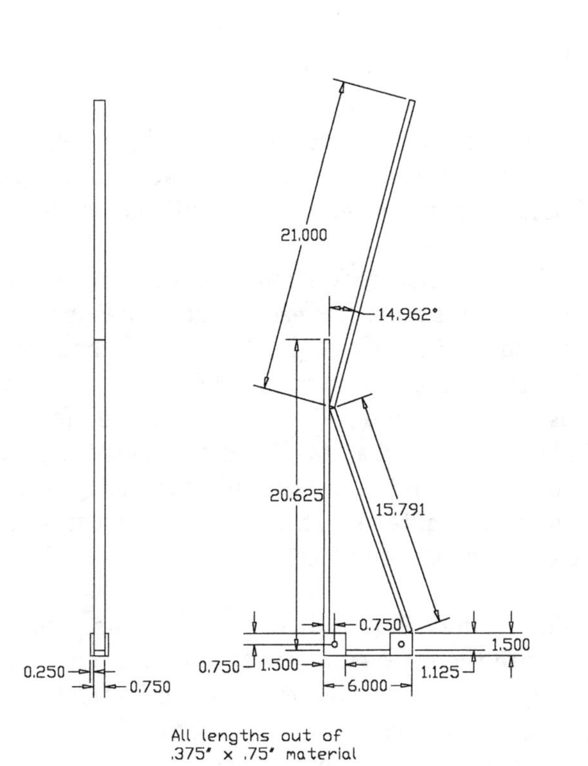

DRAWING B

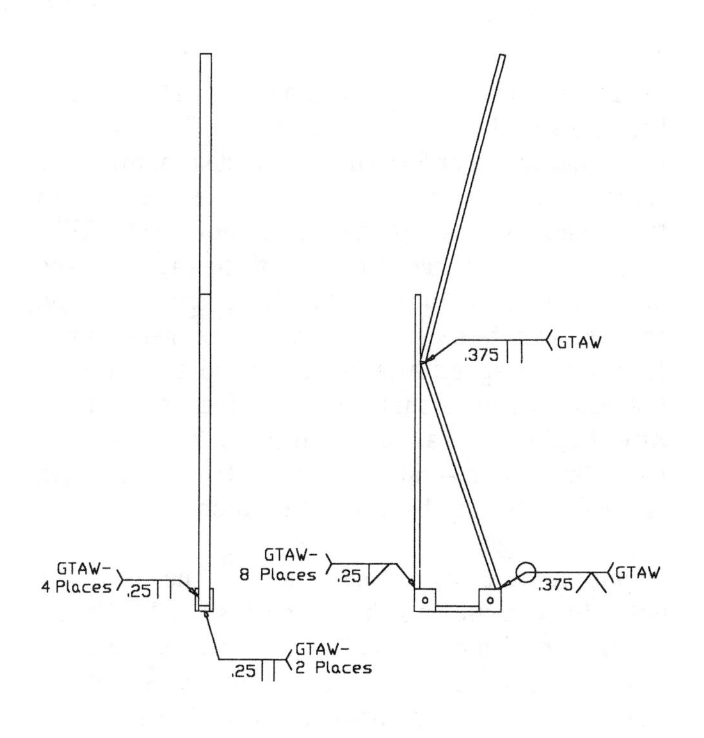

DRAWING C

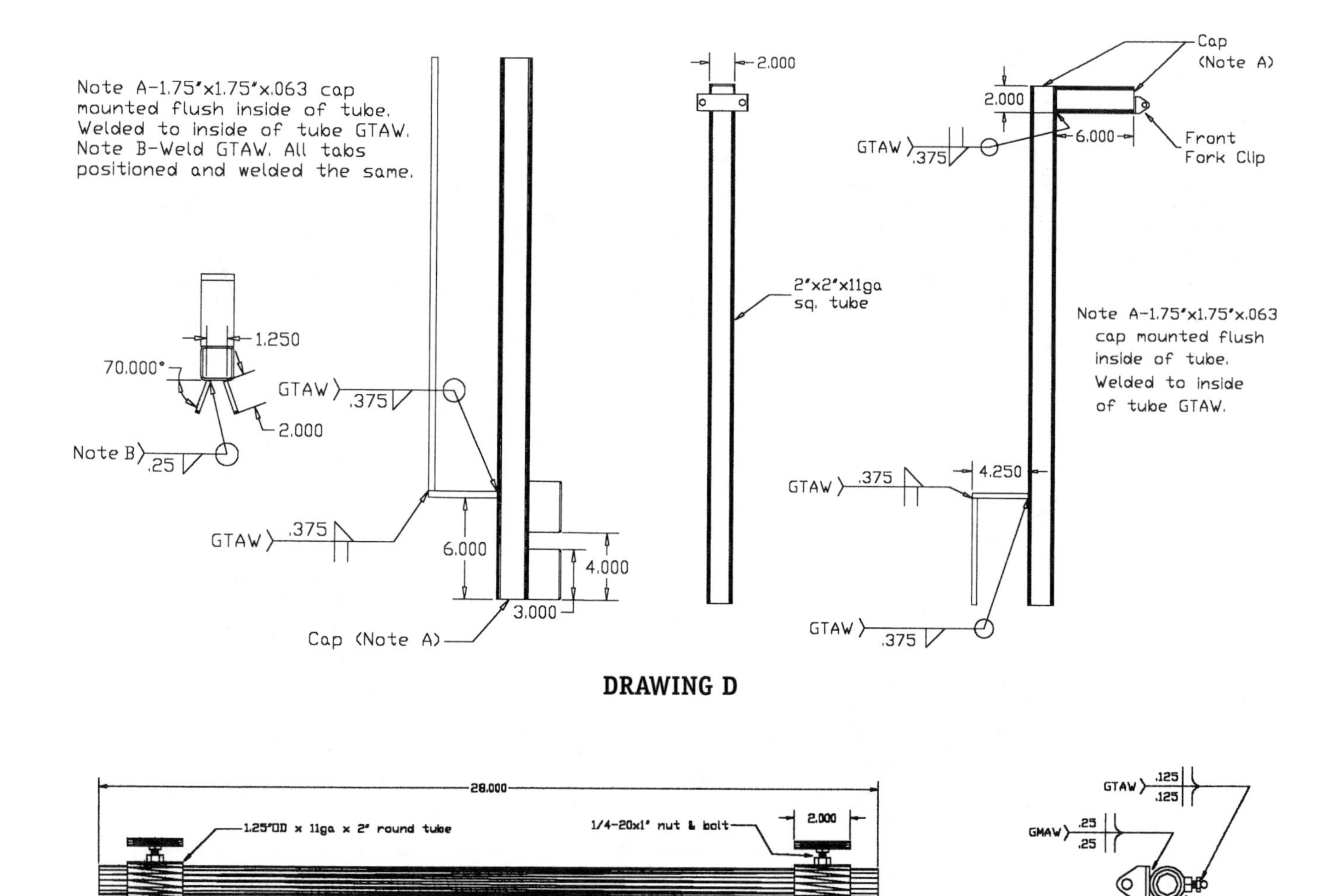

DRAWING D

DRAWING E

Weld three caps onto this main pole: one at the top of the pole and at the bottom, and one at the end of the piece that the front fork clip gets welded to. Grind the corners off from the pieces being used as caps so these pieces will be able to sit inside of the tube. Then position a magnet across the top of the cap piece and on the end of the tube, holding the cap in a position that will facilitate welding. Tack the cap piece on and remove the magnet. After removing the magnet, weld the cap entirely around to the inside of the tube. After welding the cap on, grind the welds, mostly for looks, but also to allow for the piece that has the fork clip welded to it to be in a flat position.

To put on the 6" piece that mounts to the top of the pole, the fork clip must be put on first. Weld the clip on the end with cap in it. This piece was put into the design to allow for clearance of the fender on the front tire. The pieces put onto the back are used to help give clearance to all the bikes and also some flex in the supports. These three pieces should be welded together first, with the long piece being welded in between the shorter two. The two shorter pieces should then be welded to the main pole of the rack.

4. The side bike fork mounts and rear tire mounts were made similar to each other (Drawing E and Drawing F). The first step is the same on both pieces. First the pieces (four total) sliding onto the poles must have a clearance hole drilled in them for a 1/4-20 bolt. Then weld a 1/4-20 nut around the hole. Weld a handle to the top of the nut, allowing for the bolt to be hand tightened. These bolts act as a setscrew to hold the mounts in place.

Weld the fork clip and rear tire mounts into place opposite the setscrews after they have been finished, though the tire mounts must be fabricated first. Use the same method as for the rear tire mount on the main pole. At this point, these two pieces can be welded to the main pole assembly.

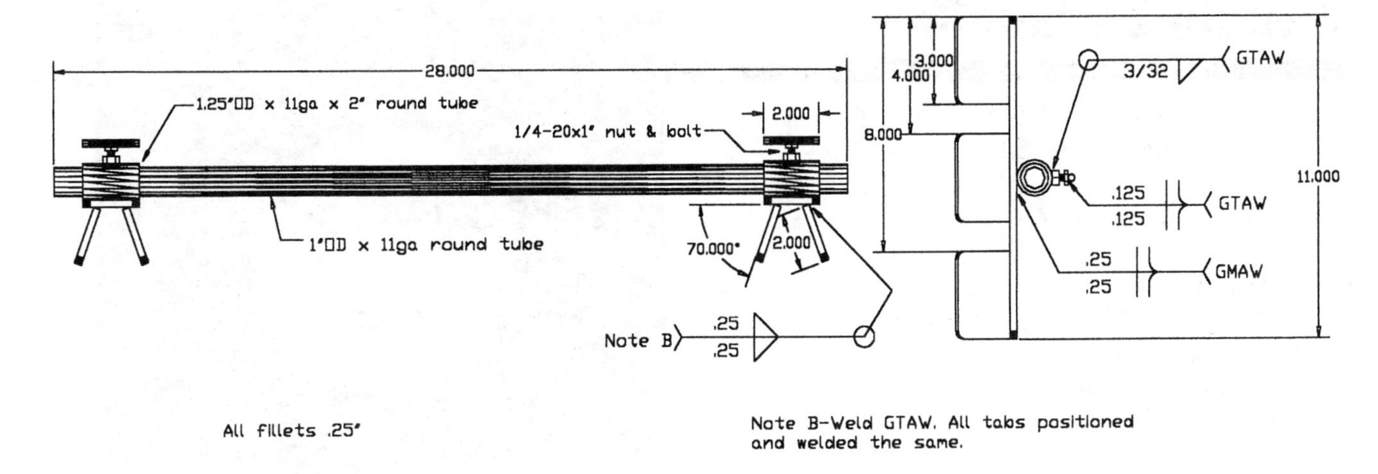

DRAWING F

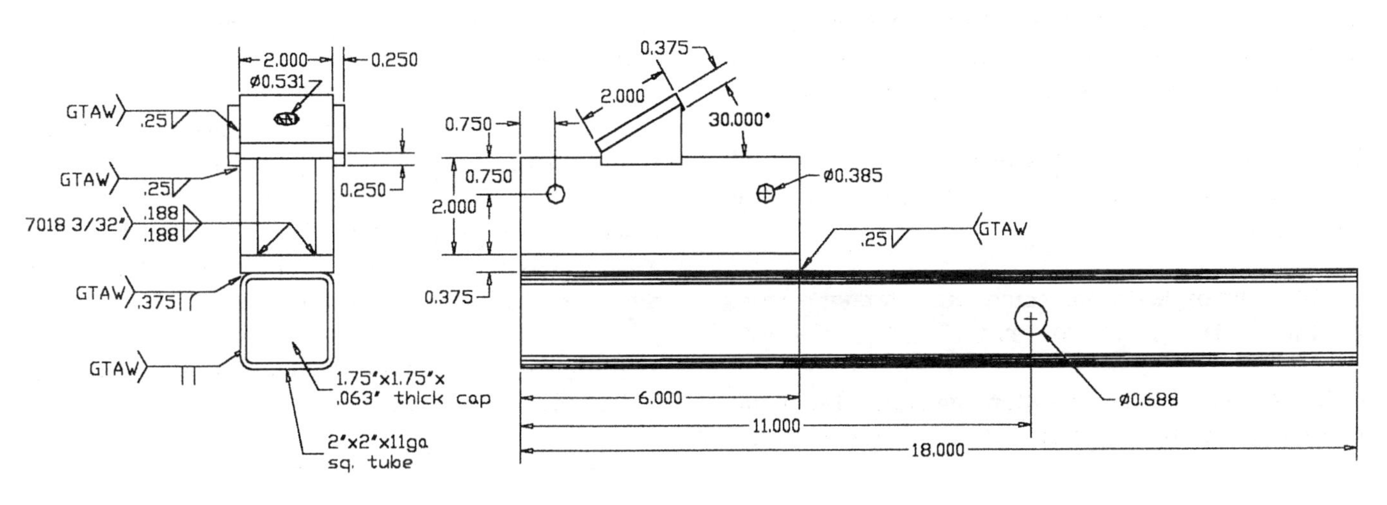

DRAWING G

5. The hitch piece is the final subassembly (Drawing G). The hitch receiver is for a 2″ hitch so the same 2″ sq. tubing will be used for this. First, put a hole to clear a 5/8″ pin in the tube. Put a cap was put onto the end furthest from the hole in the same manor that the caps were put on previously. Then put the 6″ pieces of 2″ x 0.375″ thick flat bar without the holes in it onto that same end with the ends meeting flush. Tack on the other two pieces of the same size. Insert the support subassembly between these two pieces and secure their positions with a locating pin. Then remove the support assembly and place all the welds onto these pieces. After all these welds are done, put the support assembly back between the just welded pieces and reinsert the pins. The support subassembly must be in the position described at this point or it will not be able to be put back in later.

Next, put the pivot stopping mechanism on. The two sidepieces must be cut on the shear at this point. Weld them onto the 2″ piece of 2″ x 0.375″ material with a hole already drilled in it for a 1/2″ bolt clearance. Then weld this piece onto the rest of the hitch subassembly.

6. At this point, the support and hitch subassemblies are ready to be welded to the rest of the assembly. Do this by positioning the support piece onto the back of the other assembly and welding the two contact points into place.

7. Clean up the rack using a scotch-brite pad and mineral spirits to remove any grease or dirt that may be on the parts. Wipe the rack clean of mineral spirits using a rag and paint it.

Smiling Sunshine

Author: Chris Chatterton
Instructor: Nicholas Regets
School: William D. Ford Vo-Tech Center
City & State: Westland, MI

Materials List

30″ x 30″ piece of 1/8″ thick steel plate
10 ft. of 3/16″ steel rod
5 ft. of 1/8″ steel rod
Ball bearings
Paint marker
Paint in almond, aqua, gray, dark blue, olive green, violet
Clear Coat

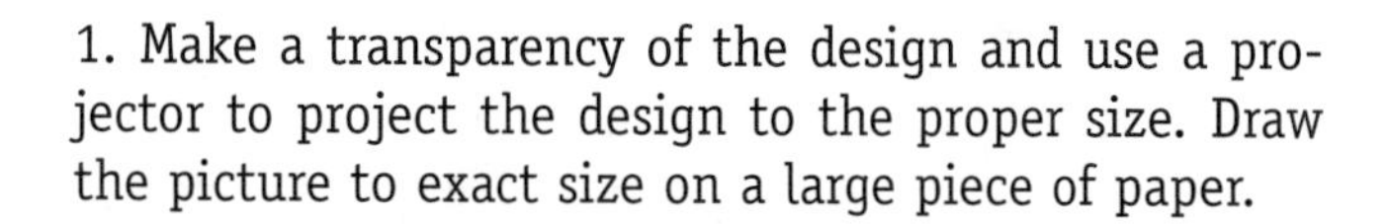

1. Make a transparency of the design and use a projector to project the design to the proper size. Draw the picture to exact size on a large piece of paper.

2. Cut out template and place on 30″ x 30″ piece of 1/8″ thick steel plate. Trace the template with the paint marker.

3. Using an oxyacetylene torch and a straight edge, cut out the triangle at the sun, then cut out the rest.

4. Use a hand grinder to soften the edges. Use a file to smooth out rough parts. Cut the pieces of 3/16″ steel rod to size, and using an anvil and a hammer pound them into shape.

5. Using a hand grinder, bevel the edges and clean up the inside corners.

6. Tack on the pieces of 3/16″ rod for the face using a TIG welder. Cut and tack together ball bearings and rods, then tack to the Sunshine.

7. Paint the Sunshine in all different colors. After it is dry, add a clear coat.

Western Lamp

Author: Kevin Duran
Instructor: Curtis Willems
School: Highland High School
City & State: Gilbert, AZ

Materials Used

1 Light Receptacle
1 Pipe Nipple
6′ Electrical Cord
6 Horse (Pony) Shoes
1 1-1/2″ Black Pipe
1 Lamp Shade
1 Cord Cap
Paint

Procedure

Make a drawing to determine dimensions.

Wire brush the pony shoes thoroughly.

Place the pony shoes on a metal slab and lay them out in their proper measurements.

Place three of the pony shoes facing outwards to make the base. Use C-clamps to hold down each shoe so they won't bow up with the heat.

Weld the base together and then use a hand-held grinder to smooth out the welds and wire brush them again.

Next, take the other three shoes that will be welded to the round shaft that will hold the electrical cord, and lay them out, making sure to keep them straight.

Place C-clamps on each shoe and weld them together. Repeat the grinding and wire brushing steps noted above.

Take an 11-1/2″ piece of black pipe and wire brush it. Thread it with a 3/8″ national pipe thread tap so that you can attach the pipe nipple and light receptacle.

After threading the pipe, weld it to the back of the three standing horse shoes. Attach the three standing shoes to the base.

Using a handheld wire brush, clean the metal and prepare it for painting.

The electrical can be purchased as a kit, with 6′ of electrical cord, one cord cap and one light receptacle.

First, twist the nipple into the pipe to make sure that the threads are good. Then attach the electrical cord to the light receptacle and thread it down through the pipe.

Once it is all the way through, attach the cord cap and plug it in with a light bulb to make sure you did not make any mistakes in wiring the light.

Remove the bulb and tape around the areas you don't want paint on. Paint the lamp base. After the paint is dry, install the bulb and the lamp shade.

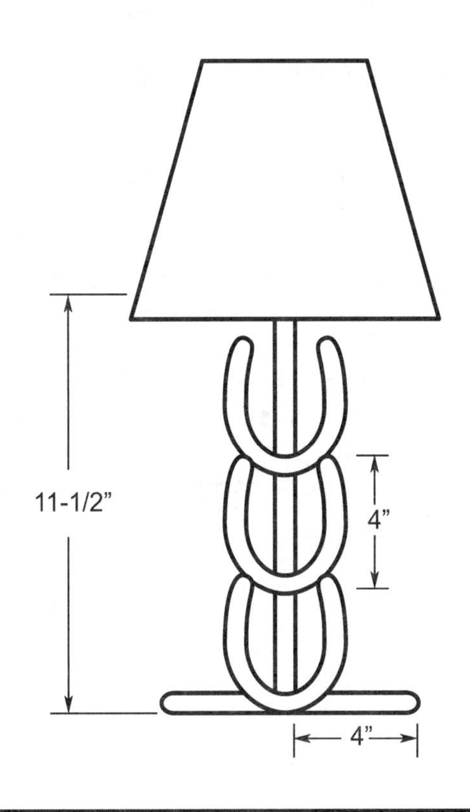

Rolling Flower Cart

Author: James A. Perkins
Instructor: Cliff Guyette
School: Franklin County Technical School
City & State: Turners Falls, MA

Materials

3/16" x 3/4" flat stock
Expandable metal for bottom and sides
Spray paint

Step-By-Step Procedure

1. For the main frame corner pieces: tack four pairs or eight pieces 4" long to form 90 degree corner angles.

2. For main cart frame: fabricate two rectangular frames, 8" x 12". Fabricate one frame on top of the other using clamps and short pieces of steel to maintain alignment. The frames need to be matched pairs.

3. Align, square and tack these two rectangular frames to corner angle pieces. Lay out and shear main frame bottom support pieces. Align and tack weld into position.

4. Shear expandable metal to fit inside of main frame. The expandable metal must be placed 3/8" below the top edge to eliminate sharp edges. Tack the sections in place.

5. Lay out, shear and bend axle support brackets.

6. Carefully align and tack axle supports into position on the bottom of the cart.

7. Using a chop saw, cut two pieces of round stock suitable for axle bushings.

8. Drill these bushings slightly over the size of the actual axle diameter. Loose fit.

9. Fit axles through bushings and align on axle supports. Carefully tack weld bushings into position.

10. Fabricate pull handle components. Align and tack weld into position.

11. Using hand rolls, fabricate two wheels. The wheels are 4-1/2" in diameter. The circumference formula is diameter X 3.1416.

12. Lay out wheel spokes, five spokes per wheel.

13. Lay out the wheels and spokes using a sheet metal template. Lay out a square that fits your wheel, then draw a line corner to corner. This is used to establish a center line or hub location. Using a pair of dividers, walk them around the wheel circumference pattern until you have them divided into five equal spoke locations. Use this template to carefully tack your wheel components into place. Then tack wheel assemblies to main axle.

14. Using a flat surface, raise your cart up on its wheels until it is perfectly level. Measure from the bottom of the cart to the flat surface. This will be the height of the cart support brackets. Fabricate and install these parts to the required height dimension.

15. Carefully tune in your welder on scrap material. Weld the flower cart together, taking care to maintain alignment and to minimize distortion.

16. Check project for any sharp edges. Clean and prepare cart for painting. Spray paint the flower cart.

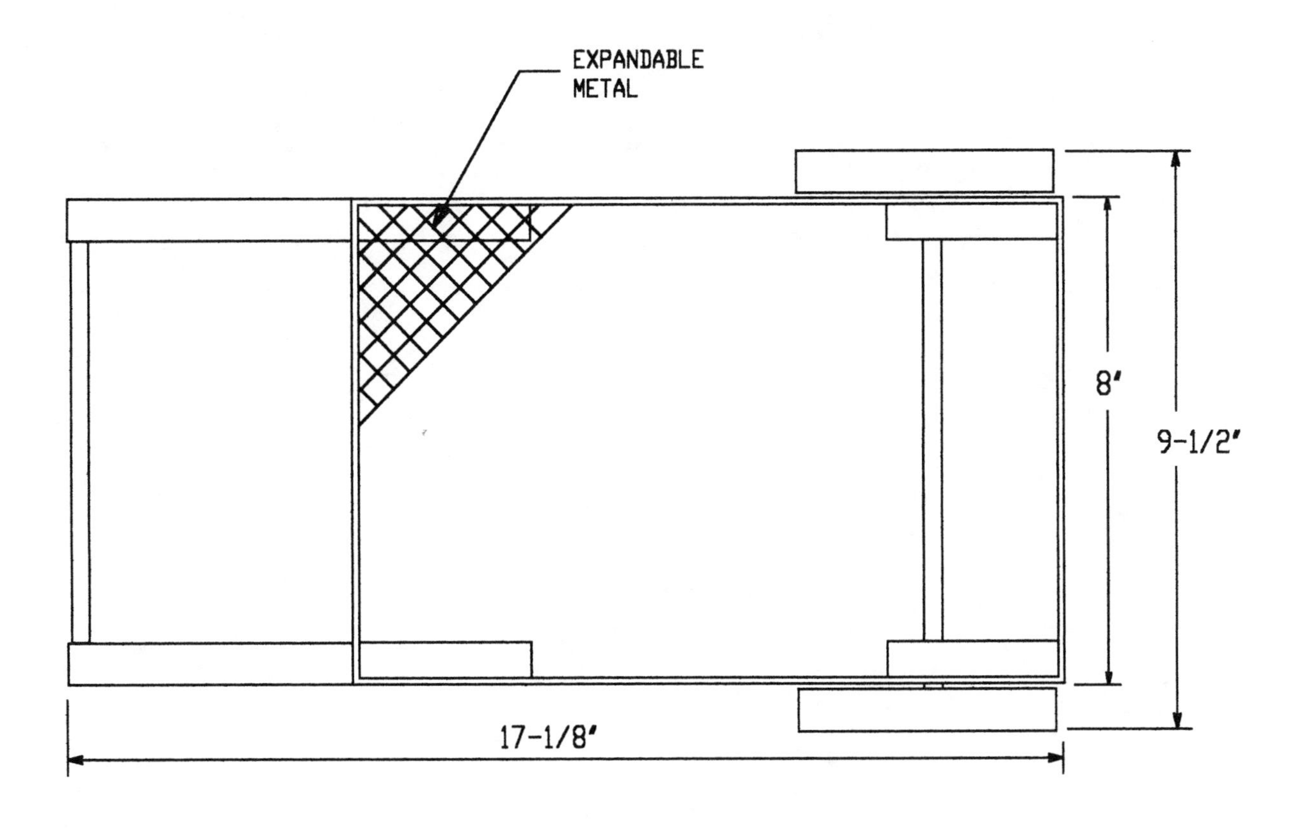
EXPANDABLE
METAL
8'
9-1/2"
17-1/8"

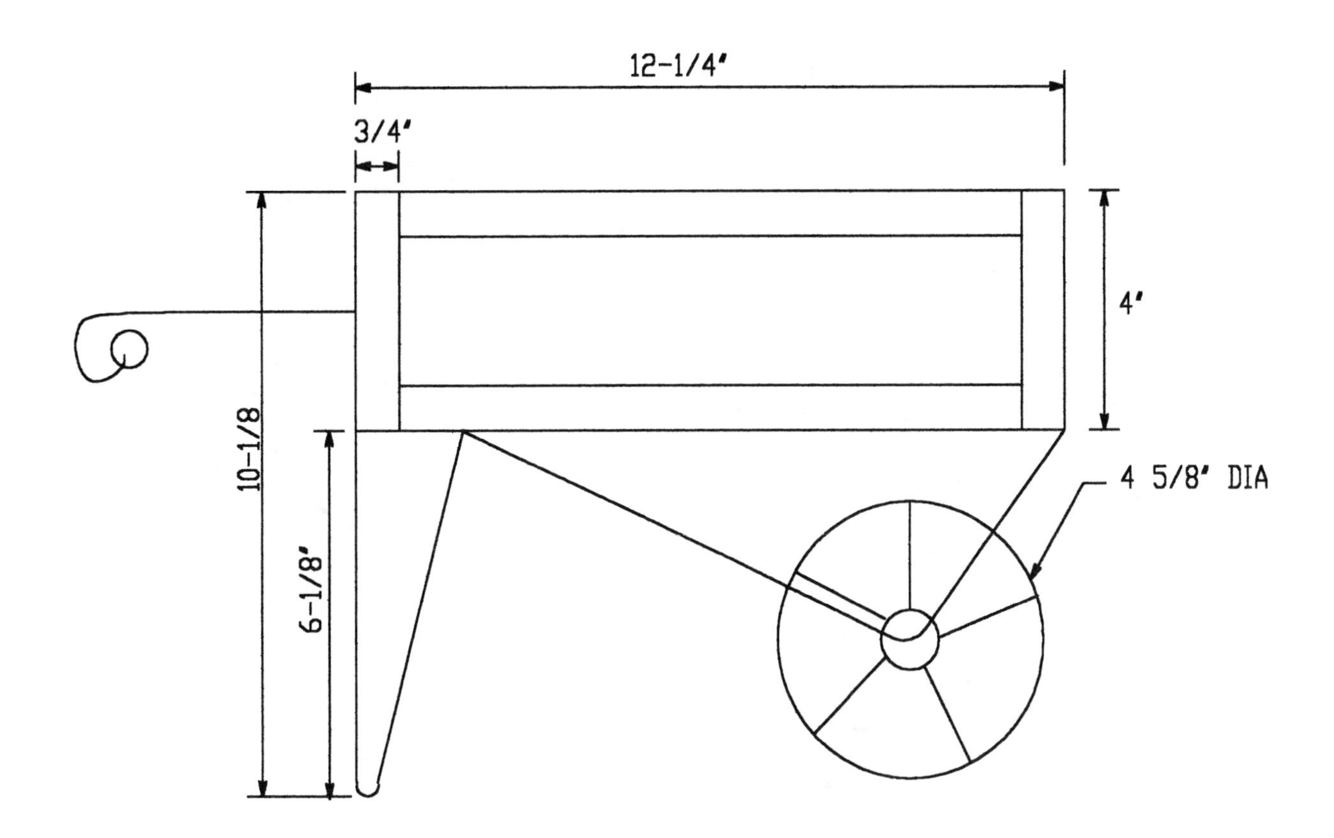
12-1/4"
3/4"
4"
10-1/8
6-1/8"
4 5/8" DIA

Miscellaneous

Table of Contents

Eight Foot Snow Pusher

Author: Mitchell Reker
Instructor: Martin Burns
School: Adrian High School
City & State: Adrian, MN

Materials

Two 4'x8'x1/4" sheets of flat steel for sides and back
20' of 1-1/2" angle iron for side edges
24' of 4" channel for support on back of blade
Quick-tach coupler to mount with skid loader
6' of 2"x2" square tubing for side support
8' of rubber cutting edge
8' of 3"x1/4" flat steel for holding on the cutting edge
7' of 5"x3/4" steel for skids
(Extra steel from the sides and back was used for skid frames, frog support, and capping the ends of the exposed angle iron.)

Concept

The objective is to build a skid-steer snow pusher with a blade big enough to move a lot of snow in a small amount of time, but not big enough to overwhelm the power output of a given skid-loader. This indicates an 8'x3'x3' pusher with a quick-tach coupler.

Fabrication Process

The first step on the project is to cut the sides and back to the right size out of the two sheets of flat steel. To make the sides, a lot of measuring is required to make the angles match so that the rolled steel blade will fit correctly on the sides. After drawing the outline of the first side piece, use a guide and a plasma cutter to cut it out.

Make the first side, then lay it directly on the next available piece of steel and trace around it to make sure both sides will match. Once you have the outline, move the guides and begin cutting the second side. When both are done, chip off the excess slag and grind the rough edges smooth with a hand grinder. Now that the sides are cut, mark which side to weld on the angle iron.

Cut the back piece at 40" by 8' using a plasma cutter and a guide to make sure the bottom edge will be straight. Chip and grind the edge until smooth. You will have to find access to a heavy steel roller to roll this piece at an 8" bend.

After the steel is rolled and the sides cut, begin cutting metal to be welded onto the sides for edge support to make the corners more rigid. Use 1-1/2" angle iron and a sliding T-bevel to determine the angles for each piece of angle iron. Draw approximate lines on the steel and then cut them with a plasma cutter. If any modifications are needed for a better fit, use a bench grinder to trim the edges.

Once all the sections are cut, tack them in place using the GMAW process. With all the pieces in place, clamp a large piece of I-beam onto the bottom to help prevent warping. To further prevent warping, weld only a small amount at a time, and go from side to side to reduce the amount of heat in any one area. It is a good idea to do the side pieces first so that all the welding can be done on the table, and the first welds do not have to be in the vertical position.

Cut the end pieces at angles and fill with a small piece of 1/4" flat steel to dress up the sides and keep water and snow from building up inside the angle iron, causing it to rust. On the very edge, weld over the flange on the side to cover the plasma cut and also to give the angle iron more support.

With the sides completed, the next step is to weld the rolled back to the two side pieces. This is not easy. It requires a forklift to hold the back in the air while you make measurements on both the top and bottom to determine where to place the back. When the back is in place, tack weld it in several places to insure the blade won't fall apart when the chain is removed.

After unlocking the chain, use a piece of channel to attach to the front of the sides to keep them from pulling when the back is being welded on. With all warping problems taken care of, it is appropriate to weld long beads with high heat to insure deep penetration. This weld is made in both the flat and vertical position because of the curve in the blade.

Before removing the channel, cut a piece of 2" square tubing at 3', then cut angles in the steel to fit the shape of the blade and back. Attach it to the side and back to hold the side pieces at a right angle to the back. Weld the support, then cut an identical match to weld to the opposite side of the pusher. Weld the second support tubing on, remove the channel, and flip the pusher over.

Weld the channel on the back to allow more support and create a place to mount the quick-tach coupler. Measure from side to side, then cut the 4" channel 1/8" short to make it fit in easily. Measure 3/4" down from the top of the back and tack weld the channel on to keep it in place. Weld on alternate side to reduce heat, and also take breaks to prevent warping of the blade. Once the first piece of channel is welded on, measure down 11-1/2" and tack weld the next piece on. Weld it to the back and repeat the same procedure for the third piece of channel.

When all the channel is welded on, weld on the purchased quick-tach coupler using 1/2" by 4" steel to support the bottom of the coupler, making multiple passes to ensure a good, strong weld.

Drill the holes for the rubber cutting edge. Make a plate of 3" x 1/4" steel to go on the front of the rubber edge to hold it in place, and drill holes to fasten it with 1/2" bolts and nuts. Once the edge is mounted, make gussets between every bolt hole to prevent the blade from bending if it hits a large object. Leftover 1/4" steel can be used for the gussets.

Make the side skids using 5" x 3/4" abrasion-resistant steel that has been bent to be welded onto a piece of 1/4" steel that will be welded to the sides of the snow pusher. The steel will have to be pressed at a machine shop with a press heavy enough to bend the thick AR steel. Weld and bolt the skids on, and paint the snow pusher.

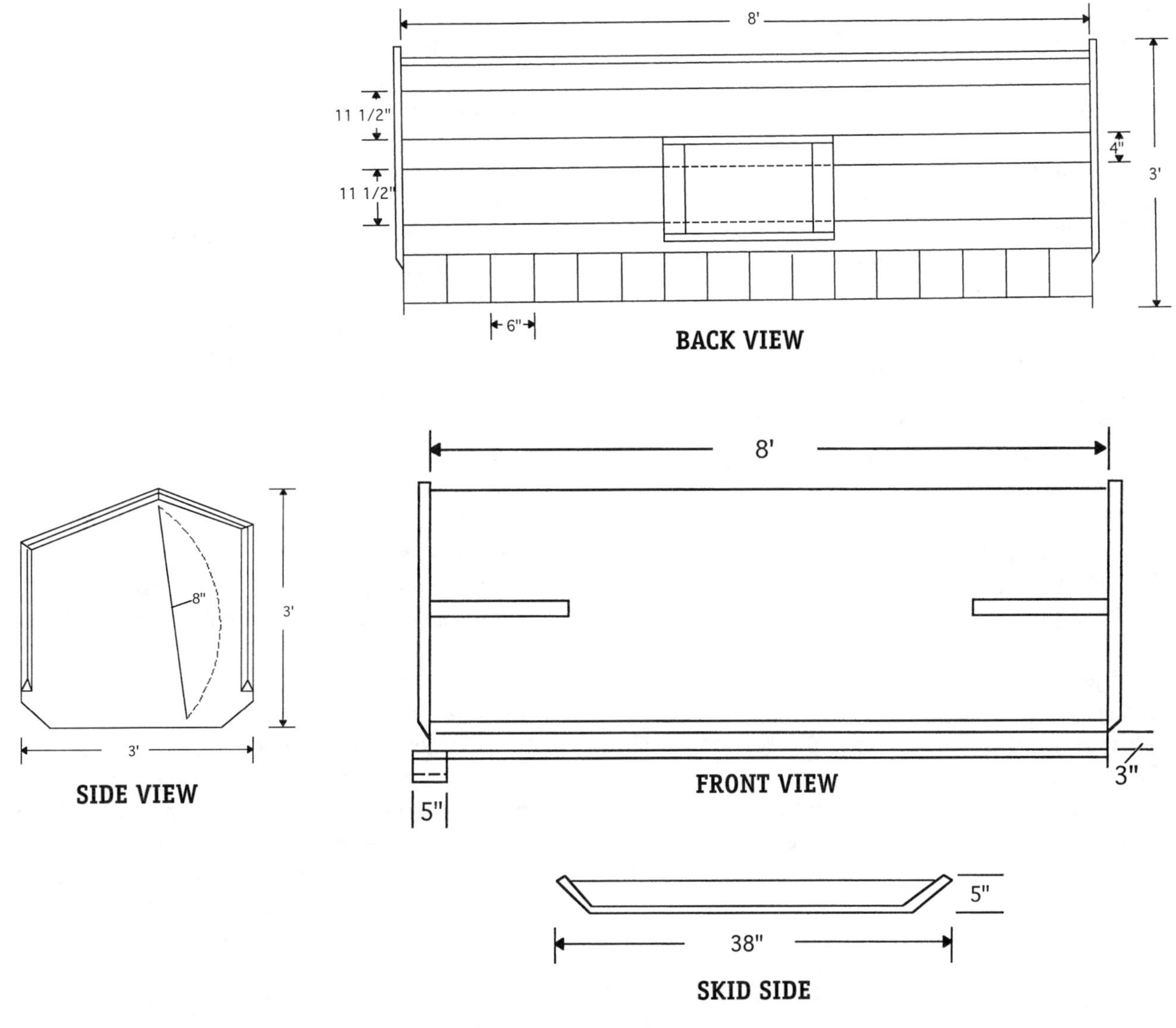

Three-Point Hitch Forks

Bill of Materials

Piece	*Qty.*	*Description*	*Length*
A	2	2"x 3" x 1/4" tube steel	64" ea.
B	2	2" x 2" x 3/16" tube steel	32" ea.
C	2	2" x 2" x 3/16" tube steel	10" ea.
D	2	2" x 2" x 3/16" tube steel	10" ea.
E	1	2" x 2" x 3/16" tube steel	30"
F	1	2" x 2" x 3/16" tube steel	17"
G	2	1/2" x 2" bar	31-3/16" ea.
H	2	1/2" x 4" bar	12" ea.
I	1	1/2" x 2" bar	16"
J	2	1/4" x 2" bar	6-3/4" ea.
K	2	1/4" x 2" bar	3" ea.
L	1	1/4" x 2" bar	2"
	2	7/8" pins	

Welding Procedure Specification #1

Job Title	1G Groove Weld
Welding Process	FCAW, Semiautomatic
Material	Mild Steel
Material thickness (T)	3/16"–1/2"
Joint design	Groove
Position	1G
Root opening	1/16"
Groove angle	NA
Land width	NA
Leg size	NA

Joint Design

Backing type	NA
Electrode # 1	E-71-T-1, Size: 0.45", Amperage: 185
Wire feed speed	280 ipm
Voltage range	24
Polarity	DCEP
Electrode stickout	1/2"
Shielding gas (type)	CO_2
Backing gas	NA

Technique

Stringer	
Travel angle	Push 70° Work angle 90°
Single pass	
Initial cleaning	wire brush
Interpass cleaning	chip, wire brush

Welding Procedure Specification #2

Job Title	2F Fillet
Welding Process	FCAW, Semiautomatic
Material	Mild Steel
Material thickness (T)	3/16"–1/2"
Joint design	Fillet
Position	2F
Root opening	NA
Groove angle	NA
Land width	NA
Leg size	Match material thickness

Joint Design

Backing type	NA
Electrode # 1	E-71-T-1, Size: 0.45",
Amperage: 180	
Wire feed speed	280 ipm
Voltage range	24
Polarity	DCEP
Electrode stickout	1/2"
Shielding gas (type)	CO_2
Backing gas	NA

Technique

Stringer	
Travel angle	Push 70° Work angle 90°
Single pass	
Initial cleaning	wire brush
Interpass cleaning	chip, wire brush

Introduction

These forks were designed to be used at a nursery where the ground gets muddy and wet, so a regular forklift cannot be used. They connect to the back of a tractor, using the tractor's 3-point hitch to turn it into a forklift, to permit hauling and lifting pallets.

Layout and Fabrication

- Measure and cut all steel to length on a band saw
- Flame cut a 60 degree section out of piece A
- Cap all open ends on pieces A and F
- Make layout lines on the floor with a chalk line (using the 3-4-5 formula)
- Layout pieces A,B,C on the floor; line them up with the lines; check for squareness
- Tack pieces together and recheck squareness
- Weld sections together as specified on drawings and WPSs
- Layout pieces D,E,F and clamp them down on a flat surface
- Tack and check for squareness
- Weld sections together as specified on drawings and WPSs
- Use the port-power to spread apart pieces D,E,F to the proper width as specified in the drawing
- Cut piece G to length
- Make layout lines on each piece according to the drawings
- Put both pieces in a vise and line the layout lines up with the top of the vise
- Apply heat to both pieces right above the vise
- Bend both pieces to a 45-degree angle
- Do exactly the same to the other end of each piece, except bend each end in the opposite direction
- Cut the 45-degree angle off of one corner on the band saw
- Drill holes as specified in the drawings
- Tack pieces H and G to pieces D,E,F and check for squareness
- Weld sections together as specified in drawings and WPSs
- Tack pieces D,E,F,H,G to pieces A,B,C and check for squareness
- Weld sections together as specified in drawings and WPSs
- Cut out piece I as specified in the drawings
- Tack piece I to the main body and check for squareness
- Weld piece I as specified in the drawings and WPSs

Finishing

- Wipe the project down with acetone
- Spray paint prepared surfaces with a primer and wait for primer to dry
- Spray paint with finish paint

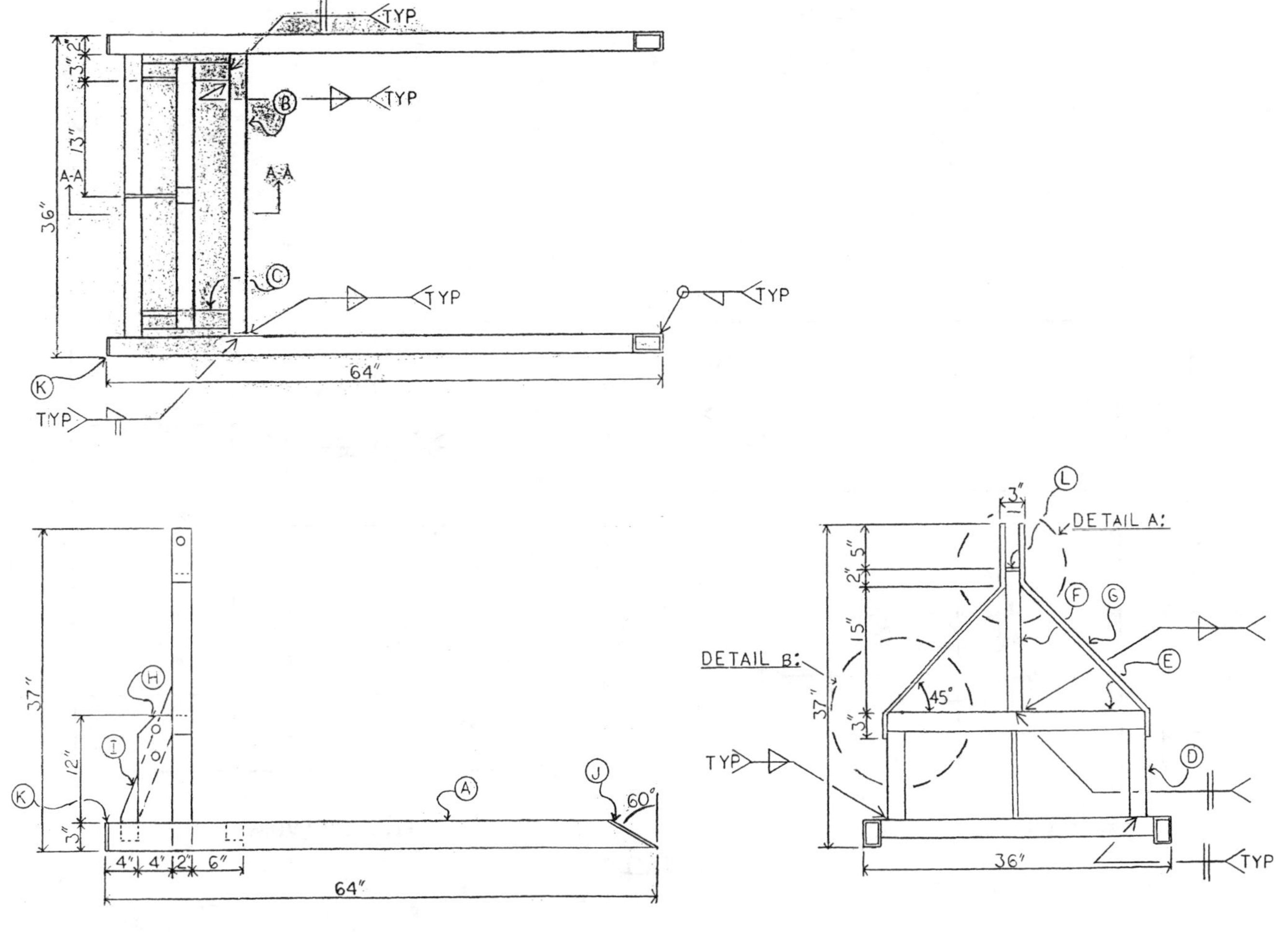

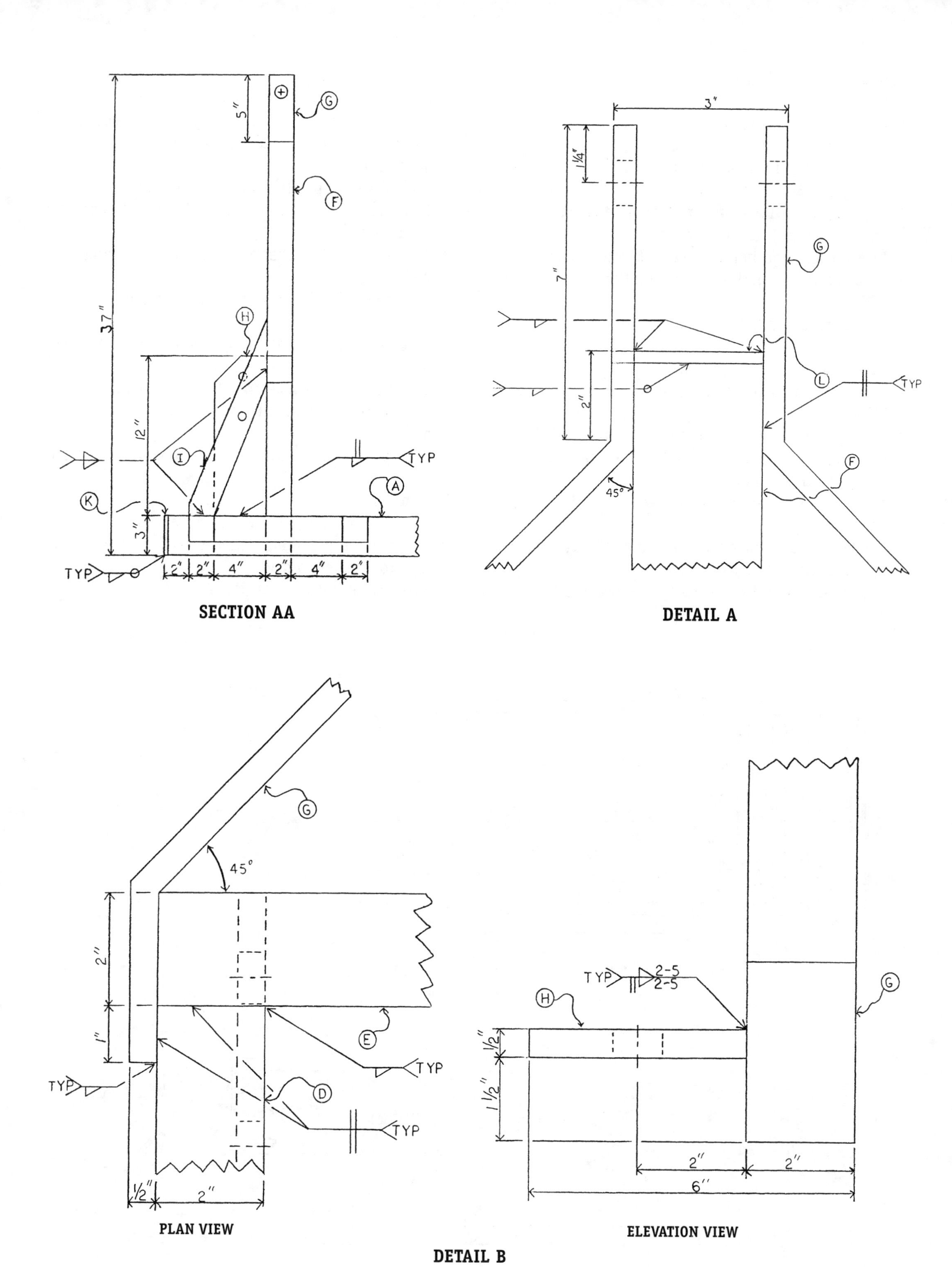

SECTION AA

DETAIL A

PLAN VIEW

ELEVATION VIEW

DETAIL B

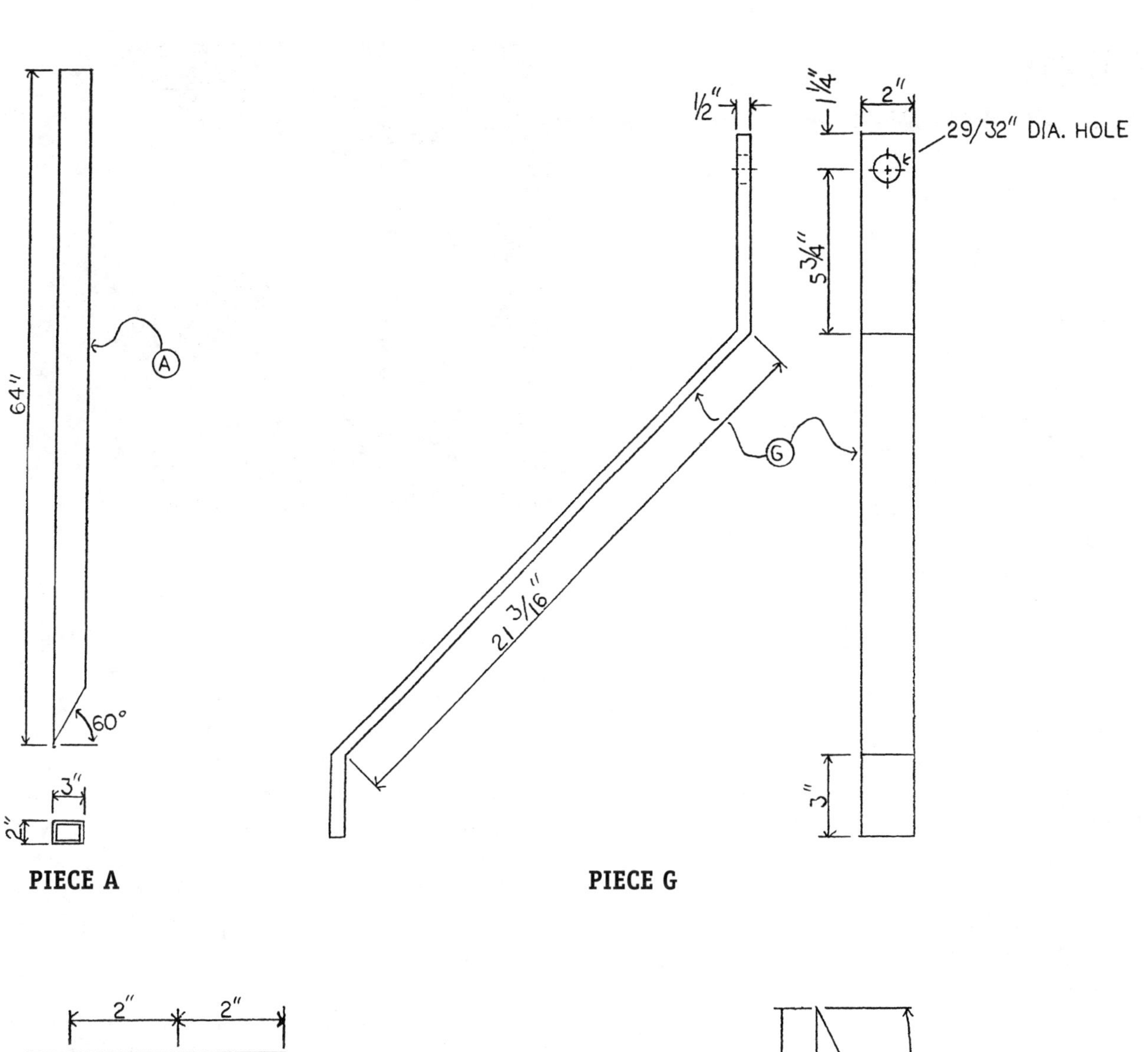

PIECE A

PIECE G

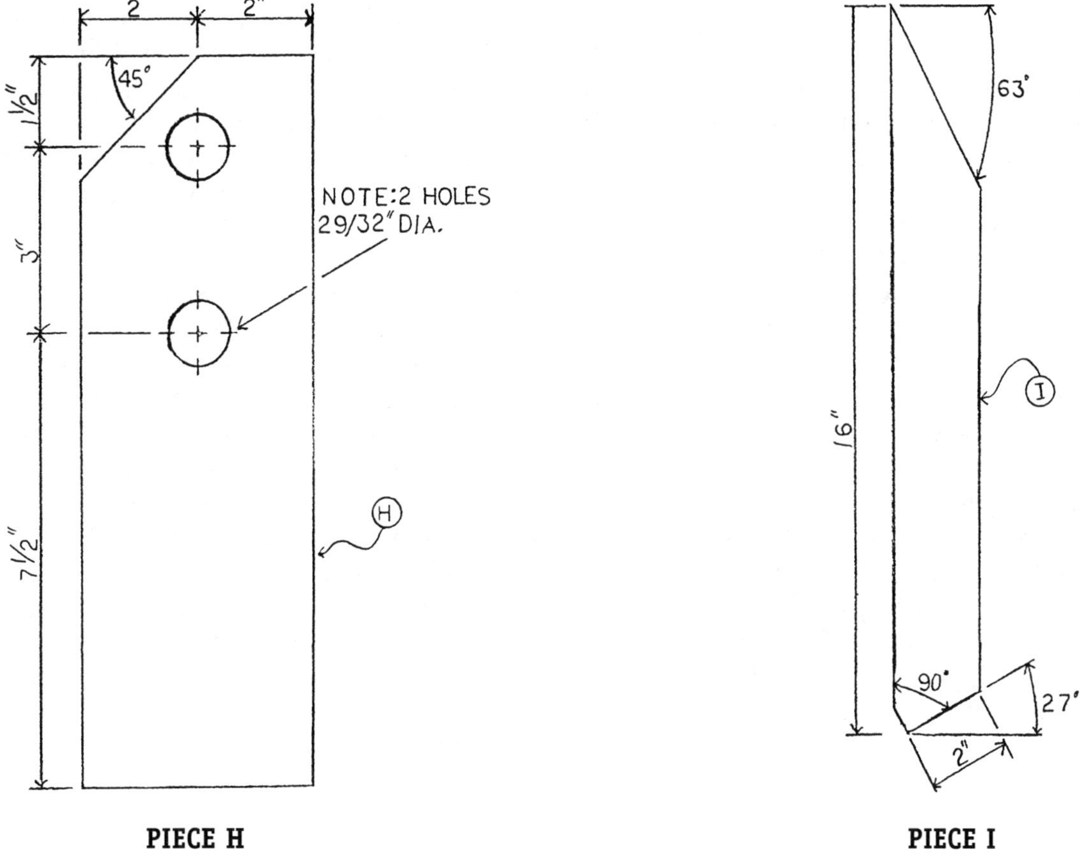

PIECE H

PIECE I

Pit Cart

Author: Jason Plantenberg
Instructor: Thayer David
School: Tomahawk High
City & State: Tomahawk, WI

Materials List

Qty	Description	Size
5	1″ x 1″ x 15 ga. Square Tubing	24′
1	1″ x 1″ x 14 ga. Square Tubing	24′
2	2″ x 2″ x 1/8″ Angle Iron	20′
2	1-1/2″ x 1-1/2″ Aluminum Angle	24′
1	Aluminum Diamond Plate	4′ x 8′
4	Pneumatic Tires	10-1/4″
2	Spindle/Bracket Sets	3-3/4″
2	13″ Econ Tie Rod Kits	3/8″–24″
4	CNT Hinges	1-1/2″ x 48″
4pk	Catch Zinc	
1	Pad Metal Finish	4-3
2	Aluminum Rivet Packs	
2pk	Aluminum Short Rivets	
1	Rivet Aluminum	3/16″ x 1″
1	Vise	4-1/2″
1	Vise Jaws	
3	Cans Spray Paint	
2	Cand Spray Primer	
1	Caulk Lexel Clear	
1pk	Fasteners	

Introduction
A pit cart of this type is used by motocross and other racers to take tools and equipment from their trailers down into the pits for easy access. The cart is designed to hold heavy objects such as toolboxes, rims and tires, air compressors, tables, and anything else needed in the pits. This pit cart incorporates tables to slide inside the cart from the side for easy access, and doors that fold into thirds to take up a minimum of space.

Building the Frame
Start by cutting the square tubing material into sections using a band saw. Cut eight lengths of 72″, four lengths of 32″, and 6 lengths of 31″. Begin by making the two end rectangles of the frame, with the 32″ sections for the height and the 31″ pieces for the width. Be sure everything is square and tack weld each corner of the two rectangles using a MIG welder. Now add in four of the 72″ pieces for length. Tack weld everything together and again check that it is square everywhere.

Add in two more sections of 72″ tubing, placing them 6″ from the bottom of the top section of the frame. (Dimensions may vary depending on the size of tables to be placed inside.) Measure another 6″ down and add another two sections of 72″ tubing, one on each side for another table. Now tack on the section of angle iron to the inside of the 72″ pieces that were just added. This will serve as the table tray. Now add in two sections of 31″ tubing to one end, directly perpendicular to the 72″ sections that were just added. These will serve as a back support so the tables do not slide out.

Add in the remaining sections of 31″ tubing to the bottom of the cart. Spacing will depend on the load requirements for the cart. In the example shown here, the spacing was roughly 8″. Finish by welding everything together, checking along the way to make sure that everything is square and making adjustments as necessary.

Assembling the Steering Components
The steering components include the handle bar, spindle bracket, spindle and bracket set, tie rods, axles, and pneumatic tires. To make the spindle bracket, start by cutting all the pieces needed. Sizes will depend upon the specific application, but in the example shown here, all flat steel was 3/16″ thick, and the following pieces were cut: one at 4″ x 2″, one at 2-1/2″ x 2″, one at 6-1/2″ x 2″, two at 2-1/2″ x 1-1/2″. After cutting the pieces, measure for holes to be drilled and drill them all. Be precise. Measure 1/2″ from one end of the 6-1/2″ x 2″ piece and 1/2″ in from the side, and drill a 3/8″ hole. Repeat on the other side, but do this to only one end.

On the other end, mark and drill a 5/8″ hole in the center and 1″ down from the top of that end. Take the 2-1/2″ x 2″ piece and mark and drill a 5/8″ hole in the center and 1″ down from the top of the piece. This piece will have to line up perfectly with the 5/8″ hole made in the 6-1/2″ x 2″ piece earlier. In the two pieces of 2-1/2″ x 1-1/2″, drill a 3/8″ hole in each, a half inch down from the top and a half inch in from the side.

Now that all all the holes are drilled, take the piece of 4″ x 2″ and tack weld it exactly perpendicular on the 6-1/2″ piece. Then take the section of the 2-1/2″ x 2″ with the 5/8″ hole and tack weld it at 90 degrees to the upright that was just tacked in. The 5/8″ holes must match up perfectly so that the pin can fit through them with ease. This assembly now resembles a lower case "h." To finish it, weld in the last two remaining sections to the back side of the only upright when the piece is laid flat on its long side. Tack these pieces in so that all edges match up and the holes drilled in the pieces are on the high side, allowing room for the handle bar to move up and down with ease.

Now MIG weld everything up. To finish up the piece, put it on a vertical mill and using a 3/8″ bit, remove 1/2″ to the bottom of the handle bar upright supports. This will give room for the tie rods to move more and give the cart a tighter turning radius. Finally, weld the axles from the kit and bolt up the wheel assembly.

Final Assembly

Check again that the frame is square and make adjustments as necessary. Paint the frame the desired color. Using a shear or similar tool cut up the sheet metal to size for: two floors, a back, one permanent side, one side that flips down, stainless steel trim around wheel wells, and a front consisting of two doors. Each door will fold up into thirds on piano hinges.

Use rivets and silicone sealant to secure everything in place. Start by adding in the floors first, and secure them permanently. For the side where the tables go, simply attach a hinge so that the door can fold down. To make the doors, cut the piece in half, and cut each half into thirds. Add hinges with rivets to create two doors that fold into thirds. Put a latch on each of these so that they will not fly open.

Using rivets, attach the stainless steel tabletop and add on the vise and anything else needed. The bottom tray may be left as open space, or it can hold plastic containers. Finally, add the tie rod assembly, trim pieces, tables, and handle bar. Clean up everything. Make steering adjustments to the tie rod assembly as necessary.

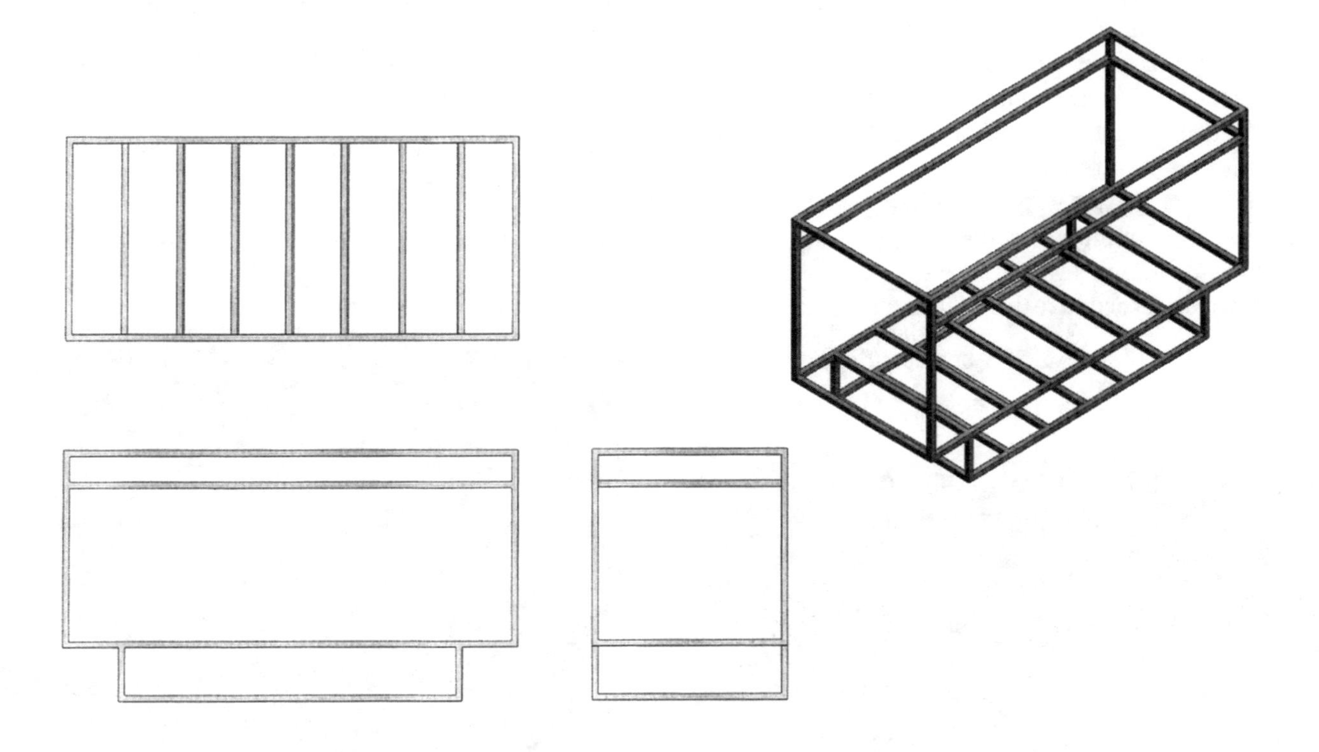

Horse-drawn Trolley

Author: Joshua D. Bishop and Andrew D. Cox
Instructor: James L. King
School: Burlington High School
City & State: Burlington, IA

Bill of Materials

Quantity	*Description*
2	8″ x 8″ I beams
7	4′ x 10′ 18 gauge sheet metal
1	4″ x 8-1/4″ gauge floor plate
40′	1″ x 1″ angle iron
56′	2 x 2 x 1/8″ Angle iron
60′	1 x 1-1/2 angle iron
48′	1″ 11 gauge sq. tube
190′	1-1/2 x 1-1/2″ 11gauge square tube
56′	2 x 2 x 3/16″ tube
24′	2 x 2 x 1/8″ tube
3′	3 x 2 1/8″ tube
230′	1 x 2 x 1/8″ tube
25′	1/2″ hot rolled rod
11′	2-1/2 x 1/4″ flat steel
10′	6 x 1/4″ flat steel
7′	4 x 1/4″ flat steel
3′	3/8 x 2″ flat steel
12′	3/16″ chain
1250	3/16″ pop rivets
15	3/16″ and 1/2″ drill bits
2	hinges
2	7/8″ pins
1	1-1/2″ king pin bolt fine thread
As needed	1/4 5/16 and 1/2″ Bolts
As needed	aluminum trim
3 gal.	white paint
4 gal.	gray paint
1 gal.	burgundy paint

Construction Procedure

Flame cut two 1/8″ 8″ x 3″ I beams 17′ long.

Place them 6′ 8″ apart.

Cut two 2″ x 2″ x 3/16″ square tubing seven feet six inches long.

Cut three 2″ x 2″ x 1/8″ square tubing seven feet six inches long.

Square up the ends of the I beam and make them flush with each other.

Measure in three feet from the front of both I beams and make a mark.

Weld one of the seven foot six 1/8″ wall pieces of square tubing to the inside of this mark. From now on, anything measuring seven feet six inches and being 2 by 22 square tubing will be referred to as a cross member.

Measure front from this mark six in. and mark.

Weld one of the 3/16″ wall cross members and just in front of this mark.

From this cross member, measure forward 1 foot and mark.

Weld one of the 3/16″ wall cross members in front of this mark.

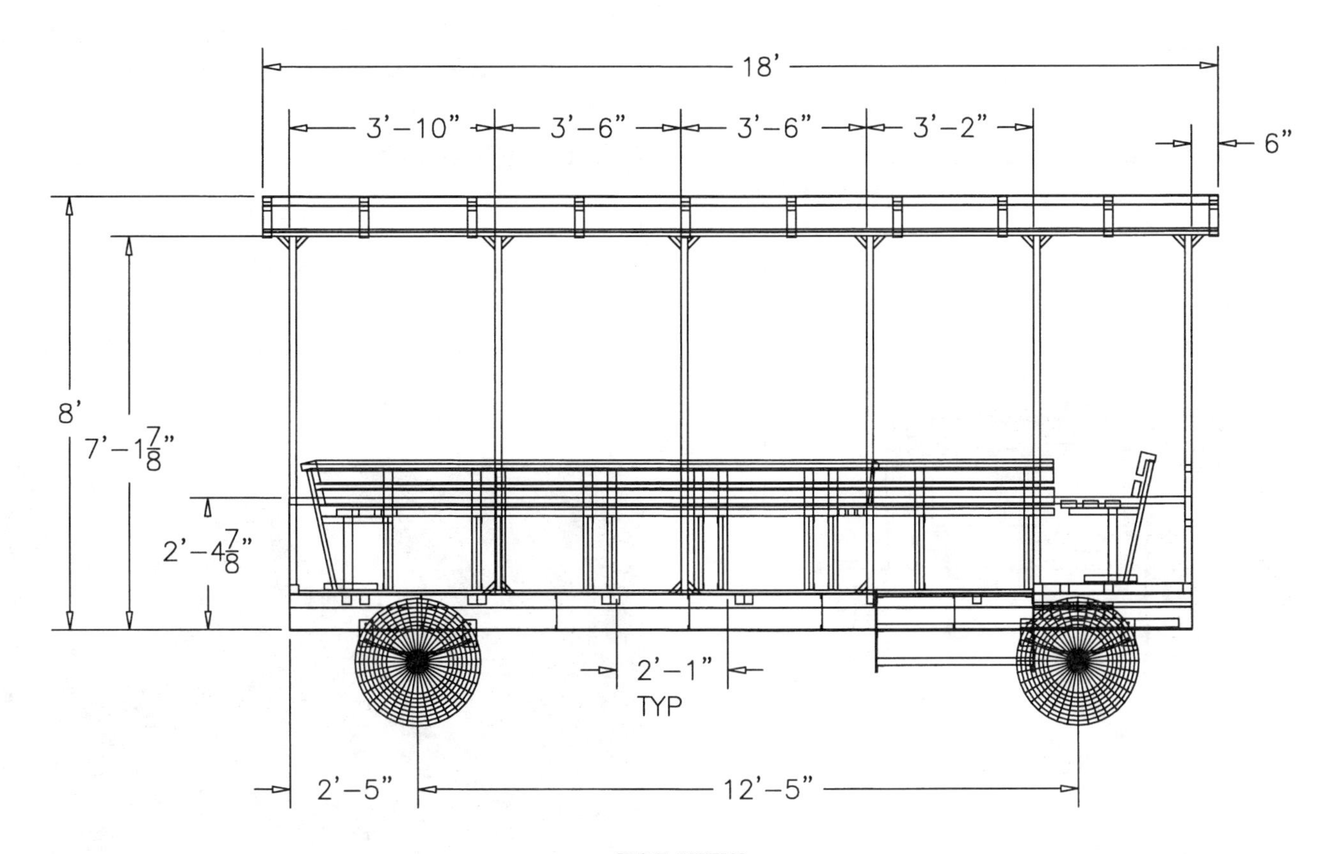

SIDE VIEW

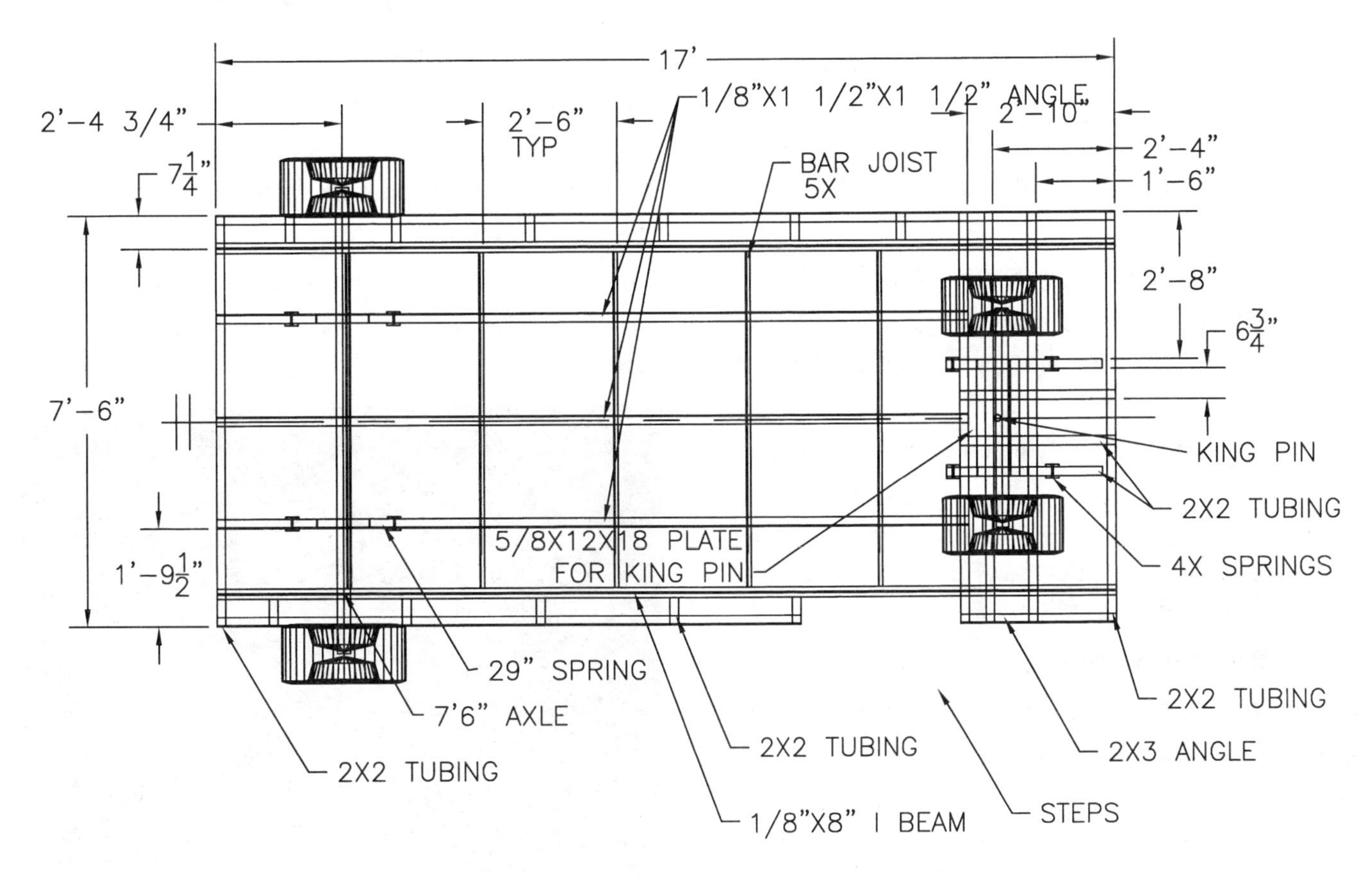

TOP VIEW

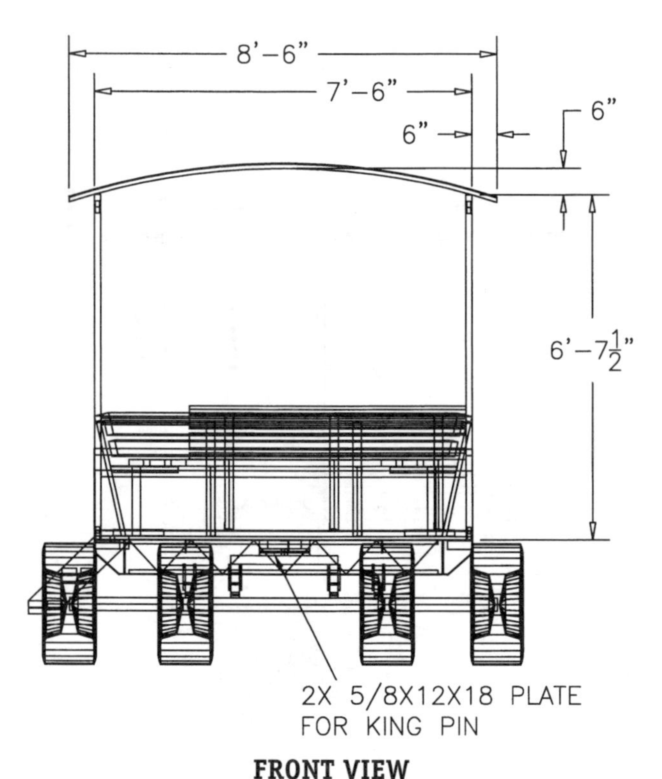

FRONT VIEW

At the very front of the I beams which should be two feet, weld the one of the 1/8" wall cross members.

At the very rear of the I beams, weld the last of the 1/8" wall cross members.

Cut eleven pieces of 3/16" wall square tubing 6 inches long.

Measure from the front of the frame back 6' on the right side of the trolley, and place the first of these pieces on.

From then on, weld the pieces on every 30 inches.

On the left side, measure back four feet, then weld a six inch 2 by 2 on. Then proceed to weld the six inch pieces on every 30 inches.

Cut two pieces of 2 by 3" angle iron seventeen feet long.

Weld one of the angle irons on each side of the frame below the cross members but on top of the 6 inch support braces. Weld to all surfaces possible. Also, have the three inch leg going down.

Cut out ten 6 foot 8 inch long pieces of 1 x 1 angle iron.

Cut out 30 pieces of 1/4 round stock 10 inches long.

Weld 1/4 inch round stock inside two of the 1 x 1 angles at 30 degree angles. These should give the appearance of an equilateral triangle. There should be six of these triangles inside each set of angle irons.

Weld the first of these braces a foot and a half from the fourth cross member. Then from there on, weld them in every 30 inches.

B. Turning Mechanism

Cut 2 pieces of 5/8 plate 18 x 12 inches.

Find the center in both plates and bore a 1 x 1/2 inch hole.

Center one of the plates on the center cross members.

Weld on.

Cut 4' out of the center of an eight foot axle.

Grind the area of the cut to a taper with a hand grinder.

Weld the axle back together with a 6010 root beat a 7018 cover bead.

Cut two one inch strips out of left over axle pipe.

Weld the resulting pieces to the outside of the short axle.

In the remaining 5/8 inch plate, drill two quarter inch holes.

Tap these holes with a tap and die set.

Thread two grease zerts into these holes.

Tack the grease zerts down to prevent them coming out.

Cut four 2 x 2 x 3/16" pieces of square tubing 25-1/2 inches long.

Weld these to the bottom plate every six inches apart.

Cut two 2" x 2" x 3/16" pieces of square tubing three feet long.

Lay the bottom plate of the turn mechanism on top of these two pieces one foot from the front.

Cut eight pieces of 2" x 2" x 3/16" square tubing 4" long.

Drill half inch holes in four of the four inch pieces.

Weld two of the blocks with holes in them flush with the back of the 3 foot long square tubing.

Weld two more blocks with the center at 29".

Shear 8 pieces of 2" wide 1/4" strap 4" long.

Shear 8 pieces of 2" wide 1/4" strap 3" long.

Drill 1/2" hole a 1/2" from the end of the 4" piece in the center. This will make a shackle for the axle.

Repeat the process mentioned in step 24 only on both ends of the 3" strap.

Bolt the short straps to the rear blocks on the turning mechanism.

Weld the long straps to the front blocks with the holes down, and have them be 29" from center to center of the shackles. Do this on both sides.

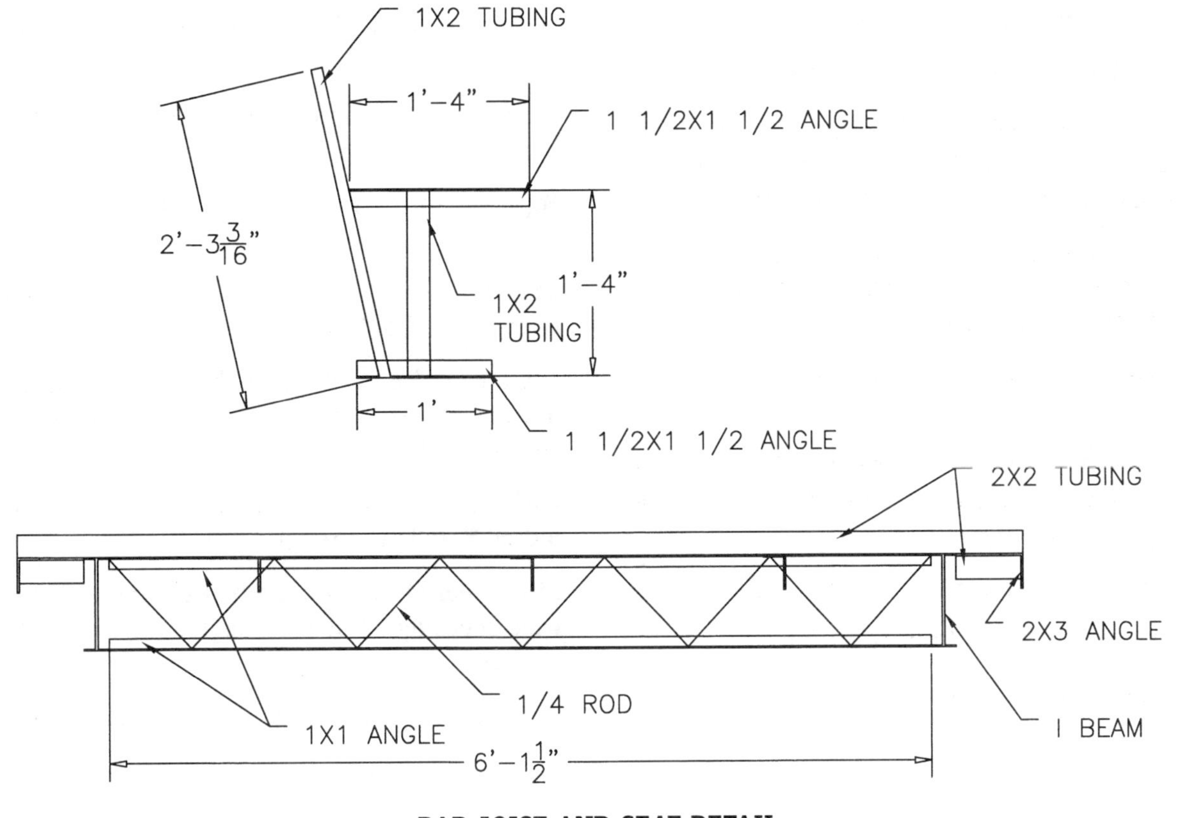

BAR JOIST AND SEAT DETAIL

Bolt the short axle to the shackles.

Roll axle and bottom plate into position and bolt on through 1-1/2" hole.

Tighten bolt and tack shut to prevent it from wiggling loose in transit.

Grease turning mechanism.

Attach the rear axle to the I-beams as well.

C. The Walls

Cut out six 6'-6" long pieces of 1" x 1/2" square, square tubing.

Cut out six 6'-4" pieces of the same material.

Cut out two pieces of the same material 18' long.

Lay all pieces out on the floor.

Weld a 6' 4" piece on six inches from the front of the 18' piece, and weld a 6' 4" post six inches from the back.

Weld another 6' 4" piece on 3' from the front post.

3' from that post, weld a 6' 6" post.

Weld two more six 6' 6" pieces on 43-1/2" apart.

Repeat these steps to make the other wall.

Cut 30 corner braces for the posts. These should be 6" long and be shaped like a trapezoid. Make them out of 1" x 2" 11 gauge square tubing.

Weld 20 of the braces to the tops of the posts.

Raise the walls and weld on. The shorter posts will sit on top of the cross members on the frame, and the long one should sit on top off the angle iron.

Weld the 10 remaining corner braces on to the bottom of the posts. Do not weld any inside the second and third posts on the right side. This is the doorway.

Cut two pieces of 1" x 2" x 11 gauge square tubing 7' 3" long.

Weld one of these pieces as a header at the top of even with the front post and flush with the top of the wall.

Repeat this process at the rear of the trolley.

Cut a 7' 2" piece of 1-1/2" square tubing.

Weld nineteen inches up from the front cross member in the frame.

Repeat this process at the rear.

Cut two 17-1/2" long supports out of 1-1/2" square tubing.

Center at the front and the back and weld in place under the bars made in process nineteen.

Sheet metal used on walls is sheet metal remaining from roof. It was sheared to 8' long and 2' wide. Use this wherever possible. We were not able to do all of the walls so some was sheared especially for the walls. Sear up a piece of 18 gauge sheet metal 7' 6" wide and 29" deep.

Rivet to the rear of the trolley starting at the 2' tall support bar.

Fit pieces of let of roof metal in where possible and rivet into place.

Shear up a 19" piece of 18 gauge sheet metal. 7' 6" inches wide.

Pop rivet to the front of the trolley beginning at the top of the cross support.

Seam up two 3' 3" pieces of sheet metal 10" wide, pop rivet to the front of the trolley.

Cut out a 2' 6" wide hole that extends up from the bottom of these two pieces. Do this with a plasma cutter.

With some scrap 18 gauge, break some 2" wide trim pieces to hide the rough edge.

At the rear of the trolley cut the corners off of the 29" wide sheet metal at a 45 degree angle. Use the plasma cutter.

Use some more 18 gauge scrap at these rough edges for trim.

D. The Roof

Cut ten pieces of 1″ x 2″ 11 gauge square tubing ten feet long.

Run through a form roller to make roof bows.

Cut all pieces to 8′ 4″.

Weld onto roof every two feet.

Cut 4 pieces of 4′ x 10′ 18 gauge sheet metal down to 4′ x 10′ pieces.

Cut one piece of 4 x 10 foot down to 2′ x 8′.

Lay sheet metal out on the roof.

Drill quarter inch holes through sheet metal and bows every four inches.

Pop rivet the sheet metal onto the bows.

E. The Floor

Cut a 4 x 10 foot piece of 14 gauge diamond plate down to 3 x 7.

Weld to the top of the four cross members at the top of the frame.

Square the ends of all rough cut oak boards and rip to varying lengths.

Run all boards through planer until they reach a 3/4 inch thickness.

Bolt all boards to the top of the I beams and through the three angle Irons on the floor joists.

Seal the floor with industrial sealant.

F. The Chairs

Cut twenty 1″x 1/2″ and 1/2″ angle irons 16″ long.

Cut twenty 1″ x 2″ 11 gauge square tubing 16″ long.

Cut twenty 1″ and 1/2″ x 1″ and 1/2″ angle irons 12″ long.

Cut twenty 1″ x 2″ 11 gauge pieces of square tubing 28-1/2″ long.

Weld the components together in a fixture made from angle irons.

Bolt chairs to floor every 25-1/2″. Bolt two chairs through the diamond deck platform 4′ apart.

Cut 12′ pine 2 x 4s to desired lengths.

Bolt pine boards to chair frame.

Stamp ends of boards with metal stamp for identification, and remove from frames for painting.

Reattach boards after painting.

G. The Steps

Cut two pieces of 8″ x 1/4″ flat steel 35-1/2″ long.

Cut a 45 degree angle at one end.

With a square, lay out a tread of 8″ and a rise of 7-1/2″.

Cut out a 3′ area of angle iron in the doorway. This is 2 x 3 in. and it is cut off the frame.

Cut two 2 x 2 x 3/16 square tubes and weld inside the door frame.

Drill a 1/2″ hole in both stair components and square tubes. Weld tubes between the 2″ x 3″ angle iron and the I beam.

Cut six pieces of 1-1/2″ x 1-1/2″ angle iron 8″ long.

Weld on the marks of the 8″ thread.

Bolt two 2 x 4s to each of the threads.

Bolt stairs to the frame.

Attach retracting chain to the door posts.

H. The Hitch

Cut two pieces of 3/8″ plate 7″ x 13″.

Drill 7/8″ hole in the center a half inch from the bottom.

Weld flush with the front of the 3′ pieces on the turning mechanism.

Cut 1 piece of 3″ x 2″ x 1/8″ square tube 2′ long.

Cut two 1/4″ straps 5-1/2″ long.

Drill a 7/8″ in both straps, centered and 3/4″ from the end.

Cut 2 pieces of 4″ x 1/4″ flat steel 37″ long.

Drill three 3/4″ holes and two 1/2″ holes in each piece of flat steel.

Cut one piece of 2″ x 2″ x 1/8″ square tube 5″ long.

Center the 5″ piece of square tube on the 3″ x 2″ square tube and weld on.

Center one of the 37″ pieces of flat steel and weld on to the block.

Center the other piece and weld it directly to the 2″ x 3″ tube.

Weld the 2″ straps to the 2″ x 3″ square tube with the holes stick opposite of the 37″ pieces.

Attach a pre-made oak tongue to the steel components with bolts.

Attach the hitch to the turning mechanism with two 7/8 pins.

Finish Schedule

Chip weld spatter off all welds with a chisel or a wire.

Grind all welds with a grinder.

Power sand and hand grind all welds.

Sandblast the project.

Apply a coat of primer paint.

Paint project.

Put aluminum trim on all sheet metal seems.

Portable Sawmill

Author: Mark Campbell
Instructor: Stanley Neal
School: Fresno City College
City & State: Fresno, CA

Bill of Materials

Qty	Size	Length	Description
2	2" x 5" x 3/16"	288"	Rectangle tube
2	2" x 5" x 3/16"	37"	Rectangle tube
8	C4 x 5.4 lbs	33"	Channel
4	2-1/2" x 2-1/2" x 1/4"	5"	Tube
4	2" x 2" x 1/4"	24"	Tube
4	6" x 1/4"	6"	Flat
2	4" x 1/4"	6"	Flat
4	1-1/4" x 1-1/4" x 0.120"	6"	Tube
4	1" x 1" x 0.120"	8"	Tube
1	3" x 3" x 0.188"	36-3/4"	Tube
1	2-1/2" x 2-1/2" x 1/4"	70"	Tube
2	1-1/2" x 1-1/2" x 1/4"	288"	Angle
4	6" x 1/4"	18"	Flat
4	1" x 1" x 0.120"	4"	Tube
2	2-1/2" x 2-1/2" x 1/4"	18"	Flat
6	4"	2"	SCH 40 pipe
2	6"	4"	SCH 40 pipe
4	1-1/2" x 1-1/2" x 0.120"	30"	Tube
2	1-1/4" x 1-1/4" x 0.120"	7"	Tube
2	1-1/2" x 1-1/2" x 0.120"	30"	Tube
4	1-1/2" x 1-1/2" x 0.120"	17"	Tube
4	2" x 4" x 1/4"	3"	Tube
1	6" diameter	4"	Stainless steel pipe
1	1-1/4" x 1-1/4" x 0.120"	2-1/2"	Stainless steel pipe
1	1" x 1" x 0.120"	16"	Tube
1	3" x 4" x 1/4"	8"	Angle
1	6" diameter	14"	Stainless steel pipe
2	1-1/2" x 1-1/2" x 1/4"	8-1/2"	Angle
2	1-1/2" x 1/4"	2-1/2"	Flat

Editor's safety note: *Be sure proper safety guards are installed along with proper warning signs.*

Sawmill Construction

Mainframe Assembly

The sawmill was assembled by welding the main frame, a 2" x 5" rectangular tubing and eight 4" channel iron cross members at 32" on center, together (see drawing part numbers 1, 2 and 3). Next, four log bunks cut out of 1/4" plate (13) were added. They are attached to cross members 2, 3, 6 and 7. (Note – the eight cross members are numbered 1-8 started with number 1 at the front.) The notch out of the center of the bunk is to keep the logs from rolling off before the log dogs are set into the log for stabilization. The log dogs (8 and 9) were the next pieces attached to cross members 3 and 6, constructed out of 1" x 1" tube that slides into 1-1/4" x 1-1/4" tube so they can adjust for different diameter logs.

Two toe board plates (7) made from 4" x 1/4" flat were welded in the center of cross members 2 and 7. These perches hold two screw jacks to compensate for logs with a taper. The two jacks were then fitted with small triangle pieces salvaged from the log bunk cutout. These pieces were sharpened so they bite into the log to prevent slipping. They were next welded to the top lifting plate of the jack. Then the modified jacks were attached to the toe board plates (7). The function of the screw jacks is to level the log so the maximum amount of lumber can be gained from the tree.

Four stabilizer legs (5) were installed on each corner of the main frame. They were made from 2" x 2" square tubing with 6" square pads (6) welded to the bottom to prevent the mill from sinking. All top ends were capped to prevent rainwater from collecting and then freezing, resulting in expansion of the tube. The stabilizer legs are adjustable as they slide through a short piece of 2-1/2" x 2-1/2" tubing (4). Holes were drilled through the side of the tubing then fitted with a 1/2" nut to accept the 1/2" set bolts, which lock the legs in any position to accommodate for uneven terrain. There was also a hole drilled through the leg at the fully retracted position so the set bolts would not have to be relied upon when transporting the mill on the highway.

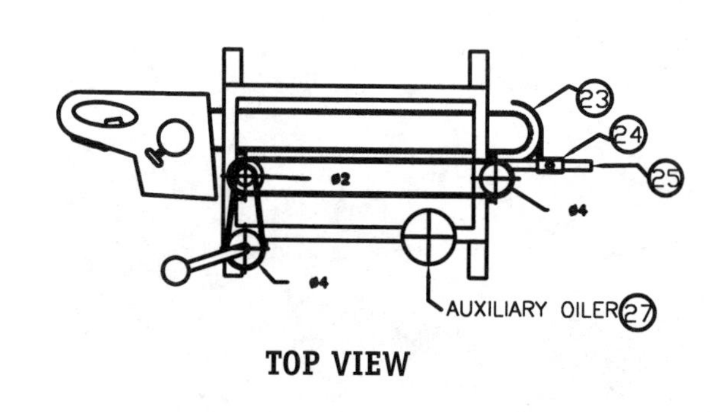

TOP VIEW

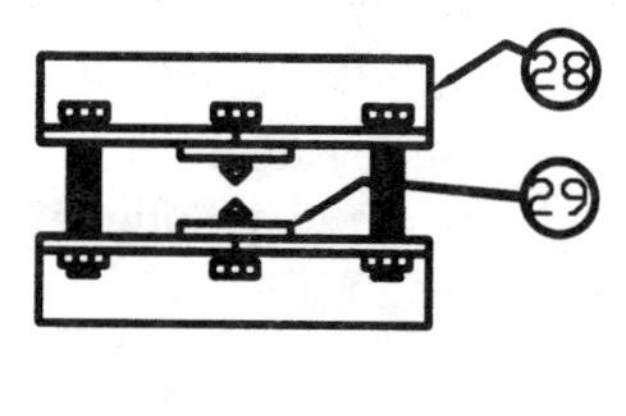

SIDE VIEW

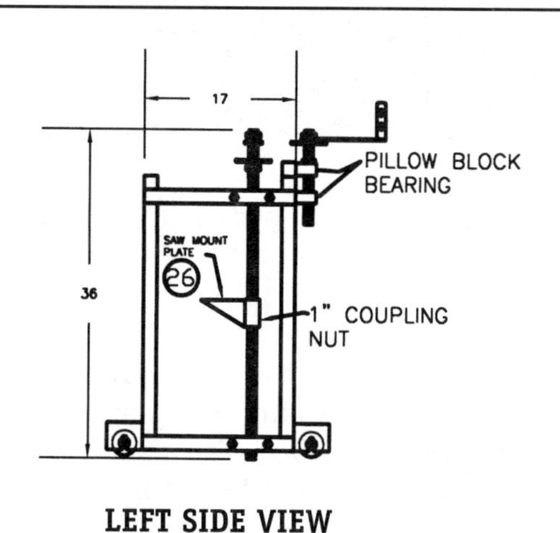

LEFT SIDE VIEW

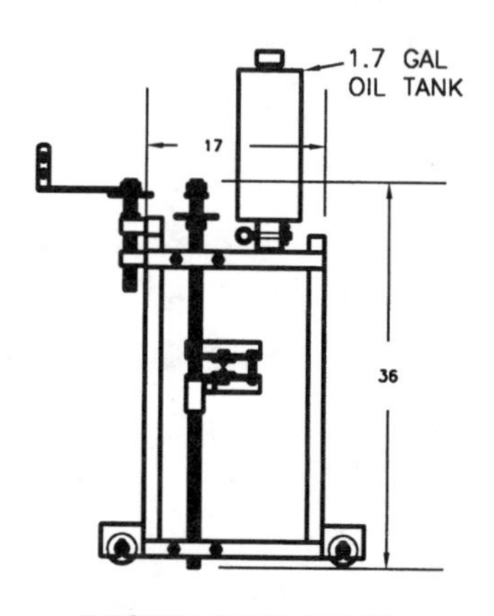

RIGHT SIDE VIEW

the saw up and down. These sprockets were also connected with #35 chain. A bicycle pedal crank was used on the top for a handle.

Oil for the saw chain is supplied from the saw to the top saw bar groove, which, under normal conditions, usually supplies enough oil. To minimize saw bar wear under adverse conditions, an auxiliary oil system was installed. By drilling a 3/8″ hole in the opposite end of the saw bar, the lower saw bar groove (or leading edge in this application) is supplied with oil through a 3/8″ bolt. The bolt was drilled through the center and one side then installed on the bar sandwiched between two copper washers. A 5/16″ vinyl hose runs from the bolt to an oil tank mounted to the top of the saw carriage. This tank was constructed from surplus 6″ stainless steel pipe with a 1/16″ wall thickness (27). The ends were capped with 3/32″ material A 2″ fitting to fill through was welded on top and oil is delivered through a 3/4″ fitting on the bottom. A hand guard was also fashioned out of the same surplus 6″ stainless steel pipe (23) attached to 1-1/2″ square tubing (24) so it can adjust for various lengths of saw bars.

Welding Process

In order to expedite construction as well as minimize clean-up while also taking into consideration the material thickness, which ranged from 0.120-0.250, the GMAW process was chosen. The machine was set on DCRP at 20 volts and 140 amps with .035 E-70s-3 wire. Shielding gas was a mixture of 75% argon and 25% carbon dioxide. The flow rate was set at 25 (cfh).

On the thicker material, edge joints were prepared with a single bevel at 60 degrees and a 1/32 gap where practical. A square butt joint with a 1/16″ gap for the thinner material was used. All material too long and too heavy to move by hand to the band saw was flame cut with an oxy-acetylene torch fitted with an 0 tip. The acetylene regulated pressure was set at 3 pounds. The oxygen was set to 30.

After all the stainless steel parts were cut with the plasma torch, they were joined by GTAW set on DCSP at 100 amps maximum. Straight argon was used for shielding gas with a flow rate of 11 (cfh). Because all the stainless steel parts were salvaged from the scrap recycle bin, their composition was unknown. ER-308-L 1/16 for filler rod was tested and worked fine.

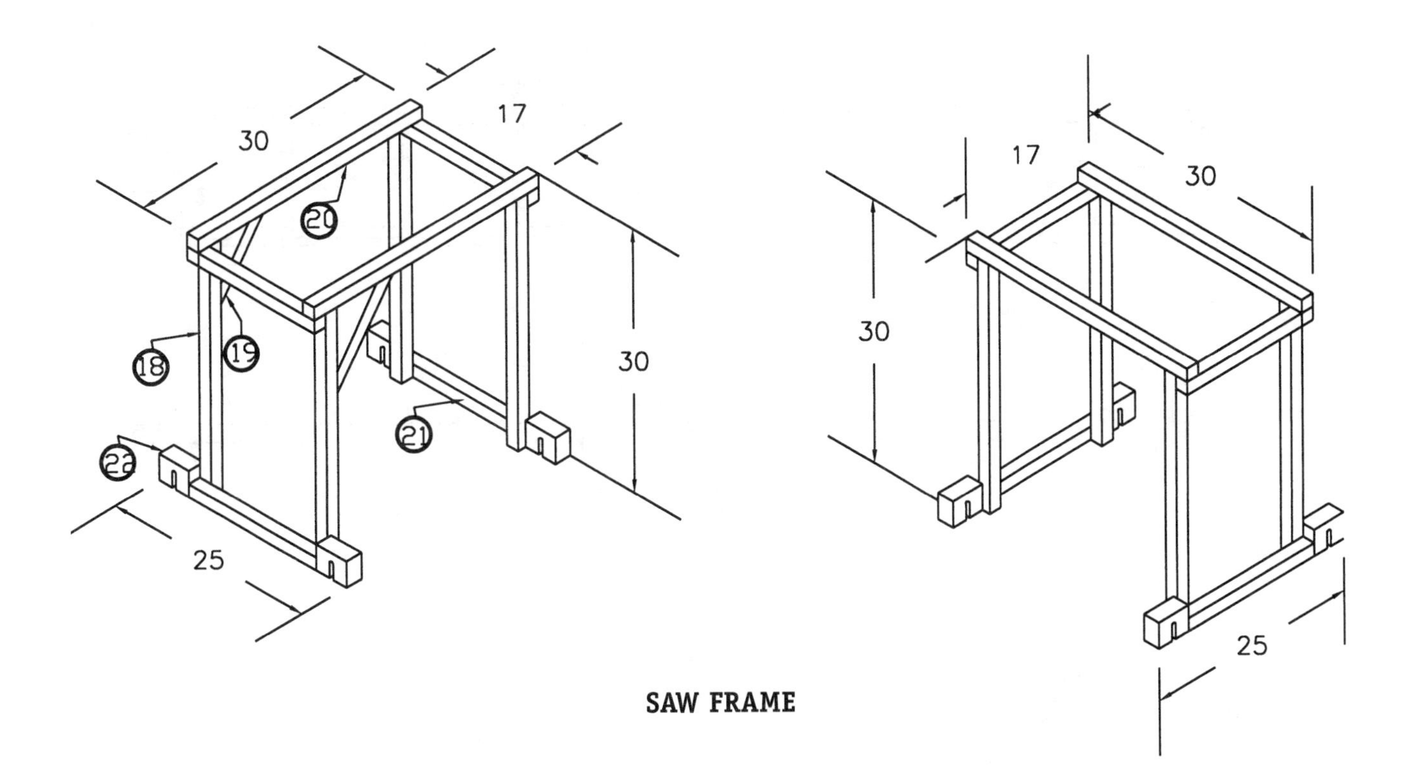

SAW FRAME

1″ diameter pillow block bearings were used to hold steady the 1″ threaded grade 8 rods that provide the vertical movement of the saw. Before these rods were attached through the pillow block bearings, 1″ x 3″ long coupling nuts are threaded onto the rods. These nuts have an extra thick wall so they can withstand having the saw mount plates welded to them without distortion.

The rear saw mount plate to hold the power head is a 3″ x 4″ angle (26). The plate was drilled with the correct spacing to accept the chain saw bar mounting studs, which are 8mm. 8 mm coupling nuts were then threaded onto the bar studs that stick through the side cover of the saw. These coupling nuts line up with the holes in the saw mount base so two 8mm x 30mm long bolts can be passed through the saw mount base and tightened into the coupling nuts to hold the saw power head securely.

On the saw bar tip end, a clamp system is used to facilitate easy removal and adjustment to accommodate different saw bars. The clamp is attached to a 1″ x 1″ tube 16″ long (25). This tube allows for different length saw bars to be used. The clamp system incorporates two 1/2″ bolts, which pass through two pieces of 1-1/2″ angle iron (28) to form a clamp. After the saw bar is clamped into position, the two sharpened bolts that pass through the saw bar pads (29) are tightened. The pointed bolts are called bar lock coupling bolts, which were salvaged out of a mechanical rebar coupler originally intended to splice concrete reinforcing bar.

Two 40 tooth sprockets connected with #35 chain are used to turn the 1″ rods. Two more sprockets, a 40 tooth and a 16 tooth for a combined ratio of 2.5, were added to minimize the number of turns needed to run

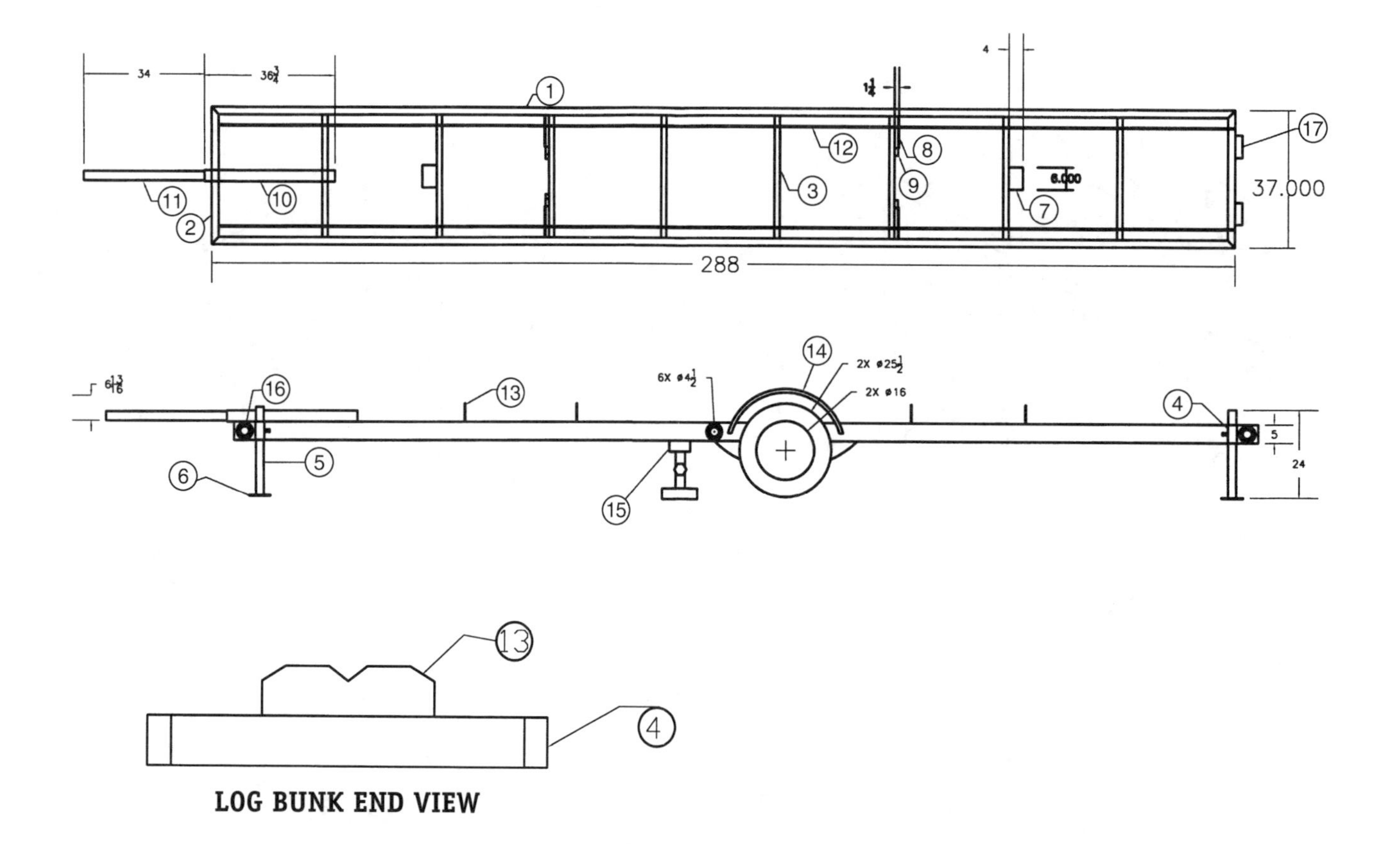

LOG BUNK END VIEW

With the stabilizers in place so the frame could be elevated off the floor, spring hangers were next welded to the frame so the axle would be set one foot past center toward the rear. An axle that was salvaged from a retired utility trailer was used after installing updated hubs, bearings and seals.

Two more screw jacks were added to the trailer main frame. These were attached by 2-1/2" square tubing (15). Their function is to take the bow out of the trailer after the stabilizer legs are lowered. With a string line attached at both ends, the jacks make it a snap to level the trailer. The jacks are also handy for tire removal. This way the mill can be lowered to the ground to make loading the logs easier.

Rails for the saw carriage to run on were made with 1-1/2" angle iron (12). The rails were tacked in place to the cross members, then bolted with 3/8" grade 8 bolts. This was done to minimize warpage from welding heat, as the rails and log bunks have to be true and on the same plane.

The fenders were made to slide on and off for easy removal with pins through a piece of 1" tube (14). The trailer tongue, a piece of 2-1/2" x 2-1/2" tubing (11), was also made this way. It slides through a piece of 3" x 3' tubing (10) so it is not in the way during sawing operations.

Removable lights were going to be used, but due to the trailer's length, permanent lights had to be installed on each corner as well as the center to comply with local traffic codes. As a result, protector rings (16 and 17) were welded in place to minimize damage to the lights while loading and unloading logs.

Saw Frame Assembly

Before the saw frame material was welded together, 1/2" diameter holes were drilled through the 1-1/2" tubing to mount the pillow block bearings. The drill press was used to ensure accuracy because the pillow block bearings need to be in perfect alignment.

The saw frame carriage was welded next using 1-1/2" 0.120 wall tubing (18, 20 and 21). Much care was taken, as this component has to be true and square. After having to cut welds apart twice, a gusset made out of 1-1/4" 0.120 wall tube (19) was added to help hold the frame square.

The carriage rolls on 2-1/2" diameter sheaves attached to the frame with 2" x 4" tubing (22). To ensure the saw carriage rolls true down the rails, 1/2" diameter slots were cut in the tube so adjustments could be made if necessary.